International Conference of the Digital Agriculture from Land to Consumers (ICDALC 2018)

IOP Conference Series: Earth and Environmental Science
Volume 335

Bogor, Indonesia
20 - 21 September 2018

Editors:

Agus Ghautsun Niam
Iin Yusliana
Harry Imantho

ISBN: 978-1-7138-0135-1
ISSN: 1755-1307

Printed from e-media with permission by:

Curran Associates, Inc.
57 Morehouse Lane
Red Hook, NY 12571

proceedings.com

Some format issues inherent in the e-media version may also appear in this print version.

Institute of Physics
Dirac House, Temple Back
Bristol BS1 6BE UK

Phone: 44 1 17 929 7481
Fax: 44 1 17 920 0979

techtracking@iop.org

Additional copies of this publication are available from:

Curran Associates, Inc.
57 Morehouse Lane
Red Hook, NY 12571 USA
Phone: 845-758-0400
Fax: 845-758-2633
Email: curran@proceedings.com
Web: www.proceedings.com

TABLE OF CONTENTS

FOOD PRODUCTION AND SURVEILLANCE SYSTEM

APPLICATIONS OF IT FOR FOOD AND AGROINDUSTRY

INTELLIGENT SYSTEM AND DEEP LEARNING FOR ENVIRONMENTAL AND NATURAL RESOURCES

IOP Conf. Series: Earth and Environmental Science **335** (2019) 011001 doi:10.1088/1755-1315/335/1/011001

PREFACE

The proceedings contain papers presented the in the International Conference on Digital Agriculture from Land to Consumers (ICDALC) 2018 in IPB University at Bogor City, West Java, Indonesia. The proceedings consist of 34 papers, which cover 5 themes: 'Smart and Precision Agriculture Solutions', 'Traceability and Logistic System', 'Food Production and Surveillance Systems', 'Applications of IT for Food and Agroindustry' and 'Intelligent System and Deep Learning for Environmental and Natural Resources Management'. These proceedings is an effort to build a framework for the development of digital agriculture in the future. Digital agriculture (DA) is the use of new and advanced technologies, integrated into one system, to enable farmers and other stakeholders within the agriculture value chain from land to consumer (upstream to downstream) to improve food as well as agroindustrial products development, production and delivery.

We would like to thank all authors for their efforts in preparing their papers. A great appreciation is also given to the reviewers for their assistance in reviewing the manuscripts. Special thanks to secretariat members for their assistance in formatting the layout of the proceedings.

My Best Regards,

Prof. Dr. Yandra Arkeman, M.Eng

Chairman

IOP Conf. Series: Earth and Environmental Science **335** (2019) 011001 doi:10.1088/1755-1315/335/1/011001

Organizing Commitee:

Prof Yandra Arkeman (Chair)
Prof Ono Suparno
Prof Slamet Budijanto
Prof Sutrisno
Prof Suprihatin
Dr Feri Kusnandar
Dr Nora Pandjaitan
Dr Slamet Widodo
Dr Liyantono
Dr Nugraha Edhi Suyatma
Dr Hartrisari Hardjomidjojo
Dr Irman Hermadi

Technical Editor Staff:

Dr. Agus Ghautsun Niam
Iin Yusliana, MM
Harry Imantho, MSc

Peer Review Statement

All papers published in this volume of *IOP Conference Series: Earth and Environmental Science* have been peer reviewed through processes administered by the proceedings Editors. Reviews were conducted by expert referees to the professional and scientific standards expected of a proceedings journal published by IOP Publishing.

IOP Conf. Series: Earth and Environmental Science **335** (2019) 011001 doi:10.1088/1755-1315/335/1/011001

2018 International Conference on Digital Agriculture from Land to Consumers

Bogor, September 20-21, 2018

IPB University, Bogor, Indonesia

Organizer

Faculty of Agricultural Engineering and Technology, IPB

Supported by

IOP Conf. Series: Earth and Environmental Science **335** (2019) 011002 doi:10.1088/1755-1315/335/1/011002

Peer review statement

All papers published in this volume of *IOP Conference Series: Earth and Environmental Science* have been peer reviewed through processes administered by the proceedings Editors. Reviews were conducted by expert referees to the professional and scientific standards expected of a proceedings journal published by IOP Publishing.

IOP Conf. Series: Earth and Environmental Science **335** (2019) 012001 doi:10.1088/1755-1315/335/1/012001

Intelligent Unmanned Aerial Vehicle for Agriculture and Agroindustry

R Hartanto[1], Y Arkeman[2], I Hermadi[2], S Sjaf[2], and M Kleinke[1]

[1]Rhine-Waal University of Applied Sciences, Kleve, NRW 47533, Germany, e-mail: ronny.hartanto@hochschule-rhein-waal.de
[2] Bogor Agricultural University, Bogor, Jawa Barat 16680, Indonesia

Abstract. Unmanned Aerial Vehicle has been adopted in various different areas starting from hobby photography to professional measurement techniques. This paper focuses on the potential of using UAV for agriculture and agroindustry to improve the overall productivity and reducing the overall product cost. It provides brief overview of available UAV systems on the market including their sensors and capabilities. More important is the software part and required algorithms for controlling the AUV in such way that it becomes intelligent. The more advanced the software the more intelligent is the AUV. Some of available algorithms and applications for this purpose are discussed in this paper.

Keywords: intelligent, Unmanned Aerial Vehicle, agroindustry, algorithms.

1. Introduction

Globalization on the food markets stimulates the international competition in the agriculture. In order to survive on the market production costs must be reduced and resources have to be utilized optimally. The UN/DESA published the growth prediction of the world population in [1] where the global population will increase from 6.5 billion to 8.9 billion in year 2030. On the other hand, the total arable land will only increase marginally [2], which means that the arable land per capita will shrink by a quarter. Therefore, the agriculture is required to utilize the limited resources better amongst others approaches. This can be achieved by precision farming which performs tasks more intelligently and efficiently.

In industrial nations, agriculture has completely changed within three generations. In the twentieth century was the era of mechanization of the agriculture sectors. Farmers are no longer tether little oxen or goats to their ploughs but using heavy machinery instead. Today, many industries have applied the fourth industrial revolution which known also as “Industry 4.0” on their daily business. This revolution is not only affecting the manufacturing industry but also agriculture and agroindustry. Information technology and data driven processes have changed the processes in the plantation and cattle industry. Machines are no longer stand-alone, they are connected to the cloud and other machines and equipped with special sensors to perform precision farming [3, 4].

In past decades, many agricultural machine industries improved the productivity by increasing the size of the machine and adding more power to the machines. However, this approach reached its limit where the size of the machine cannot be increased anymore due to the traffic limitation where nowadays those machines reached the maximum allowed size to operate on the road. Additionally, increasing the size of the machines will have negative impact of the soil, where the heavier the machine the more

IOP Conf. Series: Earth and Environmental Science **335** (2019) 012001 doi:10.1088/1755-1315/335/1/012001

damage occurs to the soil compaction. Large and heavy machines can only be operated on the large flat fields but not for small fields and mountain side crops. Knowledge about the environment, soil, and crops health can help farmers react accordingly in certain situations, for example using adequate fertilizer for certain part of the soils depending on the soil data [5]. In order to receive this knowledge, measurements using special sensors have to be conducted or using telemetry information from the satellite imagery.

Recent development in unmanned aerial vehicles (UAVs) or also knowns as drones has become an interesting game changing in the agriculture and agroindustry. The UAVs cost only a fraction of large or heavy agricultural machine and can be equipped with special sensors to perform some measurements on the field gaining additional or specific information from the crops. Therefore, the UAV is one of the many different approaches in today's farming techniques known also as smart farming. This paper focuses on the UAV technologies and their application supporting agriculture and agroindustry [6]. Two applications of using UAVs for supporting activities in agriculture and agroindustry are briefly discussed in this paper.

The following section will discuss smart farming and digital agriculture to see how far the technology and digitalization have been used in the agriculture and agroindustry. Next, the fundamental information about unmanned aerial vehicles from the classification to the autonomy is then presented. Two application of using UAVs in agriculture and agroindustry is presented in the following section. Challenges and further development are discussed in section 5. Finally, section 6 concludes the paper.

2. Smart Farming and Digital Agriculture

Smart Farming and digital agriculture are the two new trends in the agriculture technologies which change the way how farmers manage or operate their business. Information technology and digitalisation are not only change in the industrial sector but also influencing the agriculture and agroindustry. Applications of these technologies range from data management, precision farming, robotics as well as farm management to agro-engineering including sensors network in agriculture.

Experts in Germany reported that the pace of digitalization in agriculture and agroindustry is higher than in the factories. In term of involved human workers on the field, 60 years ago a farm of 100 hectares employ 25 people but today three people can do the same work on the same field size. In term of throughput, what harvested on the same area increased from 300 to 900 tonnes. Therefore, it is not surprising when the result of a poll conducted by the German Farmers Association (DBV) reported that 53% of agricultural business use already digital solution, e.g. cow milking robots, drones mapping fields and sensors measure the required nutrient of plants. 28% of agricultural companies are currently investing specifically in education and training on digital agriculture based on survey on 850 farm managers. The last AgriTechnica exhibition fair in Hanover, one whole hall was dedicated only to the digital agriculture [7].

Modern agriculture not only producing livestock and crops but also energy. *Smart energy* is the terminology used for this purpose where modern farms produce electricity, heat, and even hydrogen energy. Those energy are produced from solar and geothermal as well as biomasses, which are side products of the farms. A modern barn resembles an industrial production facility where instead of pitchfork machines and robots are used. *Livestock 4.0* is the terminology for this kind of barn which includes automatic feed dispensing, networked cattle, smart milking robots, animals' fitness trackers, video surveillance system and automated cleaning system being used in the barn. In order to deal with limited agricultural space in the future, a new approach of farming has been introduced where the crops are vertically planted. This technique is known as *vertical farming* that uses high-tech production of plant and animal products in multi-story buildings which are also known as farm-scrapers.

In digital agriculture, the farmers are required to use information technology infrastructure and have necessary knowledge about it. They have to control and monitor their daily activities on the farm using this technology. These are mostly done by the precision farming machines where heavy machines not only simple machines by an intelligent one. Most of the modern machines have been equipped with autonomous driving capability, where the farmers just sit on the control cockpit and over-watch the overall process. Some machines still require the drivers to drive the machine to move from one lane to another. Such machines which able to perform tasks autonomously are also known as *agricultural robotics*. Moreover, in some cases multiple agricultural robots are working together in swarm with a common goal. These are necessary to further optimize the infield transport logistic chain by minimizing the process time of the machines [8]. Applications of agricultural robots vary from sowing, fertilizing, and harvesting fields.

Further optimization in the agriculture can be achieved using *smart fertilizing* which reduces the cost as well as environmentally friendly. For this approach, it requires special sensors that can measure the soil quality and the amount required fertilizer by the plants, even more in certain situation individual need of a particular plant. The required fertilizer will be dosed exactly and given to the plants or individual plant. Injecting individual fertilizer can be done with the support of agricultural robots. Moreover, support from the air can also found in the modern agriculture, where agricultural drones or UAVs are used to support farming processes. The UAVs deliver not only aerial imagery but equipped with special sensors they can also detect further data such as humus composition, soil moist, or nitrogen level. Some UAVs can even fertilize or sow the field from the air. The rest of the paper is dedicated for the UAV technology including sensors and algorithms for data processing and analyzing.

3. Intelligent Unmanned Aerial Vehicle

Unmanned aerial vehicle (UAV) can also be seen as mobile robot which can move freely in three-dimensional space. It has one more degree of freedom in comparison with mobile robot which move on land, but it introduces additional challenges namely the energy required to maintain certain position on the air. A mobile robot can stop in a fixed position without drawing any energy from its battery and performing other tasks such as analyzing image or detecting object. However, a UAV will continuously draw energy from its battery to maintain its position as well draw additional energy to perform other tasks. Further challenge is on the control dynamic of a UAV which is higher than one in a mobile robot. Due to the gravitational force applied on the UAV and disturbances from wind, the UAV must be able to adjust its control command to avoid any collision. Researches and developments on the UAV technology have develop UAVs that can deal with those challenges and still have reasonable flying time to perform further tasks. Today, such UAVs are available on the market as freely owned product in affordable prices. As a result, many private persons acquiring UAVs for various purposes, e.g. hobby photography. Applications on the agriculture and agroindustry have also been coming as well with various different kind of researches and commercial application in this area. What are the requirements needed in order to make such UAV not just a flying drone but an intelligent tool for agricultural purposes?

The classification of UAVs is done based on different criteria such as the operation mode, the flying range, climb rate mass, and endurance or duration of flight. There is no one standard on the UAVs classification. Looking from the size factor, the UAVs can be classified into very small UAVs (micro or nano UAVs), small UAVs (mini UAVs), medium UAVs, and large UAVs. According to US Department of Defense, classification of UAVs based on their travel range and their endurance resulting in very low-cost close-range UAVs, close-range UAVs, short-range UAVs, mid-range UAVs, and endurance UAVs. US DoD classified the UAVs into five categories as shown in the Table 1. [9]

The tactic group of a non-profit organization ITHACA, Information Technology for Assistance, Cooperation and Action, classified the UAVs in three categories with respect to their possible usage. Each type is subdivided into subcategories based on their features and performance, travel range, maximum climb rate, endurance and weight. Table 2 shows this classification [10].

Table 1. UAVs classification according to the US Department of Defense [9].

[9]Category	Size	Maximum Gross Takeoff Weight (kg)	Normal Operating Altitude (m)	Airspeed (km/h)
Group 1	Small	< 9	< 366 above ground level (AGL)	< 185
Group 2	Medium	9.5 - 25	< 1067 AGL	< 463
Group 3	Large	< 600	< 5486 mean sea level (MSL)	< 463
Group 4	Larger	> 600	< 5486 MSL	any airspeed
Group 5	Largest	> 600	> 5486 MSL	any airspeed

Table 2. ITHACA's UAVs classification [10].

UAV Categories	Acronym	Range (km)	Climb rate (m)	Endurance (hours)	Mass (kg)
Micro	Micro	< 10	250	1	< 5
Mini	Mini	< 10	150 to 300	< 2	150
Close Range	CR	10 - 30	3000	2 to 4	150
Short Range	SR	30 - 70	3000	3 to 6	200
Medium Range	MR	70 - 200	5000	6 to 10	1250
Medium Range Endurance	MRE	> 500	8000	10 to 18	1250
Low Altitude Deep Penetration	LADP	> 250	50 - 9000	0.5 to 1	350
Low Altitude Long Endurance	LALE	> 500	3000	> 24	< 30
Medium Altitude Long Endurance	MALE	> 500	14000	24 to 48	1500

Available UAVs on the market are typically group 1 according to U.S. DoD classification or micro to mini categories according to ITHACA. These kinds of UAVs are designed to carry small payload, which are typically camera system. More professional drones will go toward group 2 and can carry heavier payload, such as fertilizer or herbicide. The higher the group the more restriction will apply to the UAVs, because such drones are able to fly in relatively high altitude and thus must follow the airline traffic regulation or civil aviation department of the corresponding country. These regulations vary from one country to another. Therefore, one must read and follow the valid regulation before operate any UAVs for agricultural or agrobusiness purposes.

3.1. UAV Locomotion

The locomotion of UAVs depends on the underlying flying principles. [11] classified UAVs based on their flying principles. Figure 1 shows this categorization, where first classified by their weight in comparison with air and followed by wing-based or rotor-based locomotion. The lighter-than-air UAVs rely on buoyancy force such as helium gas or heat air to fly. The heavier-than-air UAVs depend on aerodynamic and propulsive thrust to fly. Wing-based UAVs use their wings to generate aerodynamic lift. The fixed-wing UAVs using the same principles like aero planes. Rotor-based UAVs rely on the rotors and propellers pointing upwards to generate propulsive thrust. Wing-based UAVs have longer range and fly faster than rotor-based ones. Therefore, in agriculture application wing-based UAVs can be used for fast sweeping an area to gather some aerial imagery or throwing fertilizer or herbicide over certain area. The rotor-based fly slower than wing-based but can be controlled easily to move in 3D space in 6 degree-of-freedom (DoF).

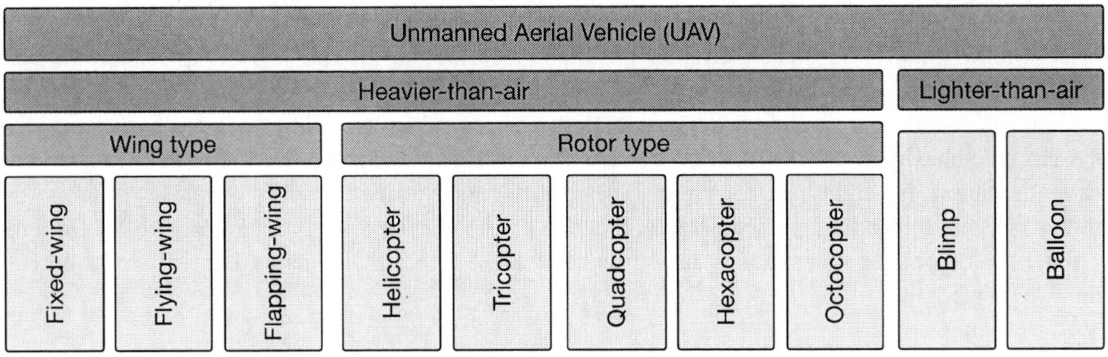

Figure 1. Type of UAVs based on locomotion based on [11].

Figure 2 shows the graph of endurance or flight time versus weight or mass. As shown in the graph, fixed-wing is heavy but still has longer flight time in comparison with rotor-based UAVs. Flapping-wing and lighter-than-air types, such as blimp or balloon have typically smaller mass. Blimp and balloon can fly longer than flapping-wing one. This is due to the required flapping action to produce enough thrust to fly but the weight must remain small. As a result, the flapping actions consume energy but the drone itself cannot carry large battery. The graph shows that the rotor-based UAVs are heavy and have less endurance than fixed-wing, but still these types are quite popular due to the ability of hovering on a certain position, which allow this kind of UAVs to perform additional tasks in addition to take aerial imagery, such as object manipulation, taking image with different time exposures, on-system object detection, etc. This capability offers various potential applications in the agriculture or agrobusiness domain. Therefore, the remaining sections focus on this type of UAVs.

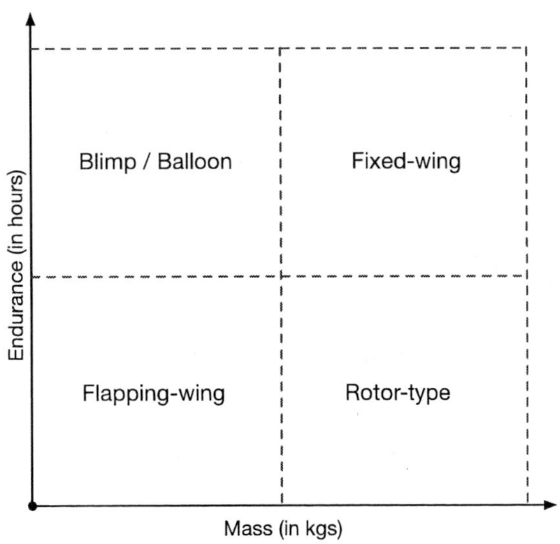

Figure 2. Endurance versus mass based on [11].

3.2. Flight Control System

UAVs are equipped with various sensors and actuators to perform the given tasks. Some of the sensors and actuators are belong to the flight control system, which is responsible for controlling the UAV, and some are used for the telemetry system. Typical sensors for the flight control system are inertial measurement unit (IMU) to estimate the position and orientation of the UAV, global positioning system

(GPS), barometer to estimate the altitude or sometimes sensors for measuring distance to ground, e.g. laser or ultrasound. The system itself must be able to measure the speed of the UAVs' motors or wind speed and direction.

Figure 3 shows an overall guidance, navigation, and control structure implemented in flight control system. The blue boxes are the proprioceptive sensors used for navigating the UAV. The orange boxes show the automatic flight control system. The control signal to the actuators can be override by the ground station. Depending on the operation mode, the UAV can be directly remote controlled from the ground station or a set of waypoints can be sent to the UAV for an automatic flight control following the waypoints.

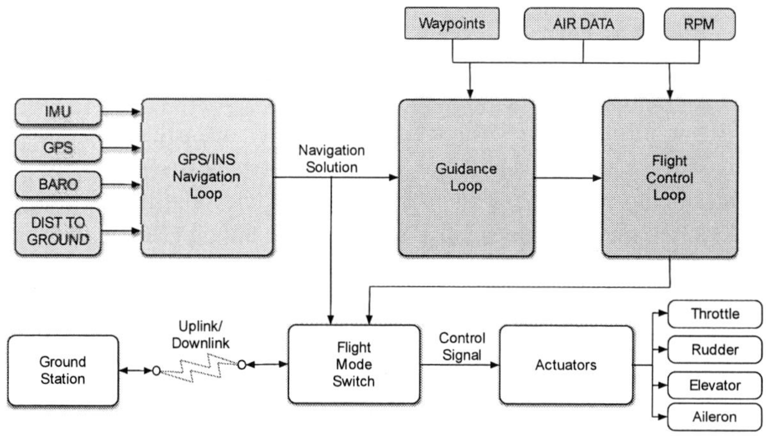

Figure 3. The structure of navigation, guidance and control loop in UAV based on [12].

An automatic flight control does not make the UAV intelligent. Table 3 shows classification of the autonomy levels of UAVs system [13]. As shown in the table, UAVs with automatic flight control can only be categorized into level 2. There are still features and capabilities required by the UAV in order to make it into full autonomous system. Achieving level 10 will enable the UAV to cooperate with other UAVs and perform a collaborative mission.

Table 3. List of autonomy levels [13].

Autonomy Level	Description
1	Remote Control
2	Automatic Flight Control
3	System Fault Adaptive
4	GPS Assisted Navigation
5	Path Planning & Execution
6	Real Time Path Planning
7	Dynamic Mission Planning
8	Real Time Collaborative Mission Planning
9	Swarm Group Decision Making
10	Full Autonomous

3.3. Telemetry Sensors

In addition to the proprioceptive sensors for the flight control, telemetry sensors are mounted on the UAVs as payloads. There are various different kind of sensors for different purposes. The most typical sensor is camera with different lenses and resolutions. Depending on the flying height of the UAV the area covered by one single snapshot can vary. The higher the drone flies the larger the coverage area for one single shot and the smaller the object appear on the image. Therefore, the flying height of the drone must be adapted to the purpose of the measurement.

Special camera such as infrared camera, multi-spectral camera, and hyper-spectral camera can be used to extract more information from the imagery. Hyper-spectral camera works on wider electromagnetic spectrum. Therefore, the image taken from the hyper-spectral camera provides more information which can help finding certain objects by identifying specific material on those objects. Infrared camera works on the small band of the spectrum namely the infrared spectrum. This camera can be used to detect living animal on the field to help field harvesting process by detecting wild animal on the field. Multi- and hyper-spectral cameras can give much more information than normal or infrared camera but the post-processing algorithms on the imagery taken from those cameras are much more complex than the one from infrared camera.

Small light detection and ranging or LiDAR can be mounted on the UAV for telemetry purpose as well. LiDAR usually uses for reconstruction of 3D imaging from either multiple 2-dimensional or 3-dimensional scan of an area. The 3D imaging provides the structure of the area, such as the height of the plants and their forms. A combination of LiDAR with camera or hyper-spectral camera can provide rich information of the scanned area for various kind of purposes. One major challenge is when the overall weight of all telemetry sensors carried by the drone. Each additional payload which a drone still can carry will reduce the flight time or the endurance.

3.4. Data Processing Algorithms

A large area is typically assembled from a number of images captured from overlapping snapshots of the sensors. Those snapshots are then combined using image stitching algorithms to create one single image forming a larger area. Before combining those images, the scaling factor resulting from snapshot from different height must be first corrected. The corrected images are known as orthophoto, orthophotograph or orthoimage. The final image is known as orthomosaic which comprises from individual orthoimages.

Further algorithms are required to process the orthomosaic image to extract various different features from it. There are numerous different image processing algorithms starting from simple color segmentation to trained deep learning algorithm for detecting specific species of plant on the image. These algorithms require typically CPU or GPU to process a large area to produce meaningful result. Once the results are generated, it can be combined with other data such as satellite imagery in order to deduce further knowledge upon the area.

Due to these two steps before one can get results from the image taken from an UAV, a real-time or online data processing is difficult to achieve using small or mini UAVs. Equipped with GPU accelerated processor on the UAV might enable the system to perform onboard orthomosaic while flying but further extraction of the image will still be challenging and might consume adequate energy which reduces the endurance of the UAVs. This is also the reason why specific purpose UAV uses specialized sensors instead of rich feature ones which can deliver the target object immediately from the sensor without or with less post processing, e.g. infrared camera for detecting wild animal.

4. Example Applications

In this section, two applications using UAVs in agriculture and agroindustry are presented. The first application is for counting the number of trees in a plantation from the aerial imagery. The second one is toward the agroindustry by evaluating the development process of a village, which can be used for further decision making and optimizing the agroindustry business.

IOP Conf. Series: Earth and Environmental Science **335** (2019) 012001 doi:10.1088/1755-1315/335/1/012001

4.1. Automated Tree Counting in Palm Plantations

The number of trees in a plantation correlated directly with the expected outcome of the final product. Having the number of trees, one can predict the outcome by averaging the number of fruits could be produced from each tree. This information can be retrieved from the previous harvesting data or rule of thumb from the expert person. Typically, this can be done quite easily while planting the trees the farmers can count the number of trees being planted. However, the interesting task is to determine how many of these trees are survive and finally grown up. Some of the trees might be damage due to various different factors. Therefore, the farmers must regularly maintain their data by counting the number of trees in their plantation. This is one activity which can be fully supported by using UAVs technology to automatically extracting this number from the aerial imagery [14-16].

Figure 4. shows the result of applying the automated tree counting algorithm on aerial imagery of a palm plantation. On the left image, it shows the colour separated image using green channel to focus on the leaves of the trees and removing any background information from the image [17]. The next step is to segment the individual palm trees. This can be done quite straight forward because the palm plantation has good structure which allow clear separation between trees. The right image shows the original image overlaid with the counted trees. From the image, it can be seen that the algorithm work quite well on the palm plantation for counting the number of palm trees.

Figure 4. Automated palm-tree counting from the aerial imagery (source: PSP3-IPB).

In some cases, the fruit can be distinguished quite easily form the tree because they have special form or colour that can be separated from the leaves. In this case, an automated process can be developed further in order to count the number of fruits on a tree. Of course, for this purpose the image has to be taken from shorter distance than the one for counting the number of trees. The drone has to be able to fly into the object of interest on the certain distance and take some image from that position. This action must not be done to every single tree but it should be enough using adequate statistical approach for sampling a certain area on the plantation. One major benefit of having this approach is that more precise approximation can be done directly from the observed plantation instead from previous year data or expert knowledge.

Moreover, not only the number of trees on the plantation is interesting for agriculture or agroindustry, there are various different parameters or indices which could be extracted from the aerial imagery. One of the aspects which also part of the research project is to determine the sustainability index of such plantation form those images. This of course presents further challenges as one of the goals is to determine the number of different species living on a palm plantation. For this purpose, the drone must fly underneath the palm canopy to take images below it. Furthermore, an human expert is needed in the beginning of the process to analyze the image and create a ground truth data from those images. This data will be used later to measure how good the algorithm can determine the sustainability index of the plantation.

4.2. Village Drone Project

In 2017, Bogor Agricultural University (IPB) team was able to extract data which can be used to manage village data more accurately by using UAV within the village drone project, or so-called "drone desa". Drones were used to capture aerial imagery of a village which will be used later for calculating village data. The research activities have produced aerial photography, village orthophoto maps, village thematic maps, and other spatial information, such as: population, education and settlement. The spatial information of the village is used for building an artificial intelligence system to support the preparation of the Village Medium Term Development Plan (MTDP), or so-called "Rencana Pembangunan Jangka Menengah (RPJM) Desa", and the Village Development Activity Plan (DAP), or so-called "Rencana Kegiatan Pembangunan (RKP) Desa". This system was needed because the past decisions to form Village RPJM or Village RKP which led to village development programs and activities did not correspond with the actual conditions and needs of the village. Therefore, one of the objectives of this research project is to collect and to update the outdated village data with the UAV to create a spatial information of the village. Furthermore, variables required to create accurate Village RPJM and Village RKP will be further investigated and developed. The main objective of this research is to develop an intelligent software framework to generate required information which is used in the preparation of the Village RPJM and the Village RKP. The framework shall use the spatial data generated from the aerial imagery captured from the drones. The resulting Village RPJM and Village RKP will then be used by the local government together with the village community as based for their decision making to improve the living quality of the community.

In order to achieve these objectives, the research study is supported by quantitative and qualitative methods. The quantitative method is done based on the questionnaire and surveys. The qualitative methods approach is performed using participatory data mining, discussions, group discussions focused on social media groups in the targeted locations. These are supported by proper dissemination of intelligence system platforms for the preparation of the Village RPJM and Village RKP to the village local community.

Qualitative and quantitative data is not enough to create an optimal result for the Village RPJM and Village RKP. Further information containing deepen knowledge of the population is required, for example people's welfare, production factors, educational level, or market demands. This information is crucial for producing more accurate data for both Village RPJM and Village RKP. Moreover, the developed system is supported by expert's decision making in the field by providing sensitive variables and strategic information used for preparing both documents.

The spatial data of the village, i.e., village area, land use, population, education, and settlement, obtained from the previous year's research study will be compared with the recent year data. This information is required in order to justify how good is the development in various different area in the previous years. The results are then fed into the developed system to predict the potential of the village. Figure 5 shows an example of fuzzy rule set designed for the Village RPJM and the Village RKP.

5. Challenges and Further Development

Although several applications using UAVs for the agriculture and agroindustry have been slowly being deployed on the market, many recent developments are still available on the research and development only. There are still many open issues which have to be further investigated in order to achieve a fully autonomous system which can collaborate with other UAVs performing a common mission. This level needs not only hardware development but software as well.

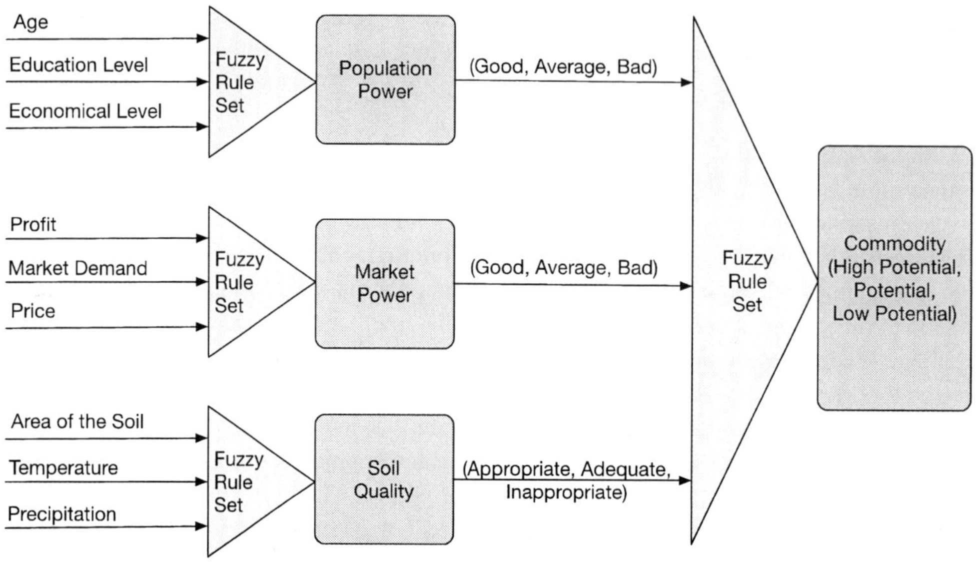

Figure 5. Fuzzy Rule-Set used in for the Village RPJM and the Village RKP.

In the aerospace system, NASA has introduced the readiness level, so called technology readiness level (TRL). This terminology has been adopted by many disciplines to define how mature the technology is. Table 4 shows the list of technology readiness levels for UAVs [13]. As one can see from the list, as long the system only passes laboratory test, it reached only level 5. A prototype system that runs on real mission can be found by a number of applications. These are the positive signs toward a fully operational system on the market.

Table 4. List of technology readiness levels (TRL) [13].

Level	Description
1	Basic Principles
2	Application Formation
3	Technology Concepts and Research
4	Technology Development and Proof of Concept
5	Low Fidelity Component Testing (Labs)
6	System Integration and Flight Tests
7	Prototype Demonstration and Operation
8	Prototype Operation in Realistic Mission Scenario
9	UAV Mission Deployment
10	Fully Operational Status

Technology readiness level is only one aspect of the hardware and software system. There are still many challenges which still have to be discussed or investigated further. One important aspect which challenging is the data ownership. As one might know that the UAVs are gathering aerial telemetry information, which can be used further to improve or make profit on a certain the industry. Who owns the information? Do the farmers have to pay for that information? How to protect the information from the third party who might create negative impact to the industry?

6. Conclusion

This paper has shown a brief review of the unmanned aerial vehicles for the agriculture and agroindustry. Together with other technological approaches such as smart farming, UAVs bring additional benefit which can complement those systems. UAVs can gather necessary data which is really crucial for the agriculture to further increase the productivity. It can help farmers predict the outcome of a certain field by gathering aerial imagery from a plantation. Having certain algorithms which can count the number of the tree on the plantation will give great benefit to the farmers. Moreover, depending on the extractable features the algorithms might also tell how many fruits can be expected from those trees.

Other example shows that UAVs can help the politician to make a wise decision. In the village drone project, the system is able to extract important information like infrastructure, number of houses, roads, etc. Further information such as educational level or ages of the population can be combined with data from the UAVs to create overview of the village. This information will then tell how or what could improve the village, which infrastructure has to be built in the next period. Do the roads good enough to transporting all goods from the sources to the silos?

The drones are still need to be further developed in order to achieve the full autonomy level and reach the highest technological readiness level. Researches and application of UAVs technology in the agricultural or agroindustry will help further development on the drones toward this goal. The drones technology is one important technological among others to improve the overall throughput in the agroindustry.

References

[1] United Nations, Department of Economic and Social Affairs, Population Division 2015 Population 2030: Demographic challenges and opportunities for sustainable development planning. ed Department of Economic and Social Affairs and P Division: (ST/ESA/SER.A/389))

[2] Bruinsma J 2009 The resource outlook to 2050: by how much do land, water, and crop yields need to increase by 2050? ed FAO: FAO: FAO Expert Meeting on 'How to feed the world in 2050')

[3] European Commission 2017 Industry 4.0 in agriculture: Focus on IoT aspects. ed D T Monitor

[4] Bayer M 2018 Digital Farming: Was Sie über Landwirtschaft 4.0 wissen müssen. In: *Computer Woche,* (Munich: IDG Business Media GmbH)

[5] Maru A, Berne D, Beer J D, Ballantyne P, Pesce V, Kalyesubula S, Fourie N, Addison C, Collett A and Chaves J 2018 Digital and data-driven agriculture: Harnessing the power of data for smallholders. (https://cgspace.cgiar.org/handle/10568/92477: Global Forum on Agricultural Research and Innovation)

[6] FAO-ITU 2018 E-Agriculture in Action: Drones for Agriculture. In: *The E-Agriculture in Action,* (http://www.fao.org/3/I8494EN/i8494en.pdf: Food and Agriculture Organization of the United Nations and the International Telecommunication Union)

[7] Ronzheimer M 2018 Digitale Landwirtschaft — Im Märzen der Bauer das IoT Anstellt. In: *VDE Dialog — Smart Farming Die Zukunft der Landwirtschaft,* pp 12-5

[8] Scheuren S, Stiene S, Hartanto R, Hertzberg J and Reinecke M 2013 Spatio-Temporally Constrained Planning for Cooperative Vehicles in a Harvesting Scenario *KI - Künstliche Intelligenz* **27** 341-6

[9] ARMY, Unmanned Aircraft Systems Center of Excellence Fort Rucker AL 2010 U.S. Army Unmanned Aircraft Systems Roadmap 2010-2035: Eyes of the Army.

[10] Bendea H, Boccardo P, Dequal S, Tonolo G, Marenchino D and Piras M 2008 Low Cost UAV for Post-Disaster Assessment *The International Archives of the Photogrammetry, Remote Sensing and Spatial Information Sciences* **XXXVII** 1373-9

[11] Liew F, DeLatte D, Takeishi N and Yairi T 2017 Recent Developments in Aerial Robotics: A Survey and Prototypes Overview

[12] Kim J-H, Sukkarieh S and Wishart S 2006 *Field and Service Robotics: Recent Advances in Reserch and Applications,* ed S i Yuta, *et al.* (Berlin, Heidelberg: Springer Berlin Heidelberg) pp 299-309

[13] Vidyadharan A, Philpott R, Kwasa J B and Bloebaum L C 2017 Analysis of Autonomous Unmanned Aerial Systems Based on Operational Scenarios Using Value Modelling *Drones* **1**

[14] Yang L, Wu X, Praun E and Ma X 2009 Tree detection from aerial imagery. In: *Proceedings of the 17th ACM SIGSPATIAL International Conference on Advances in Geographic Information Systems,* (Seattle, Washington: ACM) pp 131-7
[15] Zakharova M 2017 Automated Coconut Tree Detection In Aerial Imagery Using Deep Learning. In: *Faculty of Engineering Technology,* (Leuven: KU Leuven)
[16] Firdaus, Arkeman Y, Buono A and Hermadi I 2017 Satellite image processing for precision agriculture and agroindustry using convolutional neural network and genetic algorithm *IOP Conference Series: Earth and Environmental Science* **54** 012102
[17] Srestasathiern P and Rakwatin P 2014 Oil Palm Tree Detection with High Resolution Multi-Spectral Satellite Imagery *Remote Sensing* **6** 9749

IOP Conf. Series: Earth and Environmental Science **335** (2019) 012002 doi:10.1088/1755-1315/335/1/012002

Genetic Algorithms Optimization for water management in irrigated paddy fields

C Arif[1,4], B I Setiawan[1], M Mizoguchi[2] and B D A Nugroho[3]

[1] Department of Civil and Environmental Engineering, Bogor Agricultural University, Indonesia
e-mail: chusnul_arif@apps.ipb.ac.id
[2] Department of Global Agricultural Sciences, The University of Tokyo, Japan
[3] Department of Agricultural and Biosystem Engineering, Gadjah Mada University, Indonesia

Abstract. Water management in paddy fields is main key to produce more yield and mitigate greenhouse gas (GHG) emissions at the same time. Commonly, Indonesian farmers apply continuous flooding irrigation to combat weed growth and gain maximum yield as well. However, this method is less efficient in water use and releases more GHG emissions as represented by Global Warming Potential (GWP) value. The System of Rice Intensification (SRI), as alternative rice farming, applies intermittent irrigation that has possibility produce more yields with minimum GHG emissions. However, optimum water management for these purposes is not clear yet. The objective of this study was to search optimal soil moisture and water depth in each growth stage using genetic algorithms (GA) model for SRI rice farming. GA model was developed based on one rice season experiment that was conducted during January to May 2018 with three water management regimes, i.e., flooded (FL), moderate (MD) and dry (DR) regimes, respectively. Based on the experiment, MD regime produced highest yield by 5.26% and 10.89% higher than those FL and DR regimes, respectively. So this was the best regime among others. However, this regime release more GHG emissions than that DR regime in which its GWP value was 87.85% higher than that DR regime. So, the GA model was used to find the better regime than that MD regime. Based on those empirical data, GA model found optimal soil moisture and water depth in four growth stages. Based on GA optimal scenario, the yield can be increased up to 1.31% higher than that MD regime and GWP can be reduced up to 8.62% lower than that MD regime. More field experiments are needed to validate the model under various climate conditions.

Keywords: genetic algorithms, water management, irrigated paddy fields

1. Introduction

In the tropical paddy fields such as in Indonesia, most farmers used more water irrigation to cultivate rice by keeping water depth 2-5 cm above soil surface continuously until the late season growth stage. This anaerobic condition will promote more activities of methanogens bacteria in reducing CO2 and produce more methane emission during soil organic matter decomposition [1]. Thus, paddy field with flooded water is known as major source of methane emission. Methane is one of greenhouse gas that has contribution in global warming of 25 times higher than that carbon dioxide (CO2) [2]. In addition, flooded water is less efficient in water use since more water is loss through deep percolation, seepage and runoff [3].

The two main reasons adopted flooded waters are to combat weed growth and avoid lack water when water resource is limited during cultivation period. Also, sometime it can reduce nitrous oxide (N2O) emission during nitrification and denitrification process in the soil [4] or even absorb nitrous oxide [5].

Nitrous oxide is also one of greenhouse gas that has paying attention since it effect to global warming is 298 times higher than that carbon dioxide [2] However, flooded water continuously is not always associated with higher yield or even reduce yield [6]. The possibility of reducing yield in flooded water is less tiller numbers developments in vegetative stage and then following by less panicles development in the next stage [7].

The System of Rice Intensification (SRI) is wished as alternative rice cultivation with possibility produces more rice by less water and low greenhouse gas emission in total that is represented by global warming potential (GWP) value [8]. The SRI has at least six basic components that basically different with ordinary rice farming in Indonesia. The components are application of young seed while 2-3 leaf developed during 7-14 days, single transplanted seed in one hill, wider spacing between the hill, intermittent irrigation application, weeding regularly and compost/organic fertilizers application as much as possible to enhance soil organic matter [9]. Regarding application of intermittent irrigation, saturated soil or thin water layer depth is applied during particular time and then water is drained in others particular time. Following Arif et al. 2014 [10], it has been recommended that optimal water management for SRI is by maintained saturated soil (wet condition) during initial and vegetative/crop development growth stages, and then water level can be reduced by kept soil moisture at medium level in mid-season growth stage and finally drained water in late season growth stage to conditioned soil moisture at dry level. This scenario can increase yield 4.4% and save water input up to 12.28% compared to continuously saturated soil condition water regime.

However, that optimal water management didn't consider greenhouse gas emission effect. Thus, optimization water management in paddy fields by considering greenhouse gas emission is important especially to minimize GWP from paddy fields. Sometimes in the developing optimization model for irrigation application, there are some constraints in integrating many aspects such as weather conditions, plant growth condition and soil condition as well [11]. Here, Genetic Algorithms (GA) model can be proposed as alternative optimization method since it can search optimum global value by searching the entire population instead of moving from one point to the next as the traditional methods [12].

The main objective of this research is to find optimal irrigation water for the SRI that is represented by optimal soil moisture and water depth in four growth stages, i.e., initial, crop development, mid-season and late season stages to maximize yield and minimize GWP value. In addition, land productivity and water use efficiency are also considered in the developed model.

2. Materials and methods

2.1. Field experiments and water regime applications

Achieving the objectives, one rice planting season was conducted during 20 Jan to 13 May 2018 in wet-dry season in the field laboratory of Dept. of Civil and Environmental Engineering, West Java Indonesia. Here, rice (Oriza sativa L) variety of Pertiwi by SRI components are single transplanting, young seed (14 days after sowing), single planting in one hill with 30 cm x 30 cm interval were used. However, chemical fertilizers with doses of 275 kg/ha for urea, 100 kg for SP-36 and 150 kg/ha for KCl, respectively were used instead of organic fertilizers. Soil properties both chemical and physical properties can be shown in Table 1. Crop growth stages were divided into four stages, i.e., initial (0-27 days after transplanting (DAT)), crop development (28-72 DAT), mid-season (73-96 DAT) and late season stages (97-112 DAT) based on FAO's standard [13].

The list of authors should be indented 25 mm to match the abstract. The style for the names is initials then surname, with a comma after all but the last two names, which are separated by 'and'. Initials should not have full stops—for example **A J Smith** and *not* **A. J. Smith**. First names in full may be used if desired. If an author has additional information to appear as a footnote, such as a permanent address or to indicate that they are the corresponding author, the footnote should be entered after the surname.

Table 1. Soil properties of field location.

Soil properties	Parameters	Value

Chemical	pH (H_2O)	7.5
	C (%)	3.84
	N (%)	0.2
	C/N	19
Physical	Soil texture:	
	Sand (%)	23
	Silt (%)	34
	Clay (%)	43
	Permeability (cm/h)	8.17
	Bulk Density (g/cm^3)	1.98

For water management, there were three regimes, i.e., flooded (FL), moderate (MD), and dry (DR) regimes. We prepared 2 m x 2 m plot for each water regime. In FL regime, water depth 2-5 cm above soil surface was expected in kept during first to third growth stages, and then water was drained in the last stage. MD regime was set as moderate application in which water depth at nearly soil surface (0 cm water depth) was expected during first to third growth stages, and then water was drained in the last stage. The last regime, DR regime, was set as extreme water application with water stress condition in which water depth at nearly soil surface was expected only during 0-20 days after transplanting (DAT), after root developed well then water depth was set at 5 cm below soil surface until the end of third growth stage, and then in the last stage water was drained as the same two regimes before.

2.2. *Field Measurement Parameters*

To perform optimization model, some parameters in the fields consisted of plant growth performance, meteorological and soil parameters were measured as well as greenhouse gas emissions using some instrumentations. For growth performance, plant height, tiller and panicle numbers were measured manually every 3 days in which 5 hill samples were selected in each plot. For meteorological parameters, air temperature, relative humidity, precipitation, solar radiation, wind speed and direction were measured using automatic weather station (AWS) equipped with particular sensors with 30 minutes interval. The meteorological were used to determine crop evapotranspiration and water analysis.

Soil parameters were measured using specific sensors consisted of 5-TE sensor (Decagon Device Ltd) to measure soil moisture, soil temperature and soil Electrical Conductivity (EC) with 30 minutes interval, pHmeter (pH 3310 SET 2 incl. SenTix® 41) to measure soil pH and ORP meter (WTW Sentix) to measure soil redox potential (Eh) once a week. In particular soil Eh measurement, the output of ORP meter should be converted first by adding the redox value according to a SHE (Standard Hydrogen Electrode) reference electrode on the redox value measured by the instrument. Soil Eh values range from -300 mV to 700 mV, in which its value lower than 300 mV shows an anaerobic condition [14].

Greenhouse gas emissions were measured using a static closed chamber method [15] with the modification. This method was widely applied in greenhouse gas emission studies such as Kudo et al. 2014 [16]. The chamber was rectangular acrylic with size was 30 cm length x 30 cm width x 120 cm height equipped with mini fan to circulate air during measurement (Figure 1). The gas samples were collected at noon time using syringe four times at 0, 10, 20 and 30 minutes after the chamber placed over the single plant. Gas sample in the syringe was immediately transferred into 200-ml tedlar bag, and then collect 10-ml gas sample from that tedlar bag and transferred into 10-ml vial bottle. Each gas sample in the vial bottle then was analyzed using a gas chromatograph (Micro GC CP 4900) with flame ionization detector (FID) in the lab. From four samples, four different concentration of greenhouse gas were collected and then the flux was calculated by the following equation:

$$E = \frac{\delta C}{\delta t} \times \frac{V_{ch}}{A_{ch}} \times \frac{mW}{mV} \times \frac{273.2}{273.2+T'} \quad (1)$$

Where E is the flux of each gas (in mg/m2/min), δC/δt is slope of four different concentration during sampling time (in ppm/min), Vch is volume of chamber (in m3), Ach is area of chamber (in m2), mW is mass of molecule (in g), mV is volume of molecule (in liter), and T is temperature inside chamber during measurement (in °C).

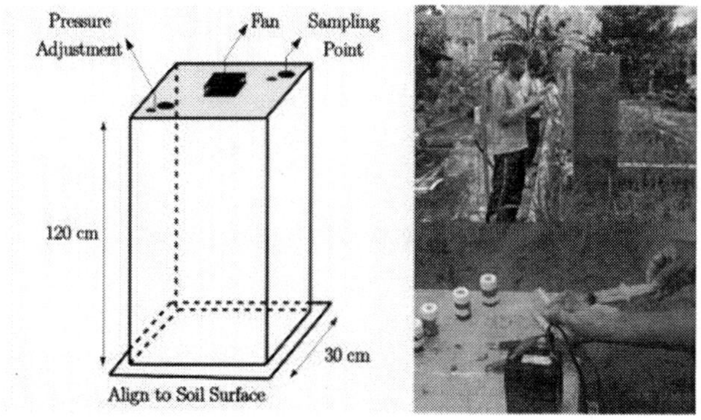

Figure 1. Chamber box schema for greenhouse gas measurement in the fields.

2.3. Optimization model development

Genetic algorithms (GA) model was developed with maximizing the following objective function:

$$F(Y, GWP) = aY - bGWP \tag{2}$$

Where Y is yield (ton/ha) and GWP is global warming potential, a and b are coefficient of those two parameters given by 0.65 and 0.35, respectively. GWP was calculated based on CH_4 and N_2O emissions total in one season with the following equation:

$$GWP = 25 \times CH_4 + 298 \times N_2O \tag{3}$$

The constraints of the objective function were given as follow:

$$WL_{min} \leq WL_1, WL_2, WL_3, WL_4 \leq WL_{max} \tag{4}$$

$$SM_{min} \leq SM_1, SM_2, SM_3, SM_4 \leq SM_{max} \tag{5}$$

Where WL_{min} and WL_{max} are minimum and maximum water depth, SM_{min} and SM_{max} are minimum and maximum soil moisture based on empirical data in each growth stage, WL_1, WL_2, WL_3, WL_4 are optimum water depth in the initial, crop development, mid-season and late season growth stages, respectively and SM_1, SM_3, SM_3, SM_4 are for optimum soil moisture in those consecutively growth stages.

To search optimal parameters, chromosome structure of the GA model was represented by an individual that was coded as a set of six-bit binary strings with the fitness function was the same as the objective function (Equation 2). The operators of the model consisted of crossover and mutation in which the rates were 60% and 5% for crossover and mutation, respectively.

3. Results and Discussions

3.1. Effect of water regimes on plant growth performances

Figure 2 shows effects of water regimes on the plant growth performances in term of plant height, tillers and panicles numbers. All those parameters were comparable among the regimes and not significant different. FL and MD regimes have higher plant height than that DR regime. However, higher plant height in FL regime was not associated to more tillers and panicles numbers. Comparing to MD regime, tillers and panicles numbers in FL regime was lower. It was indicated that MD regime has the best growth performance among others. In the harvesting day, total numbers of tillers and panicles in this regime were 25 and 24, respectively. Meanwhile, total numbers of tillers for FL and DR regime were 22 and 23, respectively, while total numbers of panicles in both regimes were same numbers, 22. The best performance in MD regime showed that non-flooded water promoted more tiller development particularly in the vegetative stage.

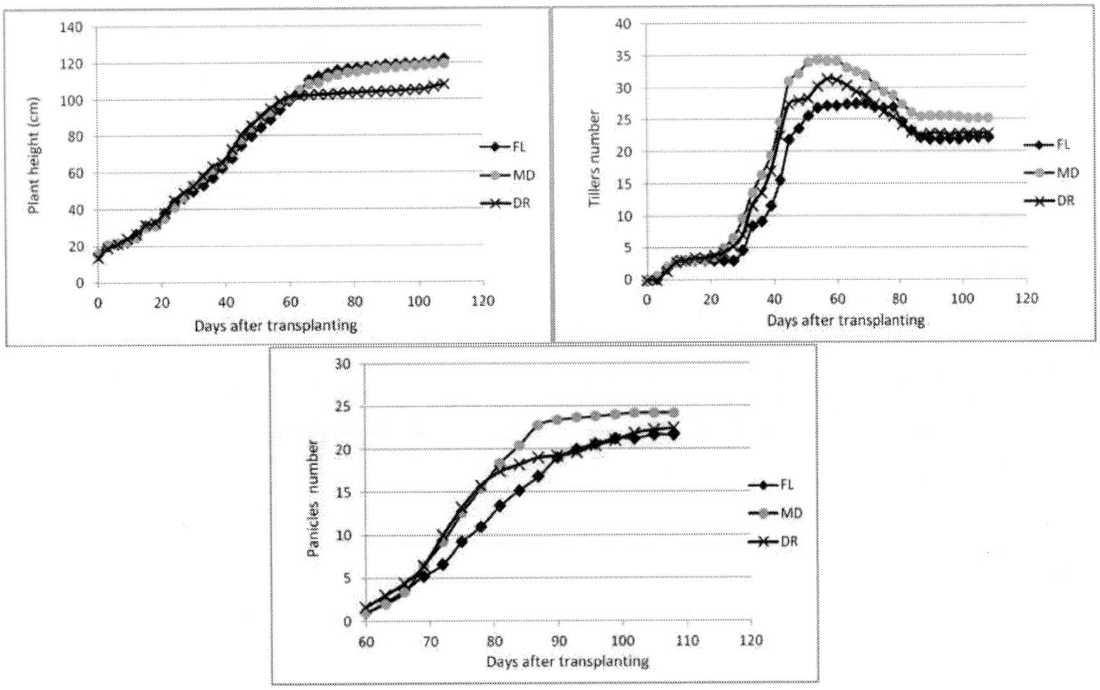

Figure 2. Plant growth performances under different water regimes.

3.2. Actual water depths and soil moisture condition

The weekly actual water depth of each regime is shown in Figure 3. The FL regime was intended to be flooded continuously, but the water depth was fluctuated and it has highest water depth in which water flooded was occurred during vegetative growth stage particularly from 2-10 weeks after transplanting. It is indicated that water loss through evapotranspiration and percolation higher than that its irrigation rate. On the other hand, the actual water depth in both MD and DR regimes were lower than soil surface during planting period. The highest water depths in both regimes were during vegetative growth stages in 1-4 weeks after transplanting. Although weekly average water depths were lower than soil surface, MD regime has the best growth performance as previously presented. It was probably caused by more oxygen availability under non-flooded, thus enhanced more tiller development. This reason is dealing with the previous finding that aerobic condition promoted more activities particularly root activity [17] and even for shoot activity under intermittent irrigation [18].

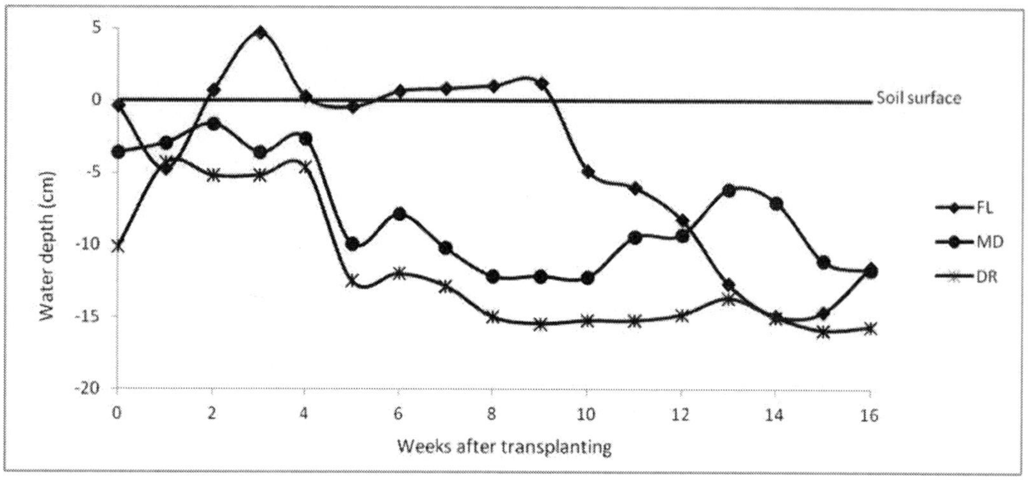

Figure 3. Actual water depth in each water regime

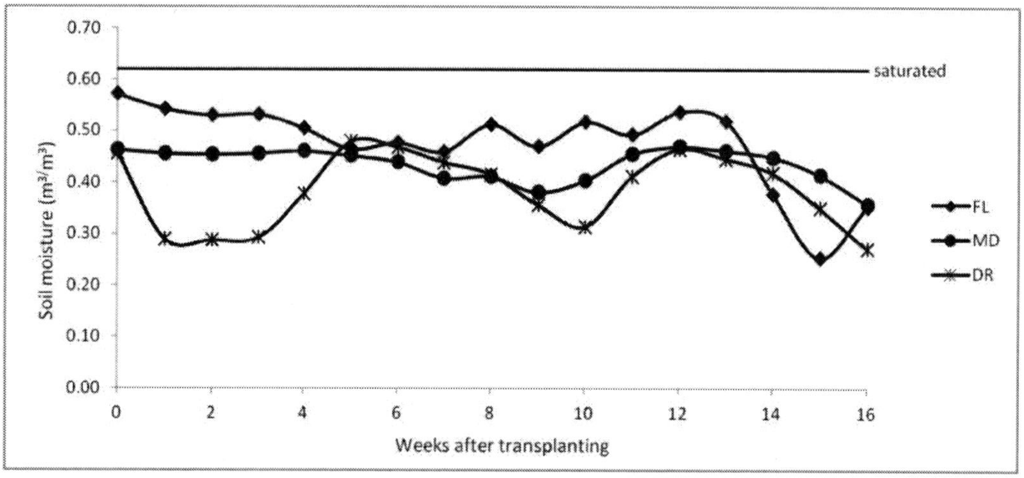

Figure 4. Actual soil moisture in each water regime.

The higher water depth in the FL regime was corresponded to higher soil moisture as shown in Figure 4. The soil moisture of FL regime was higher than that two others regimes almost in planting period except in three weeks before harvesting. It was represented that the FL regime was supplied more irrigation than that two other regimes. The average soil moisture in vegetative (initial and crop development) growth stage was 0.51 m^3/m^3, 0.44 m^3/m^3, 0.38 m^3/m^3 for the FL, MD and DR regimes, respectively. Meanwhile, in the generative (mid-season and late season) growth stages, the average soil moisture was respectively 0.41 m^3/m^3, 0.43 m^3/m^3, 0.39 m^3/m^3 for the FL, MD and DR regimes.

3.3. Effect water regimes on greenhouse gas emissions

Figure 5 shows seasonal CH_4 flux during planting period under different regimes. The peak of CH_4 emissions occurred during 3 – 9 weeks after transplanting particularly in the FL and MD regimes. On the other hand, CH_4 emission in the DR regime was lower than those the FL and MD regimes almost during planting season except in 2 weeks after transplanting. The FL regime released highest CH_4 emission in total as represented by more CH_4 emission during 3 – 9 weeks after transplanting. The average of CH_4 emission during that time was 353, 316 and -0.73 $mg/m^2/d$ for the FL, MD and DR regimes, respectively. More CH_4 released in the FL regime during that time when water depth and soil moisture higher than those others regimes (Figures 3- 4). It was indicated that more water in the field enabled methanogens activities and promoted CH_4 gas development during decomposing process. Also,

CH_4 emission is released more when the oxygen limited in the field [19] This finding dealing with many previous recent studies related with water management and CH_4 emission [20, 21].

On the other hand, after 10 weeks after transplanting less CH_4 emitted in all regimes or even their values were negative. It was occurred when the water was started drained particularly in the FL regime as indicated lower water depth and soil moisture (Figures 3-4). During this time CH_4 emission can be significantly decreased in which the average CH_4 emission was 16, 7 and -7 mg/m^2/d respectively for the FL, MD and DR regimes. It was indicated that under minimum water availability in the field, oxidative process restrained methanogen bacteria and enhanced metanotrof bacteria that can significantly reduce CH_4 gas development.

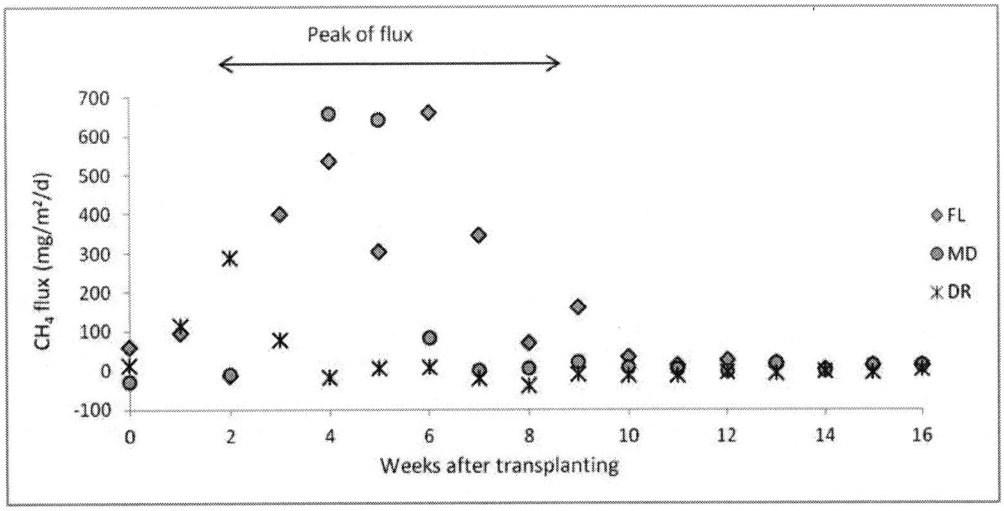

Figure 5. Weekly CH_4 flux in each water regime.

Trade off trend was shown in N2O flux as figured out in Figure 6. The DR regime released more N2O emission since this regime was driest among the regimes. The peak of N2O emission occurred 5-7 weeks after transplanting in this regime with the average was 28 mg/m2/d. This happened with the average water depth was -12 cm. For MD regime, the peak of N2O emission was in 5 weeks after transplanting when water depth was dropped from -2.63 cm to be – 9.93 cm. Meanwhile, in the FL regime, N2O emission was limited. The emission only occurred in the end of generative stage when water in the field was limited. N2O is commonly developed in aerobic soil condition when water is limited and oxygen concentration is so high. This condition triggered nitrification process that convert NO became N2O at higher pH [22].

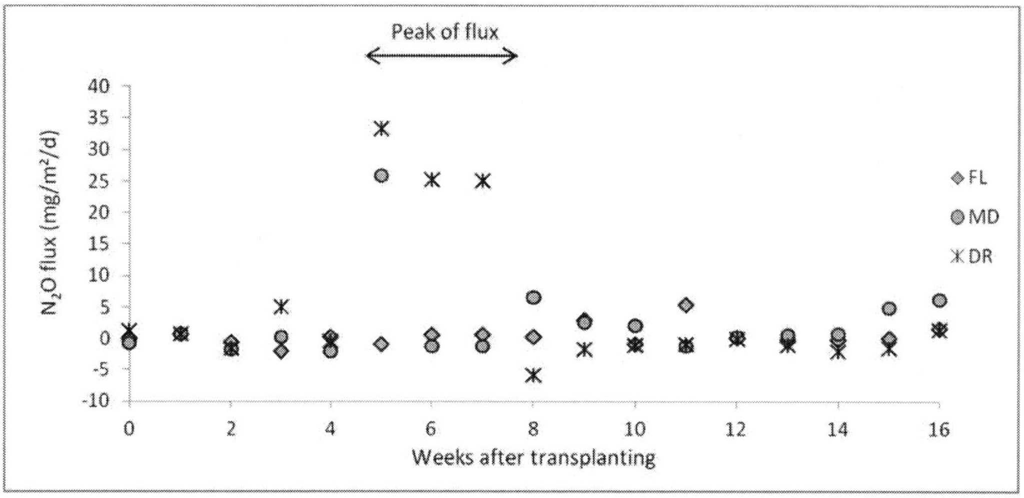

Figure 6. Weekly N_2O flux in each water regime.

3.4. Optimum water regimes

Table 2 shows the comparison between the soil moisture in each regime and optimum values generated by the GA model. The FL regime with highest soil moisture in three first growth stages produced highest CH4 emission and lowest N2O emission in total. On the other hand, the DR regime as water stress option released lowest CH4 and highest N2O emissions. As moderated option, the MD regime was proposed and it released CH4 and N2O at moderate level. Interestingly, the MD regime produced highest yield of 6.85 ton/ha. It was 5.26% and 10.89% higher than that the FL and DR regimes, respectively. However, this regime release more GHG emissions than that DR regime in which its GWP value was 87.85% higher than that DR regime. Among three empirical regimes, MD regime was the best in the point of view highest yield with moderate GWP.

Table 2. Scenario of optimum soil moisture by the GA model.

	Water Regimes			
Parameters	FL	MD	DR	Optimum
Soil Moisture in Initial stage (m^3/m^3)	0.544	0.457	0.332	0.541
Soil Moisture in Crop development (m^3/m^3)	0.483	0.422	0.411	0.331
Soil Moisture in Mid-season (m^3/m^3)	0.518	0.456	0.427	0.402
Soil Moisture in Late season (m^3/m^3)	0.331	0.429	0.429	0.429
Yield (ton/ha)	6.51	6.85	6.18	6.9
N_2O (kg/ha)	0.55	2.88	5.56	4.55
CH_4 (kg/ha)	191.31	143.86	24.44	99.26
GWP (CO_2 eq kg/ha)	4564	4166	2218	3838

Considering the empirical results, the GA model was running out and optimum soil moisture was generated properly. The main objective this model was to find better regime than that MD regime with the main subject is yield and GWP. The optimum soil moisture in the initial, crop development, mid-season and late season stages were 0.541, 0.331, 0.402, 0.429 m^3/m^3, respectively. This scenario can increased 0.73% productivity of the MD regime and also reduced 7.87% GWP of the MD regime.

Although increasing productivity insignificantly, however, the optimum soil moisture reduced GWP significantly.

Table 3. Scenario of optimum water depth by the GA model.

	Water Regimes			
Parameters	FL	MD	DR	Optimum
Water depth in Initial stage (cm) *	0.375	-2.729	-5.027	-5.027
Water depth in Crop development (cm)*	-0.802	-10.842	-13.922	-13.922
Water depth in Mid-season (cm)*	-10.408	-7.447	-14.498	-10.916
Water depth in Late season (cm)*	-13.359	-11.127	-15.746	-11.127
Yield (ton/ha)	6.51	6.85	6.18	6.94
N_2O (kg/ha)	0.55	2.88	5.56	4.86
CH_4 (kg/ha)	191.31	143.86	24.44	94.36
GWP (CO_2 eq kg/ha)	4564	4166	2218	3838

*minus values indicated below soil surface water depth

For the second scenario, the GA model was used to optimize water depth in the each growth stage with the results presented in Table 3. Based on the scenario, optimal water depth was -5.03, -3.9, -10.9 and -11.12 cm for initial, crop development, mid-season, late season growth stages, respectively. This scenario can increase the yield by 1.31% and reduce GWP by 8.36% compare to MD regime.

4. Conclusions

The developed genetic algorithms (GA) model was obtained the optimal soil moisture and water depth in each growth stage. Based on the GA model scenarios, the productivity can be increased up to 1.31% and reduced global warming potential (GWP) by up to 8.36% compare to the medium (MD) regime as the best empirical regimes. For its application in the fields, it is recommended to integrate with soil moisture and water depth control systems. However, since the model was running out based on one cropping season, thus more field experiments are needed to validate the model under various climate conditions.

References

[1] Sasai T, Nakai S, Ono K, Mano M and Miyata A, 2017 Estimation of methane emission from rice paddy soils in Japan using the diagnostic ecosystem model, *Journal of Agricultural Meteorology* 73(3): 133-139

[2] IPCC. 2007. Climate Change 2007: *The Physical Scientific Basis, Contribution of Working Group I to the Fourth Assessment Report of the Intergovernmental Panel on Climate Change*; Solomon, S., Qin,D., Manning, M., Chen, Z., Marquis, M., Averyt, K.B., Tignor, M., Miller, H.L. (Eds.).; Cambridge, UK, Cambridge University Press; 2007

[3] Bouman B A M, 2001 *Water-efficient management strategies in rice production.* International Rice Research Note 26.2

[4] Utaminingsih W and Hidayah S, 2012 Mitigasi Emisi Gas Rumah Kaca Melalui Penerapan Irigasi Intermittent di Lahan Sawah Beririgasi. *Jurnal Irigasi* 7(2): 132-141.

[5] Arif C, Setiawan B I, Munarso D T, Nugraha M D, Simarmata P W, Ardiansyah and Mizoguchi M 2016 Potensi pemanasan global dari padi sawah system of rice intensification dengan berbagai ketinggian muka air tanah *Jurnal Irigasi* 11(2): 81-90

[6] Sato S, Yamaji E and Kuroda T, 2011 Strategies and engineering adaptions to disseminate SRI methods in large-scale irrigation systems in Eastern Indonesia *Paddy Water Environment* 9:79-88

[7] Nugroho B D A, Toriyama K., Kobayashi K, Arif C, Yokoyama S and Mizoguchi M, 2018 Effect of intermittent irrigation following the system of rice intensification (SRI) on rice yield in a farmer's paddy fields in Indonesia *Paddy and Water Environment* 16(4): 715-723
[8] Uphoff N and Dazzo F B, 2016 Making Rice Production More Environmentally-Friendly. *Environments* 3: 12
[9] Uphoff N, Kassam A and Harwood R, 2011 SRI as a methodology for raising crop and water productivity: productive adaptations in rice agronomy and irrigation water management. *Paddy Water Environment* 9:3-11
[10] Arif C, Setiawan B I and Mizoguchi M, 2014. Penentuan Kelembaban Tanah Optimum Untuk Budidaya Padi Sawah SRI (*System of Rice Intensification*) Menggunakan Algoritma Genetika (Determining Optimal Soil Moisture for System of Rice Intensification Paddy Field using Genetic Algorithms). *Jurnal Irigasi* 9(1): 29-40.
[11] Zhang B, Yuan S Q, Zhang J S and Li H, 2008 *Study of corn optimization irrigation model by genetic algorithms*, IFIP International Federation for Information Processing. Springer, pp. 121-132
[12] Kuo, S F, Merkley G P and Liu C W, 2000 Decision support for irrigation project planning using a genetic algorithm. *Agricultural Water Management* 45: 243-266
[13] Allen R G, Pareira L S, Raes D and Smith M, 1998 *Crop Evapotranspiration Guidelines for computing crop water requirements*. FAO - Food and Agriculture Organization of the United Nations, Rome
[14] Inglett P W, Reddy K R and Corstanje R, 2005 *Anaerobic soils*. In: *Encyclopedia of Soils in The Environment;* Hillel D, Eds, Amsterdam, Netherland, Elsevier.
[15] Minami K and Yagi K, 1988 Method for measuring methane flux rice paddies. *J.Soil Sci. Plant Nutr* 59: 458–463
[16] Kudo Y, Noborio K, Shimoozono N and Kurihara R, 2014 The Effective Water Management Practice for Mitigating Greenhouse Gas Emissions and Maintaining Rice Yield in Central Japan. *Agriculture, Ecosystem and Environment.* 186: 77-85
[17] Uphoff N, and Kassam A, 2008 Case study: The system of rice intensification. Cornell International Institute of Food Agriculture and Development (CIIFAD) NY 14853, USA
[18] Yang J, and Zhang J, 2010 Crop management techniques to enhance harvest index in rice. *Journal of Experimental Botany*. 61(12):3177-3189.
[19] Cicerone R J, and Oremland R S, 1988 Biogeochemical aspects of atmospheric methane. *Global Biogeochem Cy* 2:229-238
[20] Pun, I and Yamaji E, 2017 Comparative study on Methane Emission and Strutural Development of Rice Plants from SRI and non-SRI methods in a Lysimeter Experiment. *International Journal of Environmental and Rural Development* 8(1): 20-26
[21] Naharia O, Setyanto P, Arsyad M, Burhan H, and Aswad M, 2018 The effect of water regime and soil management on methane (CH_4) emission of rice field. IOP *Conf. Series: Earth and Environmental Science* 157 (2018) 012012
[22] Mosier A R, Duxbury J M, Freney J R, Heinemeyer O, and Minami K, 1996 Nitrous Oxide Emissions from Agricultural Fields: Assessment, Measurement and Mitigation. *Plant Soil.* 181(1): 95-108.

Acknowledgements
We would like to thank to Kurita and AIT for generous financial support for this research by Kurita-AIT Research Grant 2017 and anonymous reviewers for constructive comments and suggestions.

Development of Jamu formula prediction system module of Ijah analytics based on pharmacology activity and particular efficacy target

N S Ramadhanti[1], W A Kusuma[1,3], and R Heryanto[2]

[1]Department of Computer Science, Bogor Agricultural University, Bogor, West Java, Indonesia
[2]Department of Chemistry, Bogor Agricultural University, Bogor, West Java, Indonesia e-mail: ananta@apps.ipb.ac.id
[3]Tropical Biopharmaca Research Center, Bogor Agricultural University, Bogor, West Java, Indonesia

Abstract. Indonesia is one of countries with abundant amount of herbal plant species. These plants have potential to be used as an ingredient of alternative medication, which is known as Jamu. There have been already many previous research related to the topic of formulating jamu efficacy using network pharmacology that construct link among plants and disease. Moreover, some researchers were more interested in predicting the interaction between chemical compound and protein to formulate Jamu. Commonly, Jamu formula contains four basics of pharmacology activities including analgesic, anti-inflammatory, antibacterial, and target pharmacology activity. Ijah has not implemented these four essential characteristics yet for predicting jamu formula. This research implemented new molecular Jamu formulation composed by some active compounds from plant with analgesic, anti-inflammatory, antibacterial and target pharmacology activity characteristic. The result of this study showed a list of plants which consisted of herbal compounds that could be used to formulate jamu as an alternative to synthetic medicine.

Keywords: Ijah, Jamu formula, pharmacology

1. Introduction

Indonesia is one of the countries in the world which has the highest number of herbal plant species. Parts of herbal plants, such as root and leave, can be processed into a concoction so that it has efficacy to maintain health and to cure many kind of illness. The concoctions are often used by people in some countries like China and Indonesia. In Indonesia this concoction is called 'Jamu'. According to Purwaningsih (2013) [1], in accordance with the policy of Indonesian Ministry of Health in 2010 [2], scientific analysis of Jamu is required to support the truth about the efficacy and safety of Jamu.

In 2010, there was a statistical research about the correlation of Jamu formula and its efficacy using bi-plot based on Robust Principal Component Analysis (PCA) by Afendi et. al. (2010) [3]. The result showed the direct effect on disease from plants that usually become the main ingredient of certain Jamu. Afendi et. al. (2012) [4] then developed a classification model of jamu formula with Partial Least Square Discriminant Analysis (PLS-DA) method with accuracy of 71.6%. Furthermore, Fitriawan (2013) [5] compared the accuracy of Afendi's worrk to the one that used Support Vector Machine (SVM) method. Based on his research, implementing data cleaning process in advance, the accuracy was improved to

IOP Conf. Series: Earth and Environmental Science **335** (2019) 012003 doi:10.1088/1755-1315/335/1/012003

95.43%. The undersampling and oversampling process can also improve the accuracy since the dataset is unbalanced. The studies conducted by Afendiet. al. (2012) [4] and Fitriawan (2013) [5] then were integrated in a Jamu Formulation Prediction System Ijah (Indonesia Jamu Herb) with a hypothesis that generally jamu is composed by plants with four particular characteristics, such as analgesic, antibacterial, anti-inflammatory, and efficacy. Next in 2016, Amir [6] conducted research about jamu formula prediction for diabetic disease with graph mining.

Conventional medicine commonly uses single component - single target approach aiming to find the most effective compound for the certain protein disease target. Furthermore, currently new diseases are exists and become more complex [7]. Therefore, we required an approach to formulate drugs that can handle the complex diseases. Gu et. al. (2011) [8] conducted research about relation of compound in drugs and protein in type 2 diabetic disease and formulate Traditional Chinese Medicine (TCM) in the molecular level with multi-component – multi-target approach using compound-protein network. Further research was conducted to find Jamu formula for the same disease. In this research, Jamu formula was formulated from the classification of plant compound and significant protein related to type 2 diabetic diseases. Characteristic for this classification was obtained from molecular docking simulation of herbal plant with target protein [9].

Barkan (2017) [10] developed Ijah Webserver that can predict relationship between plant - compound and protein - disease. Next, Kurnia (2017) [11] conducted further research which predicts link between compound and protein in IJAH using Bipartite Local Model Network Interaction-Profile Inferring (BLMNII) for compound - protein interaction network. The first version of Ijah, which developed before Ijah Webserver, is able to formulate jamu using plants with analgesic, anti-inflammatory, antibacterial, and desired efficacy characteristics while the research by Barkan (2017) [10] formulated jamu with considering the link between plant-compound and protein-diseases. This research would combine the two versions of Ijah to formulate jamu with four compounds as input. These four compounds would be adjusted with four characteristics including analgesic, anti-inflammatory, antibacterial, and desired efficacy.

2. Materials and methods

2.1 Materials

Dataset used in this research is compound data which has four characteristics, including analgesic, anti-inflammatory, antibacterial, and desired efficacy. This dataset was obtained from Ijah database. Compound and protein interaction data was obtained from Ijah database which was predicted with BLMNII based on research by Kurnia (2017) [11]. Numbers of data used in this research can be seen on table 1. Compound with certain efficacy data consists of 87 activities which were grouped into 20 sub-groups on table 2.

Table 1. Numbers of data in Ijah Database.

Data	Total
Plant	4984
Compound	17277
Protein	3334
Disease	4401
Plant and compound interaction	32461
Compound and protein interaction	7945
Protein and disease interaction	5066
Compound similarity	6425730
Protein similarity	2556150

IOP Conf. Series: Earth and Environmental Science **335** (2019) 012003 doi:10.1088/1755-1315/335/1/012003

Table 2. Efficacy and its activities numbers.

Sub-group	Activities
Bacteria, Viruses, Fungi and Parisites	7
Bites and stings	1
Bones	3
Bowles and Bladder	6
Cancer	4
Chest and Lungs	6
Ear, Nose and Throat and Eyes	3
Fevers	2
General Well Being	2
Heart and Blood	7
Infectious Diseases	1
Liver, Kidneys, and other internal bits	6
Nerves and Muscles	2
Other medicinal uses	13
Pain Relief	3
Sex and Reproduction	7
Skin, Hands and Feet	4
Stomach	5
Treatment of Wounds and Bruises	3
Whole Body	2

2.2 Methods

This research was conducted according to the sequence of processes including data collection, system design, implementations, system testing, and deployment. Firstly, data was collected by scrapping and manually taken from various sources of public domain databases, such as PubChem, DrugBank, and KEGG to get compound data which was grouped into some particular activities. Next, this dataset was combined with that of stored in Ijah database showed in table 1.The interaction data between compound and protein was also taken from Ijah as prediction results of BLMNII algorithm (figure 1).

Then, to ensure that every chosen compound has at least one connection with plants, searching was done to look for herbal compounds with high similarity score. This similarity compound data was obtained from Ijah database using similarity measure of SIMCOMP [12].

This system design includes designing input and output interface which referred to the existing of Ijah. The interface was designed with Ijah webserver template and the input component was adjusted according to the first version of Ijah which providing the option menu for chosing four plants with certain characteristics.

In the implementation, the new module was developed based on the first version of Ijah. This program code of Ijah is available on page github.com/tttor/csipb-jamu-prj. The database should also be downloaded and ran in a local server. After the program ran locally, a new page should be made to be developed as a new module.

To validate the functionality of a new module, system testing was conducted based on the black box testing technique. Afterwards, there was another test such as comparing the results to the literatures to validate the result of herbal plants prediction as candidate of Jamu formula. The list of plants and disease generated from the prediction system was compared to those of on articles, paper or any other written research results.

Finally, after testing, the application that has already run locally was placed into the server. The new data that would be used also added to the server database. Then, the server was configured so that the program run and could be accessed through Internet.

Algorithm part 1: BLMNII

```
Input: Interaction matrix (A), Compound similarity matrix (S_s), Protein similarity matrix (S_p), i, j
Output: Pre_ij
pre_i = NII_Prediction (A, S_s, S_p, i, j)
pre_j = NII_Prediction (A, S_s, S_p, i, j)
pre_ij = max(pre_i, pre_j)
```

Algorithm part 2: NII_Prediction

```
if compound-ihas an interaction:
    I = compound-i labelaccording to value in A
else
    I = compound-i labelby adding the S_s multiplication to A for each compound-i
Smith-Waterman, Gaussian and SIMCOMP kernel calculation.
Calculate kernel combination
Model = SVM_Train (Kernel, I)
if protein-j has an interaction:
Pred = SVM_Test(Model, Kernel[j])
else :
Pred = SVM_Test(Model, S_p[j])
Return Pred
```

Figure 1. BLMNII algorithm [13].

3. Results and discussion

This research yielded an additional module to improve the first version of Ijah by adding the option menu for choosing four plants with certain characteristics such as such as analgesic, anti-inflammatory, antibacterial, and desired efficacy activity. This development referred to a hypothesis that every Jamu formula should be obtained based on their pharmacology activities pattern Furthermore, as mentioned before, the dataset was collected both by scrapping and searching manually.

Scraping was done using Python programming language with Beautifulsoup4 library. Firstly, the HTML code from the pages that contain compound with particular activity was analyzed. The scraping script extracted *id* and *name of the compound.* To know the exact id and name on the HTML code, the results were inspected by conducted 'Inspect Element'. The scraping started with page parsing using lxml parser. This procedure was done to parse the document for further processing. Moreover, this information with certain characteristics is the name of compound stored in DrugBank database that obtained by analyzing the HTML code. Those lists of compound then were transformed into a data frame using pandas package and saved into CSV format.

In addition, for the uncategorized activity on DrugBank, the compounds were searched and collected manually. As an example, one of the uncategorized activities is diuretic. Collecting data manually started using keyword 'diuretic' in DrugBank search box. The id and name of the compounds listed then were copied one by one to a document.

Next, the data collected from scraping and manual process was combined into one document that consisted of activity id and compound id in CSV format. After collecting data, to synchronize the new dataset with the existing data on Ijah, compounds that did not exist on Ijah were removed. The numbers of collected data before and after removal can be seen on table 3.

Tabel 3. Numbers of collected data.

Characteristics	Total	
	Before	After
Analgesic	205	125
Anti-inflammatory	164	95
Antibacterial	205	144
Certain efficacy activity	1873	1183

Figure 2. Input interface of the first version of Ijah.

System design was initialized by designing interface for input and output of web page. The input interface was improved by combining the interface of the first version of Ijah and that of the Ijah webserver showed in figure 2. The input interface was designed to accommodate user to choose the combination of four compounds which representing different characteristics. However, because there are 87 activities, thus, activities are divided into 20 sub-groups as showed in table 2. The design of input interface could be seen on figure 3.

The formulation is represented as a network that connects plant-compound-protein-disease as showed in figure 4. On the network, there were plants which connected to herbal compounds, represented by Herbal c. Moreover, the Herbal c was connected to four selected compounds, represented by Compound 1, 2, 3, 4 which respectively have the characteristics of analgesic, antibacterial, anti-inflammatory, and activity of a particular target

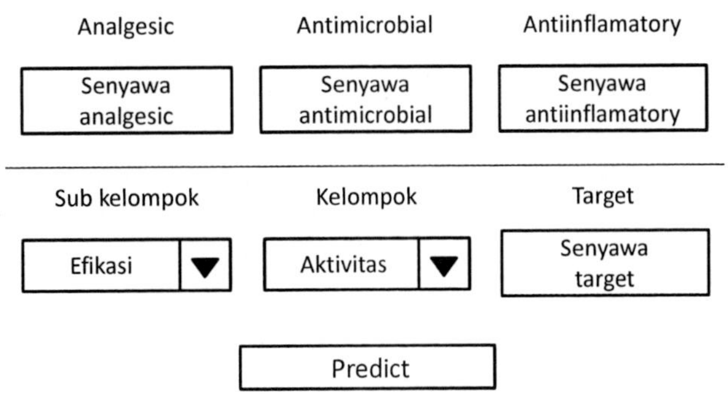

Figure 3. The input interface design.

These compounds were connected to herbal compounds if their similarity score were equal or more than 0.9. The numbers of herbal compound that connected to compound could be more than one, as shown by Compound 2 on figure 5 which connected to three herbal compounds. The connected herbal compound could also be contained in the same plant just like Herbal c 4 and 6. Next, the four selected compound would also be connected to protein and disease.

IOP Conf. Series: Earth and Environmental Science **335** (2019) 012003 doi:10.1088/1755-1315/335/1/012003

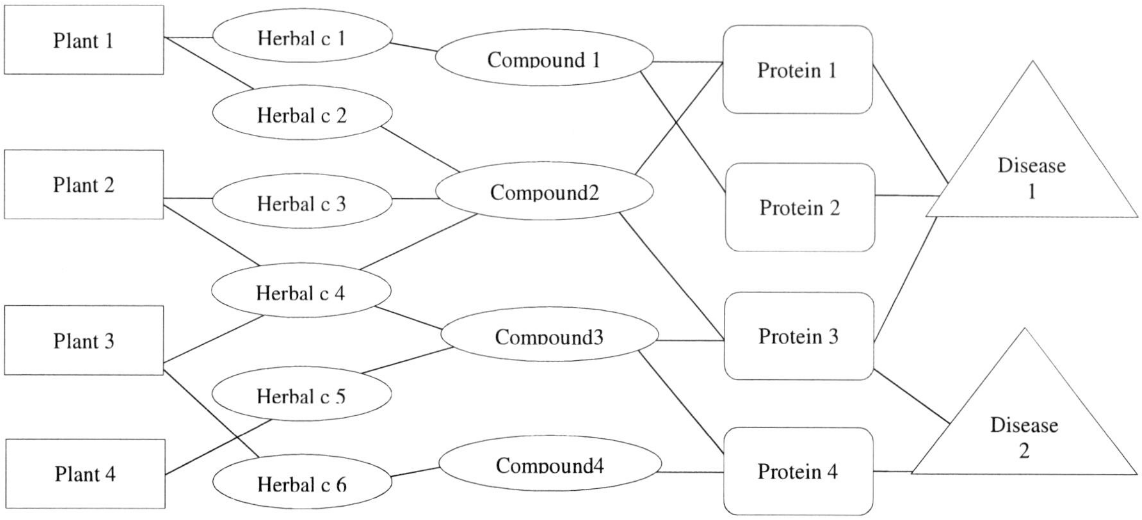

Figure 4. Connectivity among plant-compound-protein-diseases.

The implementation was conducted by modifying code of Ijah on 'webserver/src/app/ijahv1'. The first modification was conducted to change the input interface and to remove the use-case example function. The input interface was modified by adjusting according to the design shown in Figure 4. The modification result could be seen at figure 5. Input box was used to choose the compounds that belong to the analgesic, antibacterial and anti-inflammatory compound, respectively. The compound with target efficacy could be filled after selecting activity which could be selected after selecting the efficacy. Before continuing the modification, the new data of efficacy, activity, and compound-activity data,were imported to PostgreSQL. The typescript (.ts) and PHP file were modified to adjust the HTML file with the new added dataset. The adjustment was conducted to renew the selection on input box. For instance, on analgesic input box, the selection is a list of analgesic compound obtained from the collecting data.

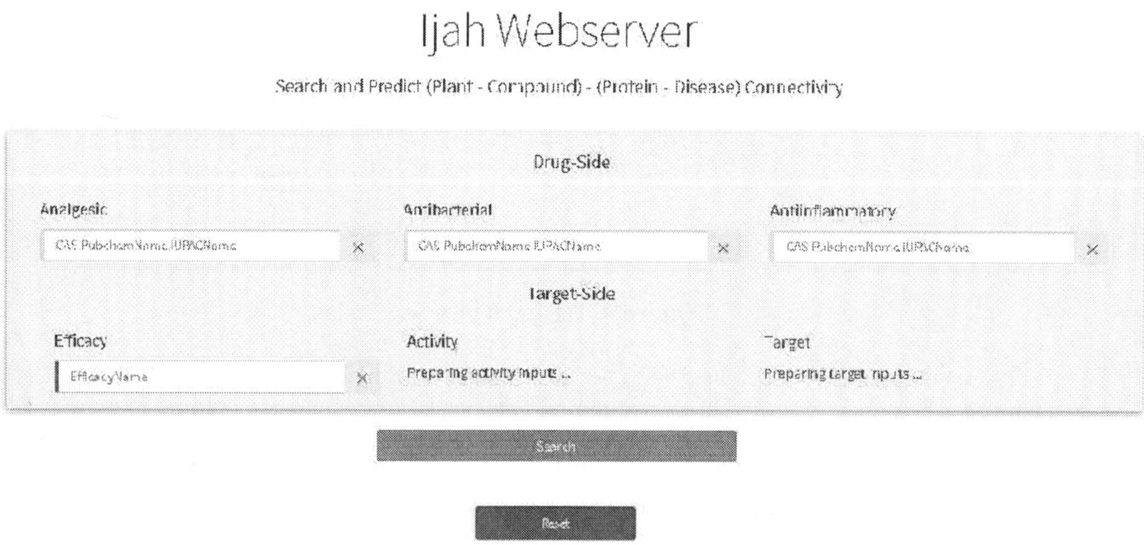

Figure 5. Implemented input interface.

Next, the output interface was designed and implemented. This interface, consist of relational network of plant, herbal compound, compound, protein, disease, was modified by modifying 'searchFromDrugSide' function and other functions related to it. The modification of this interface could be seen at figure 6.

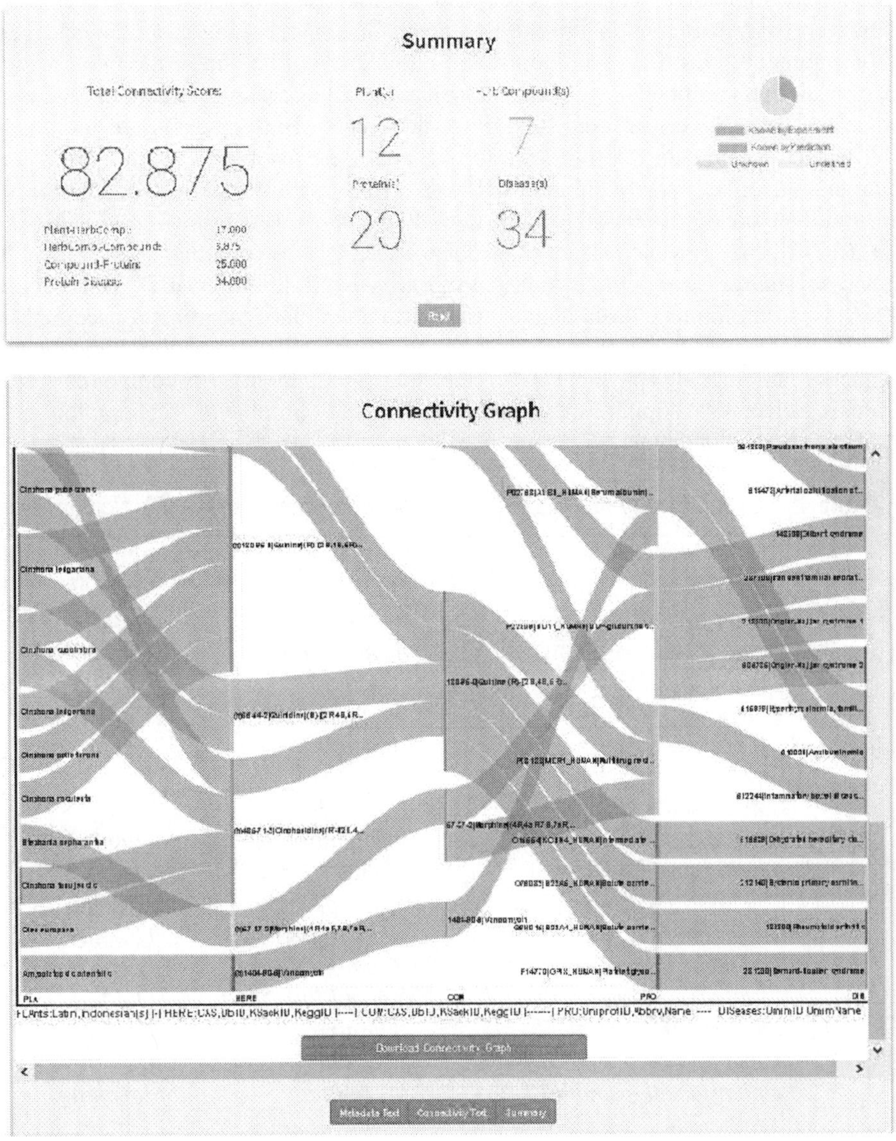

Figure 6. The output interface.

System testing was performed by using black-box testing technique to validate the functionality of the new module. System testing began with choosing compounds to be tested. Based on analysis of the collected data, the total number of compounds related to plants, including direct or indirect relationship, is 58 compounds.

This testing used the compounds of morphine, vancomycin, indomethacin, quinine as samples which have analgesic, antibacterial, anti-inflammatory, and certain characteristic, respectively. The certain efficacy activity selected was analgesic activity which belonging to the pain relief sub-group. A snip of

IOP Conf. Series: Earth and Environmental Science **335** (2019) 012003 doi:10.1088/1755-1315/335/1/012003

the generated graph could be seen on figure 6. These four compounds were chosen for the reason of that they are used as ingredients of some medicine.

System searching results showed that all four compounds were herbal compounds. The results of the connectivity data showed that there were twelve types of plants that contain herbal compounds that were similar to the four compounds selected (see table 4). Table 4 showed that all plants containing quinine compounds were only different in the species level, however belonged to the one genus, namely cinchona or commonly known as quinine, except one plant that belong to the Olea europaea species. Then the four selected compounds were also associated with 20 proteins and 34 diseases.

Two diseases were chosen in order to validate the results. Both diseases are Inflammatory Bowel Disease (IBD) and Rheumatoid Arthritis. Both of them come from proteins that are connected to the selected target compound, namely quinine. IBD is a disease with symptoms of intestine that feels hot. Therefore, quinine can be used as the main ingredient of jamu to relieve IBD. This is proofed by Sharma's (2017) [15] study which stated that Cinchona officinalis (quinine) was an ideal homeopathic drug that was natural for overcoming symptoms of excessive fluid loss in IBD. Moreover, Rheumatoid arthritis (RA) is an autoimmune disease that causes inflammation of the joints. Based on Ma and Jiang's research (2016) [16], one of the effective compounds to treat RA is quinine which can be found in quinine plants. The making of jamu with the four components of compounds could use the Amycolatopsis orientalis, Alpinia officinarum (galangal), Stephania cepharantha and Cinchona (quinine) plants showed in table 4.

Table 4. The relationship amomg plants, herbal compound, and synthetic compound.

Plant	Herbal compound	compound
Cinchona officinalis(quinine)	*Quinine*	*Quinine*
Cinchona officinalis(quinine)	*Cinchonidine*	*Quinine*
Cinchona officinalis(quinine)	*Hydroquinidine*	*Quinine*
Cinchona pubescens	*Quinine*	*Quinine*
Cinchona pubescens	*Quinidine*	*Quinine*
Amycolatopsisorientalis	*Vancomycin*	*Vancomycin*
Cinchona ledgariana	*Quinine*	*Quinine*
Cinchona ledgariana	*Quinidine*	*Quinine*
Cinchona tucujensis	*Cinchonidine*	*Quinine*
Alpiniaofficinarum(galangal)	*Indomethacin*	*Indomethacin*
Cinchona succirubra	*Quinine*	*Quinine*
Cinchona succirubra	*Cinchonidine*	*Quinine*
Cinchona ledgeriana	*Quinine*	*Quinine*
Oleaeuropaea	*Cinchonidine*	*Quinine*
Cinchona pelletierana	*Quinine*	*Quinine*
Cinchona rosulenta	*Quinine*	*Quinine*
Stephaniacepharantha	*Morphine*	*Morphine*

Further analysis on the compounds was obtained from PubChem site [14]. On PubChem [14], it is explained that beside its analgesic characteristic, morphine also has narcotic property which used to relief a severe pain. Due to its narcotic property, the usage of morphine in medicine usually is very little. On PubChem [14], vancomycin is explained as antibacterial compound with very little dose of usage. Indomethacin compound has anti-inflammatory and analgesic property. On PubChem [14], it is explained that Indomethacin is often used as cardiovascular agents, cyclooxygenase inhibitors, gout suppressants, and tocolytic agents. Aside from being a non-narcotic analgesic compound, quinine is better known as an agent used for antimalarial. In PubChem[14] it is explained that Quinine is also commonly used to overcome muscle spasms.

Finally, the new module that has been tested successfully was installed to the server of Biofarmaka with some adjustment and dependency installations to make sure that the system is running in local server as well. All dataset used in this study is also been added to the server database. The system could be accessed through IPB network on http://172.24.10.200:4200.

4. Conclusion

This study applied network pharmacology approach to the plants-compound-protein-diseases network. The compounds were characterized into four groups such as analgesic, anti-inflammatory, antibacterial, and compound with specific target activity characteristic that are generally contained in existing jamu formulas. In this research, the amount of data collected and available data was very different. This research successfully developed a new module to accommodate the four characteristics of group of compounds and integrate this module to the Ijah webserver. The black box testing showed that the functionalities of module could be implemented and could be run successfully. However, there is still drawback in this study that could be overcome in the next study. Only 0.1 % of the compounds data which were collected that have a relationship with plants. Further research could add more dataset, such as data on compounds, plants, proteins, diseases, and interaction data, so that the future research on Jamu formulations could be more comprehensive

References

[1] Purwaningsih E H 2013 Jamu, obat tradisional asli Indonesia pasang surut pemanfaatannya di Indonesia *eJKI* [internet] Available on: http://journal.ui. ac.id/index.php/eJKI/article/view/2065

[2] [Kemenkes] Kementerian Kesehatan Republik Indonesia 2010 Riset Kesehatan Dasar 2010 [internet]. Available on: http://www.diskes.baliprov.go.id/files/subdomain/content/diskes/ Januari%202015/ RISKESDAS%202010.pdf

[3] Afendi FM, Darusman L K, Hirai A, Altaf-Ul-Amin M, Takahashi H, Nakamura K and Kanaya S 2010 System biology approach for elucidating the relationship between Indonesian herbal plants and the efficacy of jamu *IEEE Int. Conf. on Data Mining Workshops* 661–668

[4] Afendi F M, Darusman L K, Morita A H, Altaf-Ul-Amin M, Takahashi H, Nakamura K, Tanaka K and Kanaya S 2012 Efficacy prediction of jamu formulations by PLS modeling *Curr. Comput. Aided Drug Des.* **9(1)** 46–59

[5] Fitriawan A 2013 Sistem klasifikasi khasiat formula jamu dengan metode support vector machine [Thesis] (Bogor: Institut Pertanian Bogor)

[6] Amir F 2016 Prediksi formula jamu untuk penyembuhan penyakit diabetes dengan teknik *graph mining* [Thesis] (Bogor: InstitutPertanian Bogor)

[7] Chen X, Yan C C, Zhang X, Zhang X, Dai F, Yin J and Zhang Y 2015 Drug–target interaction prediction: databases, webservers, and computational model *Briefings in Bioinformatics* **17(4)** 696-712

[8] Gu J, Zhang H, Chen L, Xu S, Yuan G and Xu X 2011 Drug–target network and polypharmacology studies of a Traditional Chinese Medicine for type II diabetes mellitus *Comp. Bio. Chem.* **35(5)** 293–297

[9] Wulandya S A, Afendi F M and Sumaryada T I 2017 Classification of active compounds of type 2 antidiabetic medicinal plants based on molecular docking *Int. J. Sci. Eng. Res.* **8(9)** 583-588

[10] Barkan H Z 2017 Pengembangan sistem informasi prediksi konektivitas drug-target "Ijah Webserver" dengan metode prototyping [Thesis] (Bogor: Institut Pertanian Bogor)

[11] Kurnia A 2017 Prediksi formula jamu berkhasiat menggunakan teknik link prediction dari jejaring bipartite senyawa aktif dan protein [Thesis] (Bogor: Institut Pertanian Bogor)

[12] Hattori M, Okuno Y, Goto S and Kanehisa M 2003 Heuristics for chemical compound matching. *Gen. Inf.* **14** 144–153

[13] Mei J P, Kwoh C K, Yang P, Li X L and Zheng J 2013 Drug–target interaction prediction by learning from local information and neighbors *Bioinformatics* **29(2)** 238–245

[14] [Pubchem] *Pubchem - Open Chemistry Database* 2018 *Compound Summary for* CID 3034034 [internet] Available on: https://pubchem.ncbi.nlm.nih.gov/ compound/Quinine
[15] Sharma M P 2017 A homeopathic approach to inflammatory bowel disease (crohn's and colitis) *Int. J. Rec. Adv. Sci. Tech.* **4(3)** 1-7
[16] Ma Q and Jiang J G 2016 Functional components from nature-derived drugs for the treatment of rheumatoid arthritis [review] *Curr. Drug Targets* **17(14)** 1673-1686

Acknowledgements
This research is supported by Bogor Agricultural University and Ministry of Research, Technology and Higher Education, Indonesia, under Competitive Research Grant from Directorate of Higher Education, Indonesia, 2017, contract no. 1718/IT3.11/PN/2018.

Design a fuzzy inference system to determining superior commodities for create a village medium-term development plan

W Anissudin[1], Y Arkeman[1], I Hermadi[2], S Sjaf[3]

[1*2]Department of Computer Science, Faculty of Natural Science and Mathematics, Bogor Agricultural University (IPB), Jl. Meranti Wing 20 Level V, Kampus IPB Darmaga, Bogor, Indonesia 16680 e-mail: wahid_anissudin@apps.ipb.ac.id
[3]Center for Rural and Agricultural Development Studies, Bogor Agricultural University, Bogor, West Java, Indonesia

Abstract. The Village RPJM, also known as the Village Medium Term Development Plan, is something that must be prepared very carefully. The Village RPJM, especially the Sukadamai village, currently uses intuition alone, even the data held by the village head is still not updated. So that the Village RPJM does not provide maximum results. Therefore, it is necessary to make an intelligent system that can help the village head in determining the potential commodities in Sukadamai village. The latest data with high precision was obtained with the help of the IPB PSP3 using drones and participant assistance. This intelligent system uses a Fuzzy Inference System (FIS) to determine the commodities that have potential in the town of Sukadamai. The input variables used in this study are 10. For the potential of the land and market potential, each of the 27 rules and for the potential of the citizens to produce 64 rules. In the next FIS process, it produces 27 rules to get a potential commodity, so the total rules obtained are 145 rules. Hopes that with this intelligent system, in making the Village RPJM, the village government can be more focused on developing superior commodities in the Sukadamai village.

Keyword: Commodity, Village RPJM, Fuzzy Inference System

1. Introduction

Law No. 6 of 2014 concerning Villages (Village Law) mandates the preparation of the Village Medium Term Development Plan (RPJM Desa) and the Annual Village Development Plan or what is called the Village Government Work Plan (Village RKP) to village government. The Village RPJM is a plan for village development activities for 6 (six) years, and the Village RKP as the elaboration of the Village RPJM is valid for 1 (one) year. The Village RPJM and Village RKP are the basis for rural development with the aim of making efforts to improve the quality of life and life for the highest welfare of rural communities [1].

In order for village development to work well and produce, village development must be planned, coordinated, time-bound, and following the specific conditions of the community and village area concerned. Besides, the implementation of village development involves the active role of the community, village officials, village institutions, institutions at the sub-district and district levels (or supra-village institutions), and others. It is also essential to monitor and evaluate the implementation of village development so that the direction does not deviate from the lines set in the village development planning itself [1].

Center for Rural and Agricultural Development Studies (PSP3) is part of the Institute for Research and Community Service (LPPM) IPB. PSP3 aims to develop socio-economic sciences of agriculture (in the broad sense) through community empowerment research activities [2]. PSP3 IPB researched the Sukadamai Village in Bogor regarding the Village RPJM and the completeness of the Sukadamai Village data such as the number of residents, the level of education of residents, the occupation of the residents, land area, size of the house, and others. PSP3 conducts data collection using the help of drones and participants so that the data obtained is following the current situation of Sukadamai Village.

The village head who was elected was given three months to develop the village RPJM as of the inauguration of the village head. From the time given, the Village Head must immediately be able to make a Village RPJM that is following the current village situation. Therefore we need an intelligent system that can quickly process the amount of accurate data contained in the village. Potential expert systems in helping to determining superior commodities by predicting it, so that government efforts can be carried out effectively and efficiently. The existence of an expert system can facilitate work or even replace experts, combine the knowledge and experience of several experts and provide the expertise needed by a project that does not have experts [3].

Fuzzy Logic is a branch of artificial intelligence to build intelligent systems. Fuzzy logic is often used in problem-solving that explains the system not through numbers, but linguistically, or variables that contain uncertainty/uncertainty. These uncertain values are based on reasoning that combines numerical variables, linguistic variables, and rules [4]. Fuzzy logic is a precise way to map an input space into an output space [5].

Fuzzy logic is an extension of the many values of logic in the sense of the formation of fuzzy sets and fuzzy relations as a tool into a system with many logical values [6]. The strength of FIS relies on two important characteristics. First, they are able to handle linguistic concepts. Second, they are universal approximators able to perform nonlinear mappings between inputs and outputs [7].

A research about an apple grading study using fuzzy logic. In that study, it resulted in 89% correct accuracy [8]. Apples were classified based on their surface blemishes with a classification success of 96.6% using a multi-layer feed forward neural network classifier [9]. Use of artificial neural networks provides a powerful tool for sorting operations. However, it is also associated with high computational cost and uncertainty about the working procedure of the classifier. FL, on the other hand, involves less computation and has clear implementation and working schemes.

In the expert system research, the determination of the productivity of fragrant pandanus rice using fuzzy was carried out by Arkeman, resulting in a final accuracy of 88%. This shows that fuzzy logic can be used to determine the prediction of productivity of fragrant pandan rice in the village of Cianjur Village [10].

2. Materials and Methods

2.1. Data

The data in this study were obtained from PSP3. PSP3 obtains data using drones and participants, so the data provided by PSP3 is the latest and most complete data. Other data is obtained by conducting interviews with residents and experts. Data obtained in the form of raw data is used to assist in determining input parameters and data obtained in the form of membership functions and a collection of rules.

2.2. Methods

Fuzzy logic has been used extensively in decision support systems, for strategic planning, in industrial organizations to model Fuzzy Inference System (FIS) [11]. The strength of FIS relies on two important characteristics. First, they are able to handle linguistic concepts. Second, they are universal approximators able to perform nonlinear mappings between inputs and outputs [7]. The fuzzy rules of the system make use of fuzzy linguistic terms described by membership functions. These functions capture the human expert's conception of the linguistic terms. Fuzzy rules take the form IF (conditions) THEN (actions). Where conditions and actions are linguistic values applied to input and output variables respectively [5].

The Mamdani method is often known as the Max-min method. This method was introduced by Ebrahim Mamdani in 1975. Four steps are needed to get output [5]:

1. Formation of fuzzy sets
   In the Mamdani method, both input variables, and output variables are divided into one or more fuzzy sets.
2. Application function implications
   In the Mamdani method, the function implication used is min.
3. Composition of rules
   There are three methods used in doing fuzzy system inference, namely max, additive and probabilistic OR. If all propositions have been evaluated, the output will contain a fuzzy set that reflects the contribution of each proposition.
4. Defuzzification
   The input of the defuzzification process is a fuzzy set that is obtained from the composition of fuzzy rules, while the resulting output is a number in the fuzzy set domain in a certain range, so a specific crisp value must be taken as the output. There are several defuzzification methods in the composition of the Mamdani rules, namely center of gravity, mean of max, and bisector of the area as in the figure 1 [5].

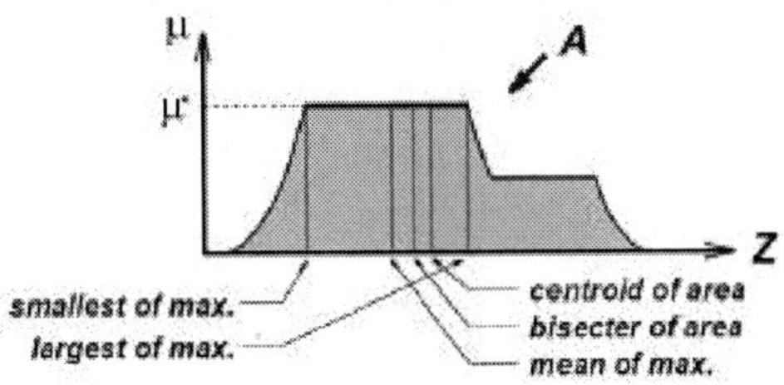

Figure 1. Defuzzification of Mamdani model [5].

In this study using the center of gravity defuzzification. In this method, the crisp solution is obtained by taking the center point (x *) of the fuzzy region.

$$x^* = \frac{\int x.\mu_c(x)dx}{\int \mu_c(x)dx} \tag{1}$$

x∗ is the x-coordinate of center of gravity. $\int x.\mu_c(x)dx$ denotes the area of the region bounded by the curve μC.

$$x^* = \frac{\sum_{i=1}^{n} x_i.\mu_c(x_i)}{\sum_{i=1}^{n} \mu_c(x_i)} \tag{2}$$

If μC is defined with a discrete membership function, then CoG can be stated as equation (2).

3. Methodology

The method used to design superior commodity intelligent system consists of several stages in full is presented in figure 2.

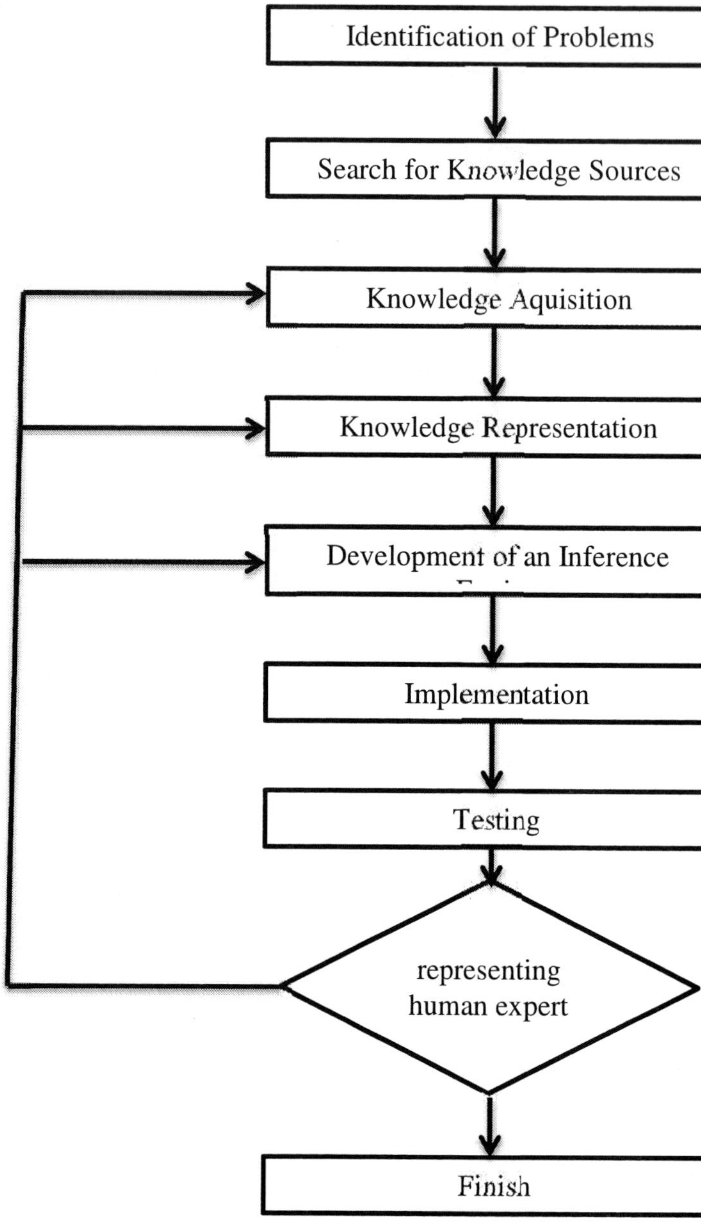

Figure 2. Expert system formation stage [3].

3.1. Knowledge Acquisition

Knowledge acquisition is needed to acquire knowledge, facts, and rules required by intelligent systems. sources of knowledge in this study were obtained from data, farmers, Sukadamai villagers, and related journals. Based on knowledge acquisition that has been done, obtained four fuzzy inference systems. The first FIS is to determine the potential of citizens, the second FIS to determine market potential, the third FIS to determine the potential of the land and the last FIS is the determination of superior commodities. Each FIS has many input parameters, inference systems, and a set of rules.

The first FIS has four inputs, and the second and third FIS have three inputs. the output of the three FISs will be input for the fourth FIS. The following are the parameters that will be used in FIS in table 1-4.

IOP Conf. Series: Earth and Environmental Science **335** (2019) 012004 doi:10.1088/1755-1315/335/1/012004

Table 1. Description of input parameters for the first FIS.

No	Parameter	Unit
1	Man	
2	Quality of Education	
3	Ilness	
4	Age	

Table 2. Description of input parameters for the second FIS.

No	Parameter	Unit
1	Profit	Rp
2	Dema nd	Kg
3	Price	Rp

Table 3. Description of input parameters for the third FIS.

No	Parameter	Unit
1	Rainfall	mmHg
2	Land Area	M^2
3	Irrigation	lt/seconds/Ha

Table 4. Description of input parameters for the fourth FIS.

No	Parameter	Unit
1	Villager Potential	Range (0-10)
2	Land Potential	Range (0-10)
3	Market Potential	Range (0-10)

Every process of FIS has the input parameters, the fuzzy inference system, a set of rules. The dependence diagram of fuzzy expert system for predicting commodity potential is presented in figure 3.

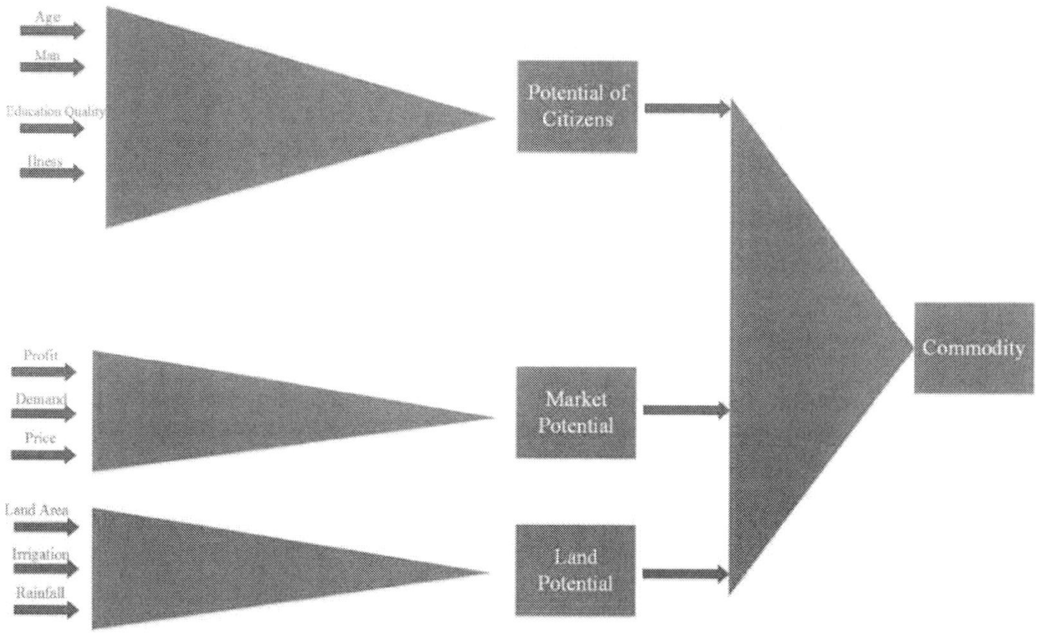

Figure 3. Dependency diagram of the intelligent system.

4. Result and Discussion

4.1. Knowledge Representation

After obtaining knowledge from various sources, then the representation technique is used to develop expert systems. The development of this expert system uses production rules to present procedural knowledge representation. The form of production rules used are if-then forms. "If" is a premise that shows a condition that will be assessed and "then" is a conclusion that shows actions that can be taken if conditions are met. The complete rules obtained are 145 rules consisting of 64 of the first FIS, and 27 rules for the second, third and fourth FIS respectively. An example of rules for the first FIS:

IF (Man = Many) AND (Age = Nature) AND (Ilness = Low) AND (Quality of Education = High) THEN (Villager Potential = Very Potential).

An example of rules for the second FIS:

IF (Profit = High) AND (Demand = High) AND (Price = Middle) THEN (Market Potential = Very Potential).

An example of rules for the third FIS:

IF (Rainfall = Middle) AND (Land Are = Middle) AND (Irrigation = Middle) THEN (Land Potential = Potential).

4.2. Development of Inference Engine

An intelligent system for determining superior commodities using the Mamdani model. For defuzzification using a center of gravity, as in the figure 4.

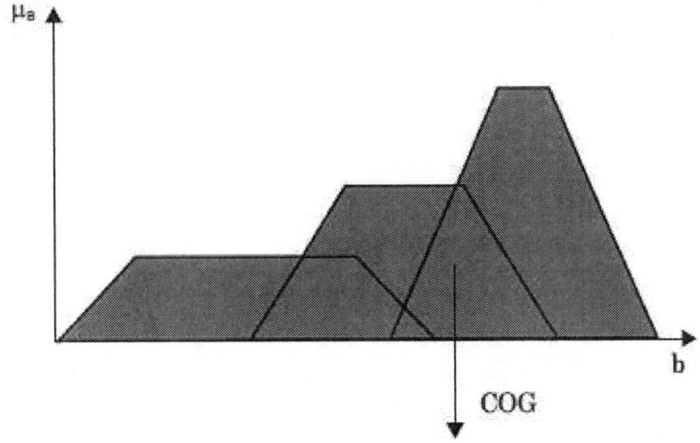

Figure 4. Defuzzification using the Center of Gravity (CoG) method [12].

The first FIS is developed for producing the output of potential of citizens which is the important factor in determining commodity potential. The fuzzification of this FIS involves four parameters of fuzzy input and one parameter of fuzzy output. Fuzzy classification, range and domain of the input and output parameter of the first FIS is presented in table 5.

Table 5. Fuzzy classification and domain of the first FIS.

Function	Variable	Category	Range	Domain
Input	Age	Young	[0 - 75]	[0 10 20]
		Nature		[10 35 60]
		Old		[50 65 75]
	Quality of Education	Low	[0 - 500]	[0 100 200]
		Middle		[125 250 375]
		High		[300 400 500]
	Man	Low	[0-500]	[0 100 200]
		Middle		[125 250 375]
		High		[300 400 500]
	Ilness	Low	[5.5 - 8.5]	[5.5 5.8 6.5]
		Middle		[5.8 7 8.2]
		High		[7.5 8.2 8.5]
Output	Potential of Citizens	Not Potential	[0 - 10]	[0 4]
		Potential		[2 5 8]
		Very Potential		[6 10]

The second FIS is developed for producing the output of market potential which is the important factor in determining commodity potential. The fuzzification of this FIS involves three parameters of fuzzy input and one parameter of fuzzy output. Fuzzy classification, range and domain of the input and output parameter of the second FIS is presented in table 6.

IOP Conf. Series: Earth and Environmental Science **335** (2019) 012004 doi:10.1088/1755-1315/335/1/012004

Table 6. Fuzzy classification and domain of the second FIS.

Function	Variable	Category	Range	Domain
Input	Profit	Low	[0 - 1500]	[0 300 600]
		Middle		[400 750 1000]
		High		[850 1250 1500]
	Demand	Low	[0 - 1000]	[0 250 600]
		Middle		[300 625 800]
		High		[700 850 1000]
	Price	Low	[0 - 5000]	[0 1250 2000]
		Middle		[1500 3250 3750]
		High		[3000 4200 5000]
Output	Market Potential	Not Potential	[0 - 10]	[0 4]
		Potential		[2 5 8]
		Very Potential		[6 10]

The third FIS is developed for producing the output of land potential which is the important factor in determining commodity potential. The fuzzification of this FIS involves three parameters of fuzzy input and one parameter of fuzzy output. Fuzzy classification, range and domain of the input and output parameter of the second FIS is presented in table 7.

Table 7. Fuzzy classification and domain of the third FIS.

Function	Variable	Category	Range	Domain
Input	Rainfall	Low	[0 - 450]	[0 150]
		Middle		[100 225 350]
		High		[300 450]
	Land Area	Low	[0 - 1500]	[0 600]
		Middle		[450 750 1100]
		High		[900 1500]
	Irrigaiton	Low	[0 - 2]	[0 0.6]
		Middle		[0.4 1 1.6]
		High		[1.4 2]

Function	Variable	Category	Range	Domain
Output	Land Potential	Not Potential	[0 - 10]	[0 4]
		Potential		[2 5 8]
		Very Potential		[6 10]

The fourth FIS is developed to produce the final output namely the commodity potential. The fuzzification of this FIS involves three parameters of fuzzy input coming from the output of first, second and third FIS and one parameter of fuzzy output. Fuzzy classification, range and domain of the input and output parameter of the third FIS is presented in table 8.

Table 8. Fuzzy classification and domain of the final FIS.

Function	Variable	Category	Range	Domain
Input	Land Potential	Not Potential	[0 - 10]	[0 4]
		Potential		[2 5 8]
		Very Potential		[6 10]
	Market Potential	Not Potential	[0 - 10]	[0 4]
		Potential		[2 5 8]
		Very Potential		[6 10]
	Villager Potential	Not Potential	[0 - 10]	[0 4]
		Potential		[2 5 8]
		Very Potential		[6 10]
Output	Superior Commodities	Not Potential	[0 - 8]	[0 3]
		Potential		[3 4 5]
		Very Potential		[5 8]

Following one of the membership function of input and output in the form of a formula:
Membership Function of each fuzzy set of villager potential parameter is as follows:

$$\mu\, Not\ Potential\ (x) = \begin{cases} 0, \ x \geq 4 \\ 1, \ x = 0 \\ (4-x)/(4-0) \quad 0 < x < 4 \end{cases} \tag{3}$$

$$\mu\, Potential\ (x) = \begin{cases} 0, \ x \leq 2 \, or \, x \geq 8 \\ (x-2)/(5-2), \ 2 < x < 5 \\ 1, \ x = 5 \\ (8-x)/(8-5), \ 5 < x < 8 \end{cases} \tag{4}$$

$$\mu\, Very\ Potential\ (x) = \begin{cases} 0, \ x \leq 6 \\ 1, \ x = 10 \\ (x-6)/(10-6) \quad 6 < x < 10 \end{cases} \tag{5}$$

Membership Function of each fuzzy set of market potential parameter is as follows:

$$\mu\, Not\ Potential\ (x) = \begin{cases} 0,\ x \geq 4 \\ 1,\ x = 0 \\ (4 - x)/(4 - 0) \quad 0 < x < 4 \end{cases} \tag{6}$$

$$\mu\, Potential\ (x) = \begin{cases} 0,\ x \leq 2\ or\ x \geq 8 \\ (x - 2)/(5 - 2),\ 2 < x < 5 \\ 1,\ x = 5 \\ (8 - x)/(8 - 5),\ 5 < x < 8 \end{cases} \tag{7}$$

$$\mu\, Very\ Potential\ (x) = \begin{cases} 0,\ x \leq 6 \\ 1,\ x = 10 \\ (x - 6)/(10 - 6) \quad 6 < x < 10 \end{cases} \tag{8}$$

Membership Function of each fuzzy set of land potential parameter is as follows:

$$\mu\, Not\ Potential\ (x) = \begin{cases} 0,\ x \geq 4 \\ 1,\ x = 0 \\ (4 - x)/(4 - 0) \quad 0 < x < 4 \end{cases} \tag{9}$$

$$\mu\, Potential\ (x) = \begin{cases} 0,\ x \leq 2\ or\ x \geq 8 \\ (x - 2)/(5 - 2),\ 2 < x < 5 \\ 1,\ x = 5 \\ (8 - x)/(8 - 5),\ 5 < x < 8 \end{cases} \tag{10}$$

$$\mu\, Very\ Potential\ (x) = \begin{cases} 0,\ x \leq 6 \\ 1,\ x = 10 \\ (x - 6)/(10 - 6) \quad 6 < x < 10 \end{cases} \tag{11}$$

5. Conclusions and Suggestions

Based on the discussion that has been discussed, some conclusions can be drawn. The input variables used in this study are 10. For the potential of the land and market potential, each of the 27 rules and for the potential of the citizens to produce 64 rules. In the next FIS process, it produces 27 rules to get a potential commodity, so the total rules obtained are 145 rules. input parameters and rules are obtained based on data, interviews with villagers (practical experts), and related journals and textbooks. Hopes that with this intelligent system, in making the Village RPJM, the village government can be more focused on developing superior commodities in the Sukadamai village. With the development of preferred commodities in the Sukadamai village, can help improve the quality of life of Sukadamai villagers.

References

[1] P Djuni 2015 *Panduan Penyusuan Rencana Pembangunan Jangka Menengah Desa* (Indonesia: Yayasan Penabulu)

[2] IPB [Internet] 2018 *Pusat Pembangunan Pertanian dan Pedesaan (PSP3) LPPM IPB* (Indonesia: PSP3 [cited 2018 Sep 10] Available from: http://lppm.ipb.ac.id

[3] Marimin 2009 *Teori dan Aplikasi Sistem Pakar dalam Teknologi Manajerial* (Bogor: IPB Press)

[4] D J Yulmaini 2015 Penggunaan metode fuzzy inference system (fis) mamdani dalam pemilihan peminatan mahasiswa untuk tugas akhir *Jurnal Informatika* **15**

[5] S Kusumadewi and H Purnomo 2010, *Aplikasi Logika Fuzzy: Untuk Pendukung Keputusan* (Yogyakarta: Graha Ilmu)

[6] G Bojadziev and M Bojadziev 2007 *Fuzzy logic for Business, Finance, and Management* (Singapore: Word Scientific)

[7] G Serge 2001 Designing fuzzy inference systems from data: interpretability oriented *Review IEEE Transactions on Fuzzy Systems* **9(3)** 426-442

[8] I Kavdir and E Daniel Guyer 2003 Apple grading using fuzzy logic turk *J. Agric. for.* **27** 75-382

[9] Q Yang 1993 Classification of apple surface features using machine vision and neural networks *Comput Electron Agriculture* **9** 112
[10] Y Arkeman, M Al Khanif, B Agus and B S Kudang 2016 A Fuzzy expert system for predicting pandan wangi paddy productivity in indonesia *Proceedings WCCA AFITA*
[11] I Rawabdeha, A Al-Refaie and H Arabiyatd Developing a fuzzy logic decision support system for strategic planning in industrial organizations advanced technology & science *IJISAE* **1(2)** 14-23
[12] F Saboya J R, M G Alves and W D Pinto 2006 Assessment of failure susceptibility of soil slopes using fuzzy logic *Eng. Geol* **86** 211-224

Acknowledgements
Thanks to Allah SWT for the grace and guidance, so that I can make this paper and get the opportunity to help one of the villages in Indonesia. Thank you to Prof. Dr. Ir. Yandra Arkeman M.Eng, Irman Hermadi, S.Kom, MS, PhD who have given direction, criticism, and enthusiasm to complete this task. Thank you to family and friends who always provide support and prayer. Thank you to Dr. Sofyan Sjaf, SPt, M.Si and friends from PSP3 who have helped to obtain accurate data.

Implementation of hybrid sampling technique for predicting active compound and protein interaction in unbalanced dataset

W A Kusuma[1,3], A S Rahmi[1], and R Heryanto[2,3]

[1] Department of Computer Science, Bogor Agricultural University, Bogor, West Java, Indonesia e-mail: ananta@apps.ipb.ac.id
[2] Department of Chemistry, Bogor Agricultural University, Bogor, West Java, Indonesia
[3] Tropical Biopharmaca Research Center, Bogor Agricultural University, Bogor, West Java, Indonesia

Abstract. Indonesia Jamu Herbs (Ijah) web server aims to predict Jamu efficacy based on interaction between active compound and disease's protein. However, the interaction between compound and protein data is unbalance since there are many unknown interactions between active compounds and protein target. Thus, the prediction result is still not optimal. In this research, the hybrid sampling technique, combining complementary fuzzy support vector machine (CMTFSVM) and synthetic minority oversampling technique (SMOTE) was used to handle imbalanced data interaction between active compound and protein for Ijah, web server to predict candidate Jamu formula for certain disease. Performance was measured using geometric mean (Gmean), area under curve (AUC), and accuracy. The evaluation results showed that the hybrid sampling technique could increase the instance of minority class three times. Moreover, the prediction model could obtain the value of 0.8346, 0.6812, and 0.5319 for accuracy, Gmean, and AUC, respectively.

Keywords: Indonesia Jamu Herbs (Ijah), efficacy, active compound, fuzzy support vector machine, SMOTE

1. Introduction

Jamu is Indonesian herbal medicine which is used to maintain health and to treat a certain disease[1]. The mixture of some plants composing Jamu can treat certain health problems. The knowledge about Jamu is derived based on local society experience and culture[2]. Study about the association between plant and disease conducted to find jamu formula using the relation between active compound in plant and protein in disease[3].

Implementation of bipartite local model network–based interaction–profile inferring (BLMNII) can predict the relation between active compound and protein to create jamu formula which is implemented to web application called Indonesia Jamu Herbs (Ijah)[4]. BLMNII uses Support Vector Machine (SVM) classiffier and the information of active compound and protein interaction to predict candidate jamu formula for a certain disease. If there are no information of protein and active compound interaction, calculation of the class label uses the most similar active compound and protein from itself. Prediction without interaction information was the worst cases in BLMNII with $O(n^2)$. The implementation in [4] used unbalanced data where the information between active compound and protein interaction was only 0.0001%. Prediction with unbalanced data in BLMNII used similarity of other data to predict Jamu candidate, this situation was considered as the worst case of BLMNII. BLMNII can not handle unbalanced data therefore there should be a method to balancing the active compound and protein interaction data in order to minimize the bias of classifier decision.

There are sampling techniques used to handle many unbalanced data cases such as undersampling, oversampling, and hybrid sampling[5,6]. Undersampling technique is conducted by decreassing the number of majority class data while oversampling technique aims to increase the number of data in minority class to achieve more balanced data. The hybrid sampling is the combination of undersampling and oversampling techniques. The Complementary Fuzzy Support Vector Machine (CMTFSVM) combined with Synthetic Minority Oversampling Technique (SMOTE) is an example of hybrid sampling technique which could handle unbalanced data effectively[6]. CMTFSVM is an undersampling technique which applies complementer (CMT) concept with truth model and falsity model. CMTFSVM uses fuzzy membership values to representing the importance of data to their class. SMOTE as oversampling technique, was combined to increase the number of minority class and to create more balanced dataset.

This study aims to implement hybrid sampling technique using CMTFSVM dan SMOTE to handle unbalanced data interaction between active compound and protein in Ijah. Implementation method referred to the research conducted by Pruengkarn[6].

2. Materials and methods

2.1. Dataset

Dataset used in this research referred to the research data used by Pruengkarn[6], Kurnia[4], and by webcrawling from PubChem. Pruengkarn's dataset was consisted of three dataset including German, Yeast3, and Glass5 dataset. All dataset was binary dataset which only had two classes: positive class (minority class) and negative class (majority class). The German dataset is used to identify credit card fraud, where the positive class represents fraud occurance. Moreover, the Yeast3 dataset cluster yeast into ME3 (positive class) and others (negative class). Lastly, the Glass5 dataset tried to identify glass source based on the chemical information. There were two classes in the Glass5 dataset, including container glass (positive class) and non container glass (negative class). Table 1 summariseschacacteristics of these three datasets. The summarization was conducted according to attributes such as the number of total instances (#instance), the majority class percentage (%major), and the minority percentage (%minor).

Table 1. Summary dataset Preuengkarn[6].

Dataset	#instance	%major	%minor
German	1 000	70	30
Yeast3	1 484	89.02	10.98
Glass5	214	95.79	4.21

Kurnia's[4] dataset consists of list of compound, list of protein, list of compound-protein interaction, and amino-acid sequence taken from Uniprot. Every compound has chemicals characteristics called compound descriptor. In this study, compound descriptor was referred to PubChem. Compound descriptors were crawled from PubChem using Compound ID (CID) taken from list of compound. Moreover, protein descriptors were obtained based on feature calculations. For instance, Amino acid composition (AAC) is one type of protein descriptors, it describe a protein as the proportion of each amino acid divided by total amino acid that construct the protein. Protein feature server (Profeat)[7] provides protein descriptor using amino acid composition (AAC) feature.

Protein descriptor and compound descriptor are the variable of list of compound-protein interaction used in Ijah dataset to predict jamu formula. The value of compound-protein interaction becomes the label of class, where the positive class represents the existence of interaction between compound and protein, the negative class, otherwise. The dataset generated by pairing of protein descriptor and compound descriptor, that labelled based on list of compound-protein interaction.

2.2. Methods

Figure 1 illustrates the methods. The experiment was divided into two part. Part (a) used the three dataset from Pruengkarn[6] with FSVM as classifier. The Geometric Mean (Gmean), Area Under Curve (AUC), and Accuracy was used to evaluate the results. The purpose of stage (a) is to validate the implementation of hybrid sampling technique which refer to Pruengkarn's study[6].

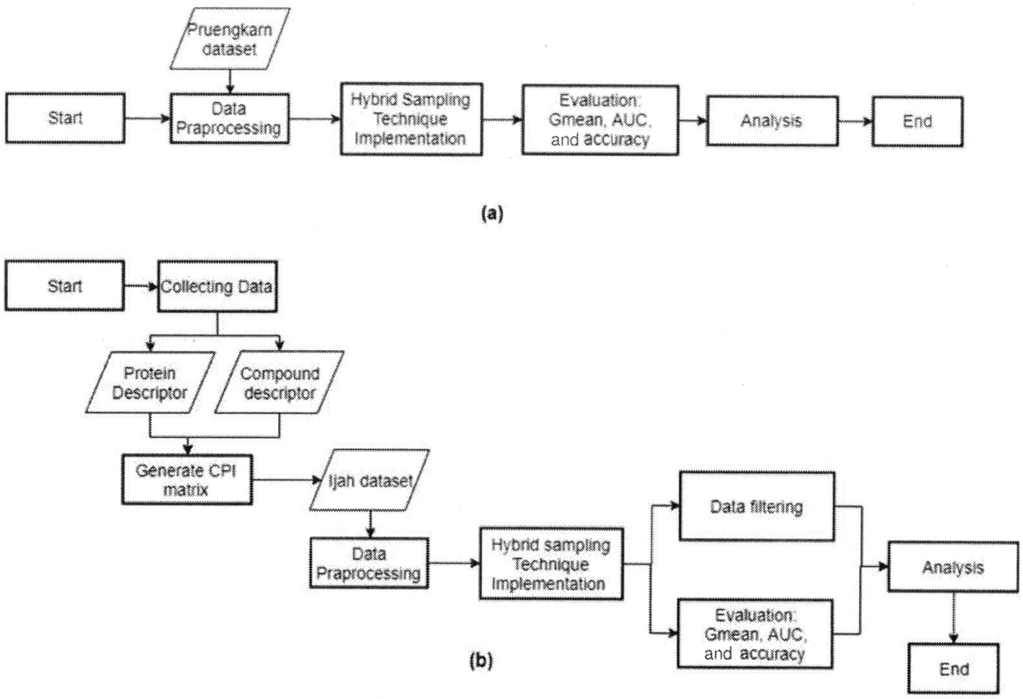

Figure 1. Experimental process.

Stage (b) is the implementation of hybrid sampling technique using Ijah dataset which used compound descriptor and protein descriptor. Ijah dataset is a matrix which pairing each protein descriptor with each compound descriptor and labelled by compound protein interaction (CPI) data. The evaluation was conducted using SVM as classifier to get geometric mean (Gmean) and the accuracy of model. The balanced dataset was taken from hybrid sampling technique filtered by CPI similarity and was used as BLMNII input.

The implementation of Hybrid sampling technique can be seen in figure 2. The hybrid sampling technique was implemented using CMTFSVM combining with SMOTE. The order of implementation was referred to study performed by Pruengkarn[6]. The evaluation was conducted by comparing Gmean and AUC of using undersampling, oversampling, and hybrid sampling technique in different order using Support Vector Machine (SVM), Fuzzy Support Vector Machine (FSVM), and Neural Network (NN) classifier. The study showed that hybrid sampling technique with CMTFSVM followed by SMOTE using FSVM classifier provided better result on imbalanced dataset.

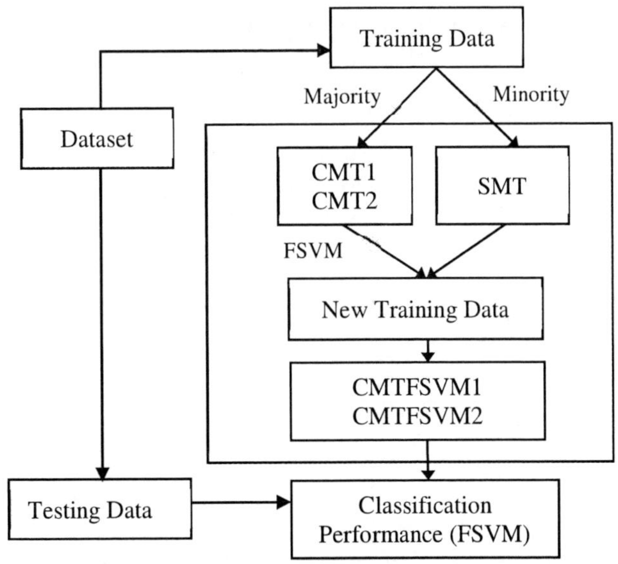

Figure 2. Hybrid sampling technique implementation.

First stage of hybrid sampling technique was undersampling using CMTFSVM for the majority class. CMTFSVM used truth model and falsity model. Truth model is the real instance of dataset. Falsity model is similar to truth model, however in the falsity model the target output is the complement of the truth model output. The truth model was trained by truth memberships, on the other hand the falsity model was trained by false membership. The membership represented the importance of data to their classes. Membership value aims to surpress the effect of noise and outlier in a class. The membership function define as (1a) and (1b). Membership value for minority class represented by $m_i^+$with x_i^+ as data example, while m_i^-represent the membership value of negative class example x_i^-. The within-class importance for each class represent by$f(x_i)$ which has value between 0 and 1. The ratio of imbalanced class was reflected by r^-which reflected the negative/majority class ratio andr^+which reflected positive/minority class ratio where$r^+>r^-$. Membership value is between [0, r_i] interval, where$r<1$ reflected negative class. In order to reflect the imbalanced class, the fuzzy membership value has to be$r^-=r$ and$r^+=1$ for training example.

$$m_i^+=f(x_i^+)\times r^+ \tag{1a}$$

$$m_i^-=f(x_i^-)\times r^- \tag{1b}$$

$F(x_i)$ is based on the distance of data to the actual hyperplane defined as (1c). Variable β is the steepness of the decay increased by 0.1 in [0,1] interval, where $d_i^{hyp}$is a functional margin of each example$x_i$which is equivalent with absolute value of the SVM decision value defined as (1d). The variation of β makes ten possible fuzzy membership values. The membership value for each example selected based on the highest value on Gmean.

$$f_{exp}^{hyp}(x_i)=\frac{2}{1+\exp\left(\beta d_i^{hyp}\right)} \tag{1c}$$

$$d_i^{hyp}=y_i(\omega\times\Phi(x_i)+b) \tag{1d}$$

There are two types of undersampling technique which are CMTFSVM1 (CMT1) and CMTFSVM2 (CMT2). Training dataset in CMTFSVM1 created by eliminating all misclassification samples by the truth and falsity model, however CMTFSVM2 eliminated samples which appear both in truth and falsity model.

IOP Conf. Series: Earth and Environmental Science **335** (2019) 012005 doi:10.1088/1755-1315/335/1/012005

The next step of hybrid sampling technique implementation was balancing dataset after CMTFSVM method using SMOTE. SMOTE replicated the minority data based on neirest neighbor to create synthetic data and made balanced dataset[8]. Synthetic data created larger data with less specific decision area. The more region created by synthetic data the larger minority class area created. The synthetics data generation process could be described as follows:

1 Selected minority data sample and k neirest neighbor
2 Calculated the different (dif) of sample with its neighbor
3 Randomized gap between 0,1
4 Multiplied dif and gap
5 Synthetics data was the addition of selected sample and result from step 4

2.3. Evaluation

The evaluation of model was conducted using 10-fold cross validation method. We used the evaluation measure of Gmean, AUC, and accuracy using FSVM classifier for all dataset. Gmean evaluation is standard technique to measure an imbalance dataset classifier[6]. Gmean is the square root of the product of true positive rate and true negative rate as shown in equation (2). While, AUC is evaluation measure which used trade off between *benefit* (TP_{rate} = TP / (TP + FN)) and *cost* (FP_{rate} = FP / (FP + TN)). True Positive (TP) and True Negative (TN) are the instance that classified correctly, on the other hand False Positive (FP) and False Negative (FN) are the instance of negative data and positive data that missclassified.

$$\text{Gmean} = \left(\frac{TP}{TP+FN} \times \frac{TN}{TN+FP}\right)^{\frac{1}{2}} \quad (2)$$

The experiment was conducted using specification of development environtment is Personal computer Asus A456U, Intel Core i5-7200U 2.5 Ghz processor with 4 GB RAM. We used the package of 'mlfromscratch', python implementation of machine learning package for SVM implementation[9] and the package of 'imblearn', python implementation of imbalance learning algorithm for SMOTE implementation[10].

3. Results and discussion

3.1. Data

German dataset was downloaded from UCI, while Glass5 dataset and Yeast3 dataset was downloaded from KEEL. Compound descriptor data were crawled from PubChem using compund ID (CID) from Kurnia's study[4] and available API. Protein descriptor was obtained from protein feature server (Profeat) using amino acid composition (AAC) feature. Profeat used FASTA file from amino acid sequences, where in this study it was collected from Kurnia's study[4].

Compound descriptor had fourteen variables that described the compound as numerical value. On the other hand, protein descriptor had 20 variables represented the numeric description of protein. The variables were the proportion of each amino acid divided by the total of amino acid in the protein. Example of compound descriptor and protein descriptor are listed in table 2 and table 3.

Table 2. Compound descriptor.

Compound ID	1005	1001	100004	10001388	14505
Molecular Formula	C3H5O6P	C8H11N	C20H24N2O2	C25H26O5	C6H6O2
Molecular Weight	168.041	121.183	324.424	406.478	110.112
H Bond Donor Count	3	1	1	3	0
H Bond Acceptor Count	6	1	4	5	2
Rotatable Bond Count	3	2	3	5	1
Complexity	201.0	65.0	600	713	98.7
Monoiotopic Mass	167.982	121.089	324.184	406.178	110.037

IOP Conf. Series: Earth and Environmental Science **335** (2019) 012005 doi:10.1088/1755-1315/335/1/012005

Compound ID	1005	1001	100004	10001388	14505
Exact Mass	167.982	121.089	324.184	406.178	110.037
XLogP	-1.1	1.4	3.1	6.5	0.5
Charge	0	0	0	0	0
Heavy Atom	10	9	24	30	8
Defined Bond Stereo Count	0	0	0	0	0
Undefined Bond Stereo Count	0	0	0	0	0
Isotope Atom	0	0	0	0	0
Covalent Unit	1	1	1	1	1

Table 3. Protein descriptor.

Protein ID	A0PJY2	A1A4Y4	B7U540	C9JR72	O00116
A	8.632	7.735	8.083	14.410	8.055
C	4.211	1.105	2.309	3.057	2.128
D	2.316	3.867	6.005	5.022	5.015
E	3.158	7.735	7.159	5.677	6.839
F	4.842	3.867	5.774	5.022	4.407
G	6.316	5.525	6.928	8.734	8.967
H	6.105	1.657	3.002	1.965	2.584
I	2.105	4.420	6.467	1.310	5.927
K	8.211	5.525	3.695	0.873	5.927
L	8.632	8.840	9.007	12.445	7.903
M	2.316	6.077	3.233	0.873	1.824
N	4.421	6.630	3.233	1.528	4.103
P	10.947	3.315	3.233	6.114	4.407
Q	4.211	3.315	3.464	3.275	3.495
R	4.000	3.315	7.159	7.642	5.775
S	6.947	8.287	6.005	4.367	6.231
T	5.474	7.735	4.388	5.895	5.927
V	4.842	7.182	7.159	7.205	5.927
W	0.000	1.105	1.155	1.747	1.216
Y	2.316	2.762	2.540	2.838	3.343

Ijah compound data interaction (CPI) was obtained from Kurnia's study[4]. If protein had interaction with compound, the class of CPI was positive, otherwise negative. Table 4 shown the example of CPI from Kurnia's study[4] and table 5 summarises the data that used in this study.

Table 4. Compound protein interaction.

Compound ID	Protein ID	Weight
COM00000014	PRO00000377	1
COM00000016	PRO00001875	1
COM00000020	PRO00002082	1
COM00000020	PRO00001846	1
COM00000020	PRO00000615	1

IOP Conf. Series: Earth and Environmental Science **335** (2019) 012005 doi:10.1088/1755-1315/335/1/012005

Table 5. Summary of data obtained.

Data	Source	Description
Protein	Profeat	3 335 *protein descriptor data*
Compound	PubChem	7 119 *compound descriptor data*
CPI	Ijah	3 693 compund protein interaction
Amino acid sequence	Uniprot	3 335 amino acid sequence in *.fasta

3.2. Data Praprocess

All attributes from Yeast3 dataset and Glass5 dataset were numerics. We conducted normalization to both datasets. While, data praprocessing for German dataset was conducted by transforming the nominal attribute using label encoding, scaling the ordinal attributes, and conducting normalization for numeric attributes. Label encoding create biner -1,1 data for nominal attribute. No data categorized as noise and outlier for all dataset. Ijah dataset is CPI matrix which consist compound descriptor, protein descriptor, and CPI information. CPI matrix created from each protein descriptor that crossed with each compound descriptor and labelled using CPI information from Kurnia's study[4]. There are missing Pubchem CID from list of compound in Kurnia's study[4], it caused some compound does not have compound descriptor. CPI matrix which does not have compound descriptor are deleted. After the deletion, there are 23 741 865 record data for Ijah dataset which consist 2 908 positive class. The example of Ijah dataset is listed in table 6.

Table 6. Ijah dataset.

Properties	Data-0	Data-1	Data-2	Data-3	Data-14574
A	0.00634	0.00837	0.00754	0.00587	0.0005
C	0.00078	0.00232	0.00128	0.00181	0.00024
D	0.0057	0.00604	0.00384	0.00158	0.00024
E	0.00686	0.00651	0.00901	0.00406	0.00029
F	0.00453	0.00558	0.00297	0.00316	0.00041
G	0.00621	0.00372	0.00544	0.00429	0.00036
H	0.00311	0	0.00242	0.00135	0.00021
I	0.00531	0.00465	0.00416	0.00565	0.00036
K	0.00686	0.0079	0.00768	0.00497	0.00018
L	0.00958	0.00697	0.01042	0.00948	0.0008
M	0.00272	0.00186	0.0016	0.00158	9.00E-05
N	0.00272	0.00604	0.00288	0.00271	0.00014
P	0.00492	0.00325	0.00585	0.00181	0.00039
Q	0.00401	0.00511	0.00553	0.00203	0.00012
R	0.00479	0.00465	0.00658	0.00226	0.00027
S	0.00583	0.00558	0.0074	0.00452	0.00052
T	0.00531	0.00604	0.00475	0.00429	0.00046
V	0.00608	0.00744	0.00553	0.00497	0.00042
W	0.00104	0.00139	0.00073	0.0009	0.00026
Y	0.00311	0.00279	0.00297	0.00271	0.00014
MolecularWeight	0.41919	0.40445	0.42632	0.42612	0.32879

Properties	Data-0	Data-1	Data-2	Data-3	Data-14574
HBondDonorCount	0.00192	0	0.00592	0.0007	0.00533
H BondAcceptorCount	0.00766	0.00866	0.00986	0.0056	0.00488
RotatableBondCount	0.00479	0.00673	0.00394	0.0021	0.012
Complexity	0.68695	0.71286	0.67343	0.67411	0.82167
MonoisotopicMass	0.41884	0.40419	0.42605	0.42581	0.32859
ExactMass	0.41884	0.40419	0.42605	0.42581	0.32871
XLogP	0.00297	0.00221	0.00069	0.00427	-0.00152
Record	0	1	2	3	14574
Charge	0	0	0	0	0
HeavyAtomCount	0.0297	0.02886	0.03057	0.0315	0.02311
DefinedBondStereoCount	0	0	0.00099	0	6.00E-05
UndefinedBondStereoCount	0	0	0	0	0
IsotopeAtomCount	0	0	0	0	0
CovalentUnitCount	0.00096	0.00096	0.00099	0.0007	6.00E-05
Class	0	0	0	0	1

This study used 14 575 data from Ijah dataset that classified as negative class (no interaction between protein and compound) and 2 908 data from positive class (compound and protein has interaction). Sample data which used 17 483 record with positive class proportion was 20%. After generating the sample data, all features were normalized using sklearn library.

3.3. Evaluation

The Gmean and AUC resulted in this study were compared to those of Pruengkarn[6]as described in table 7 dan table 8. This comparison showed that the results of this study were differ to those of previous research. It probably was caused by the different techniques used in the preprocessing. In addition, in Pruengkarn[6], in the oversamplingusing SMOTE, the variables used in their experiments were not explained clearly. In this study, the variable of neirest neighbor (k) was set into equal or greater than 2 ($k \geq 2$).

Table 7 shows the evaluation of undersampling, oversampling and hybrid sampling approach. CMT1 and CMT2 was represent two type of undersampling technique which are CMTFSVM1 and CMTFSVM2 respectively, SMOTE was oversampling approach, however CMTSMT1 and CMTSMT2 was hybrid sampling approach used two types of undersampling technique combined by oversampling technique, SMOTE (SMT).

Table 7. Gmean measurement comparison between this study and Pruengkarn's.[6]

Dataset	Approach	Gmean of this study	Gmean of Pruengkarn	ΔGmean
German	CMT1	0.5828	0.7244	- 0.1416
	CMT2	0.5828	0.7262	- 0.1792
	SMOTE	0.6434	0.7294	- 0.0860
	CMTSMT1	0.6389	0.7141	- 0.0752
	CMTSMT2	0.6389	0.7367	- 0.0978
Glass5	CMT1	0.7155	0.0000	+ 0.7155
	CMT2	0.7155	0.1000	+ 0.7155
	SMOTE	0.8209	0.9625	- 0.1416

Dataset	Approach	Gmean of this study	Gmean of Pruengkarn	ΔGmean
	CMTSMT1	0.8346	0.9625	- 0.1279
	CMTSMT2	0.8346	0.9638	- 0.1292
Yeast3	CMT1	0.7989	0.8244	- 0.0255
	CMT2	0.7989	0.8241	- 0.0252
	SMOTE	0.7748	0.9032	- 0.1284
	CMTSMT1	0.8027	0.9026	- 0.0999
	CMTSMT2	0.8027	0.8997	- 0.0970

Table 8. AUC measurement comparison between between this study and Pruengkarn's.[6]

Dataset	Approach	AUC of this study	AUC of Pruengkarn	ΔAUC
German	CMT1	0.6049	0.7313	- 0.1264
	CMT2	0.6049	0.7358	- 0.1309
	SMOTE	0.6477	0.7358	- 0.0881
	CMTSMT1	0.6411	0.7187	- 0.0776
	CMTSMT2	0.6411	0.7388	- 0.0977
Glass5	CMT1	0.7165	0.5000	+ 0.2165
	CMT2	0.7165	0.5500	+ 0.1665
	SMOTE	0.8652	0.9634	- 0.0982
	CMTSMT1	0.8902	0.9634	- 0.0732
	CMTSMT2	0.8902	0.9646	- 0.0744
Yeast3	CMT1	0.7949	0.8380	- 0.0431
	CMT2	0.7949	0.8380	- 0.0431
	SMOTE	0.8445	0.9043	- 0.0598
	CMTSMT1	0.8066	0.9034	- 0.0968
	CMTSMT2	0.8066	0.9006	- 0.0940

The implementation of hybrid sampling technique measured by Gmean and AUC showed slightly different value. It meant that the implementation of hybrid sampling technique in this study was quite similar to the study conducted by Pruengkarn[6]. The implementation of hybrid sampling technique then was applied to active compound and protein interaction in Ijah dataset. In order to know the effect of hybrid sampling technique in Ijah dataset, the unbalanced dataset was classified using SVM. Table 9 showed how the performance of SVM classifying the unbalanced dataset. The classification performance had 0.0000,0.5000, and 0.7677 for Gmean, AUC, and accuracy, respectively. The result showed that SVM classifier was failed to identify the minority class.

Table 9. Unbalance Ijah dataset classified using SVM.

Measurement	Gmean	0.0000
	AUC	0.5000
	Accuracy	0.7677
Minority samples	Total of minority samples	2 908
	Minority : majority proportion	0.2 : 0.8

Table 10 presents the result of hybrid sampling technique on Ijah dataset. Gmean and AUC increased up to 0.6812 and 0.0319, respectively. Table 10 showed better accuracy compared to those of Table 9 where data was imbalanced. The result showed that all classifiers could recognise both minority and majority samples. After hybrid sampling technique was implementated, the total instances were 13 116 data. Comparing the result to the original CPI matrix, filtered 9 851 data which was similar to original CPI matrix.

Table 10. Hybrid sampling technique on Ijah dataset.

Measurement	Gmean	0.6812
	AUC	0.5319
	Accuracy	0.8346
Minority samples	Total of minority samples	9 851
	Minority : majority proportion	0.4 : 0.6

4. Conclusion

This study used the combination of Complementary Fuzzy Support Vector Machine and Synthetic Minority Oversampling Technique as hybrid sampling technique to handle imbalanced data.The implementation had Gmean which was differ from -0.0752 to -0.1792. Moreover, the value of AUC also was slight difference of -0.0732 to 0.0977 compared to the the previous research conducted by Pruengkarn.The hybrid sampling technique can handle Ijah active compound and protein interaction imbalanced problem. The result showed that the data in minority class was increased up to 150% (9 581 data) from the original data that only consist of 2 908 data. Futhermore, it presented good classification result with 0.8346 and 0.5319 for Gmean and AUC, respectively.

References

[1] Torri M C 2013 *Europeanon J. Med. Plants* **3(1)** 25-39
[2] [WHO] World Health Organization 2008 Traditonal medicine fact sheet 134
[3] Amir F 2016 Prediksi formula jamu untuk penyembuhan penyakit diabetes dengan teknik graph mining[thesis] Institut Pertanian Bogor, Bogor, Indonesia
[4] Kurnia A 2017 Prediksi formula jamu berkhasiat menggunakan teknik link prediction dari jejaring bipartite senyawa aktif dan protein [Thesis] (Bogor: Bogor Agricultural University)
[5] Ali A, Shamsuddin S M and Ralescu A L 2015 *Int. J. Adv. Soft Comput. Appl.* **7(3)** 176-204
[6] Pruengkarn R, Wong K W and Fung C C 2017 *IEEE Int. Conf. Sys. Man Cybern. (Banff)*(Canada) 978-83
[7] Zhang P, Tao L, Zeng X, Qin C, Chen S Y, Zhu F, Yang S Y, Li Z R, Chen W O and Chen Y Z 2016 *J. Molecular Biology* **429(3)** 416-25
[8] Chawla N V, Bowyer K W, Hall L O, and Kegelmeyer W P 2002 *J. Artif. Intell. Res.* **16** 321-57
[9] Noren E L 2017 ML-From-Scratch *Github Repository* Available at https://github.com/eriklindernoren/ML-From-Scratch
[10] Nogueira F and Aridas C K 2017 *J. Mach. Learn. Res.* **18(17)** 1-5

Acknowledgements

This research is supported by Bogor Agricultural University and Ministry of Research, Technology and Higher Education, Indonesia, under Competitive Research Grantfrom Directorate of Higher Education, Indonesia, 2017

Analysis of farming efficiency and smart farming system development in supporting garlic self-sufficiency: A concept

B Waryanto[1), T Agustina [1) and D Tjiptodharmono [2)

[1) Ministry of Agriculture, Jl. Harson RM 3 Gd D Lt 4, Jakarta e-mail: budi_w@pertanian.go.id
[2) Agricultural practitioner, Palu, Central Sulawesi

Abstract. In Indonesia, the agricultural sector has a strategic role in sustaining the national economy. Various policies have been made with the aim of increasing the strategic role, both in the form of policies to achieve food self-sufficiency, reduce imports and even policies to encourage exports. One of the important commodities in the agricultural sector is garlic. This commodity is the government's priority to be managed and targeted for self-sufficiency by 2021. This paper tries to trace the success of garlic farming through data studies and related literature. The review of existing data shows that over the past four years (2014 to 2017) the ratio of garlic production to imports was very small just only 4.15%, indicating that Indonesia still depends on imported garlic. Amount of imported garlic grew 5.09% per year, but on the other hand, domestic production grew negatively -1.73%, meaning that domestic production is still not enough for domestic consumption. As a discouraging form towards the achievement of self-sufficiency, this paper also reviewed the literature on the efficiency concept of garlic farming at the farm level through the approach of econometric models, as well as the concept of developing smart farming systems used for product tracking functions to the consumer level, on the contrary can be traced back to products produced by farmers. Through the offering of these two thought concepts, it is expected to be contribute in supporting the government program to achieve garlic self-sufficiency.

Keywords: garlic self-sufficiency, farming efficiency, smart farming system

1. Introduction

Indonesian government has made a policy to achieve garlic self-sufficiency in 2021, with the aim to reduce imports which tend to increase. This step is very important because the average ratio between garlic production and imports from 2014 to 2017 just only 4.15%. This shows that the dependence on imported garlic is still high. The Ministry of Agriculture estimates the land which is needed for garlic planting to achieve the self-sufficiency covering 65 thousand hectares, while for the seedling land it takes 14 thousand hectares [24]. The land needs are very large because in 2017 there was only 2,146 hectares of the harvested land. Therefore, an additional 60 thousand is needed to met the land needs [23].

Regarding land availability [7], stated that the garlic-producing land in Indonesia is concentrated in the highland with altitude from 600 metres to 1,100 metres above sea level (asl), with cool and dry air. In lowland areas where the availability of land is still wide, it is rare to find garlic-producing areas. Therefore, from the needs of a large area to achieve self-sufficiency, a suitable location needs to be chosen by considering the ability of farmers as managers of garlic farming.

IOP Conf. Series: Earth and Environmental Science **335** (2019) 012006 doi:10.1088/1755-1315/335/1/012006

Farmers cultivation techniques which has an important part of the garlic farming system have not been efficient. The use of production inputs such as land, seeds and fertilizers are technically not yet efficient. Seedlings is one of the main factors that technically leads to low farm efficiency [16]. The decline in garlic productivity from year to year was caused by tuber seeds that were suspected of carrying the disease [13]. In addition, a number of local seeds are more widely available for highland such as Lumbu Hijau and Lumbu Kuning, only Lumbu Putih varieties are available for lowland [20]. Highland garlic varieties are not good if planted in the lowland.

Local garlic produced by farmers is often less competitive with imported garlic in terms of size and price. This makes the numbers of imported garlic. On the contrary, garlic planted by farmers decrease. On the other hand, domestic consumption is increasing along with the increase in population. The increase in imports actually began since Indonesia signed *Punta del Este* as the GATT system in 1986 [17]. The system requires each country to change non-tariff barriers to tariffs and reduce domestic and export subsidies for agricultural commodities.

Although there are many obstacles to achieving garlic self-sufficiency in the future, it is not necessary to think backward. The challenges ahead will be more complex, the agricultural sector is faced with global demands which are often known as the "Industrial Revolution 4.0". [22] Expressed their opinions about the challenge, that the Industrial Revolution 4.0 is a form of response to the rapid development of sensor technology, interconnection, and data analysis integrated into various industrial fields, including the agricultural industry. The challenge also affects the government's desire to achieve garlic self-sufficiency. It is not only government obligation to think about it, but also for the academics and practitioners.

By looking at wide-range obstacles and challenges to achieve garlic self-sufficiency, this article examines the aspects of garlic farming on a micro-scale based on technology. This is a form of contribution from practitioners to realizing self-sufficiency in garlic, reducing imports, and it can even be found that the concept of smart farming can meet the demands of the industrial revolution 4.0. The approach method used is a literature study of various garlic farming and trading policies, also exploring the facts of local wisdom, the concept of quantitative analysis on farming aspects and the study of the Industrial Revolution 4.0.

2. Import, production management and land availability

In 2017, Indonesia's imported garlic about 559.73 thousand tons, while production was only 19.16 thousand tons (figure 1). During the period 2014 to 2017, imported garlic grew by an average of 5.09%, whereas production in the same period decreased by 1.73%. The main import of Indonesian garlic comes from China. Indonesia is the largest garlic importer from China with a contribution of 17.9% of China's total exports, above Vietnam with 6.2% and Malaysia with 5.6%. In 2017, China's garlic stock reached a record of 3.2 million tons high, rose 1.05 million tons from the previous year [25]. Seeing the high amount of garlic imports in Indonesia, the estimated of land needs for planting to support self-sufficiency in the area of 65 thousand hectares becomes a tough job. One other way that can be done is to increase the productivity of garlic.

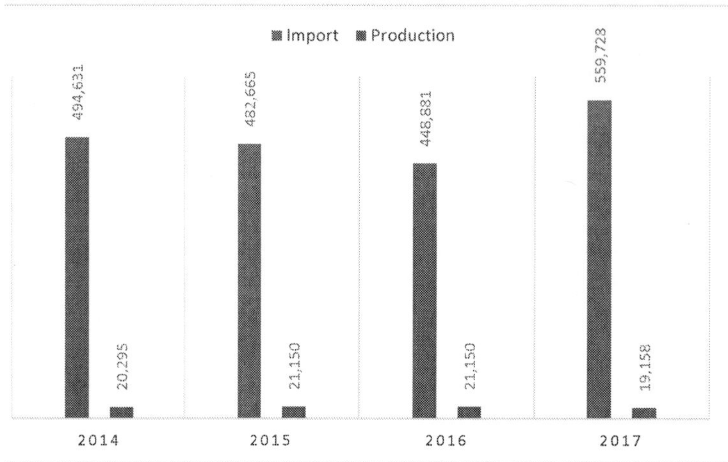

Figure 1. Import and Production of Garlic in 2014-2017 (Metric tons).

The productivity of garlic produced by Indonesian farmers was not optimal, the reason is the application of technical cultivation that is not in accordance with the SOP (Standard Operational Procedure), also the use of low-quality seeds [11]. In addition, diseases also affect tuber size, resulting in low productivity. This condition triggers an increase in import volume. In 1983, the import of garlic only reached 9.92 thousand tons, but in 2017 the number reached 559.73 thousand tons. Efforts to increase garlic productivity through the use of good seeds is believed to have a significant effect in increasing production, so that import can be reduced [9, 15]. The problem faced so far is the conventional seed production techniques which have not produced high-quality seeds, one of which is due to viruses that are always carried on the seed bulbs. Seed production through tissue culture techniques can be implemented as an alternative method. Tissue culture methods could create new plants that are virus free from plants that were originally infected [18].

To increase garlic production in the future, it is not only depending on the seed's quality, the management of irrigation and fertilization of garlic plants are also a determinant of production. Surface irrigation management combined with fertilization in production input management system can actually increase 19% of garlic production [19]. Surface drip irrigation has a noticeable effect on garlic yields than other types of irrigation. Drip irrigation technology combined with fertilization can be considered as a breakthrough than previous irrigation methods which use water flooding on garlic land.

By looking at the constraints of using garlic cultivation technology, especially the use of production inputs such as seed, irrigation, and fertilization, it is necessary to breakthrough to obtain an efficient garlic farming system. The step that needs to be done is to develop the concept of efficiency farming on new land by considering various factors. The concept of practical development and research needs to be carried out simultaneously because time to achieve the garlic self-sufficiency target in 2021 is getting closer. Therefore, the selection of development sites is a determining factor. Parigi Moutong Regency in Central Sulawesi is one of the areas which has a large potential of land, there is a land that has an altitude above 600 meters and land under 600 meters. However, agricultural land in Parigi has limitations in terms of water availability [5].

This paper describes the concept of efficient garlic farming development combined with the application of smart farming in Parigi Moutong Regency, Central Sulawesi. This region chosen with the consideration of the ease in integrating various technologies, scientific studies, and management of its implementation. Various parties such as business people, government and farmers in Parigi Moutong District have been willing to become new development areas. If this implementation is considered successful in terms of various aspects, then replication is carried out on a wide scale in other locations.

3. Efficiency concept on garlic farming

The efficiency of the farming system is basically in the concept that is intended for producers/farmers to be able to manage their farms well and ultimately increase production and profits. A rational farmer will use optimal production input as long as the added value produced is equal to or greater than the additional cost of adding the input. Efficiency is a comparison of output with inputs used in a production process [8].

The concept of efficiency into three parts, namely: 1) Technical Efficiency, 2) Price Efficiency, and 3) Economic Efficiency [2]. Technical efficiency measures production performance at certain levels of input use. Price efficiency or allocative efficiency measures the level of success of farmers in their farming to achieve maximum profit. Maximum profit can be achieved when the marginal productivity value (MP) of each given factor of production equals with the marginal cost. Economic efficiency is a combination between technical efficiency and price efficiency.

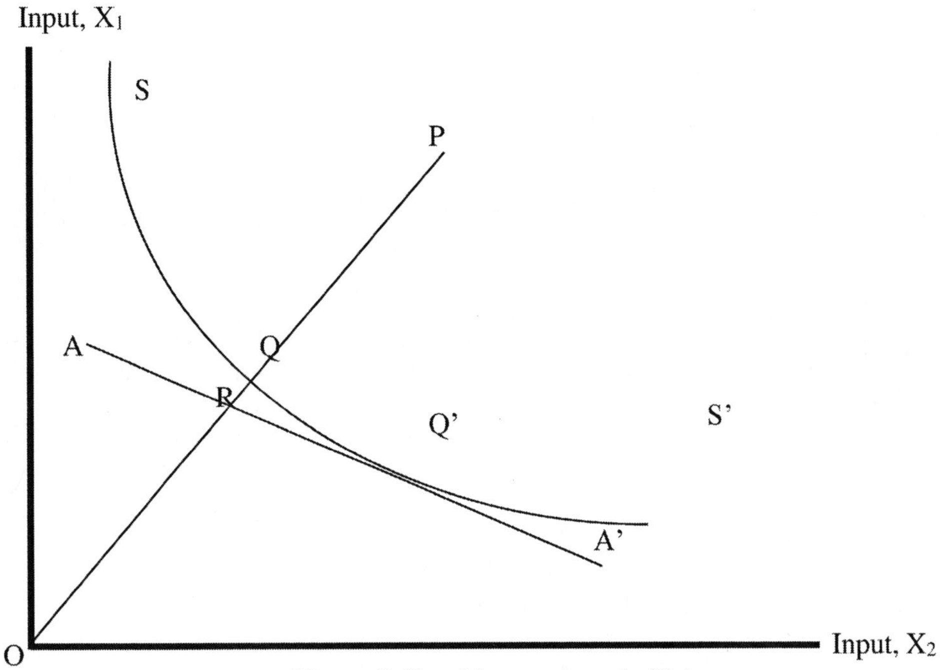

Figure 2. Farrel input-oriented efficiency measure

The concept of technical efficiency can be explained in terms of input use [1, 10] and is assumed to be in the Constant Return to Scale condition. figure 2 shows the concept of technical efficiency, where the horizontal axis represents the use of input X_2 and the vertical axis describes the use of input X_1. The SS curve is *isoquant* which represents various combinations of inputs (X_1 and X_2) which are used to produce a certain number of outputs (Y). All points on the *isoquant* reflect technical efficient production. An effort can be made to measure the efficiency of the farming system, which operates at point P. At this point (P), the on-farm is assumed to produce the same level of output (Y) as it is produced on the *isoquant* / SS'. To determine the technical efficiency of the farming system, a line can be drawn from the origin (0) to the point P. The line crosses the *isoquant* at point Q. In the case of a technically efficient farm, the total output (Y) produced by using input X_1 and X_2 are defined by input point Q. At the point P, input X_1 and X_2 are not used efficiently by farmers. So, the observed technical efficiency (TE) of a farmer is defined as the OQ / OP ratio, which shows the proportion where the combination of P inputs can be reduced.

Related to figure 2, if input prices are available, allocative efficiency (AE) or according to [1] also called price efficiency can be determined. The *isocost* line (AA') is described as offending SS' *isoquant* and cuts the OP line at point R. Point R shows the optimum input-output ratio which minimizes production costs at a certain level of output because the *isoquant* slope is the same as the *isocost* line. Point Q is technically efficient, but is allocatively inefficient because of production at point Q costs higher than point Q'. The OR-OQ distance shows a decrease in production costs if production occurs at point Q' (efficient on TE and AE). If TE and AE can be calculated, then economic efficiency (EE) can be calculated by multiplying TE and AE values.

3.1. Stochastic frontier production function and efficiency calculation

The equation of the stochastic frontier production function in summary as follows [4]:

$$Y_i = x_i\beta + (V_i - U_i), \text{ i=1,2,3,..., n} \tag{1}$$

where:

Y_i = production produced by the i-th farmer

x_i = input factor used by the i-th farmer

β = parameter vector that will be estimated

V_i = random variables related to external factors, dispersed $(V_i \approx N(0,\sigma_v^2))$

U_i = non negative random variables, assumed to affect the level of technical inefficiency and related to internal factors, dispersed $(U_i \approx |N(0,\sigma_U^2)|)$

The stochastic frontier production function equation has been used for two decades and experienced development by accommodating various things, including: assuming a general distribution for U_i such as truncated normal distribution, considering panel data and time, variations in technical efficiency, development of methodology for cost function and also estimation of the existing equation system [4]. Parameter estimation of the stochastic frontier production function can be completed using the *Frontier 4.1* software program.

In the stochastic frontier production function, the components of model error consist of V_i and Ui, as shown in equation (2). Particularly for the U_i component which describes the effects of technical inefficiency can be modeled through the following equations [3]:

$$U_i = z_i\delta + W_i \tag{2}$$

W_i is assumed truncated normal dispersed with mean 0 and variety σ^2.

3.2. Technical Efficiency and Economic Efficiency

Frontier 4.1. Software can be used to estimate the production function with a variety of options, namely: 1) presume the production function model with panel data that is not balance and have a truncated normal distribution, 2) a two-stage regression function by calculating the estimation of production functions and efficiency, then the predicted inefficiency effects value of specific variables (such as: managerial, producer characteristics, etc.) and 3) estimation of cost functions and efficiency predictions for both technical efficiency and cost function efficiency [4]. In terms of efficiency calculations [3], make general equations for technical efficiency, as follows:

$$ET_i = \exp(-U_i) = \exp(-z_i\delta - W_i) \tag{3}$$

where the value of ET_i between 0 and 1 or $0 \leq ET_i \leq 1$.

Assuming that the on-farm is intended to achieve profits, then it must allocate the minimum cost from the existing inputs, or the on-farm achieve allocative efficiency (AE). Allocative efficiency can be

IOP Conf. Series: Earth and Environmental Science **335** (2019) 012006 doi:10.1088/1755-1315/335/1/012006

calculated by dividing the value of Economic Efficiency (EE) with ET. For this reason, the EE value must be calculated first. Before calculating the EE value, from the stochastic frontier regression function, the dual frontier cost function must be derived with the following equation:

$$C_i = C(y_i, p_i, \beta_i) + u_i \quad (4)$$

where:

C_i = cost production

y_i = numbers of output

p_i = input price

β_i = parameter coefficient

u_i = error term (cost inefficiency effect)

Economic efficiency can be calculated based on the ratio of the total observed minimum production costs (C*) to the total actual production costs (C), as follows [6]:

$$EE_i = \frac{C_i^*}{C_i} = \frac{E(C_i \mid u_i = 0, y_i, p_i)}{E(C_i \mid u_i, y_i, p_i)} \quad (5)$$

4. Smart farming concept

The concept of smart farming in this paper emphasizes the garlic business process carried out by anticipating the phenomenon of Industrial Revolution 4.0. Indonesia as part of a global system that is moving towards digital business must adapt quickly. Attempt to increase garlic production is not enough. Farmers and stakeholders together with government and business people must create an integrated smart farming concept, one of which is through the application of Cyber Physical System (CPS) integrated with the Internet of Things and Services (IoT and IoS) into the industrial process [12]. CPS technology is chosen because it has a concept that combines the real world with cyberspace. This merger can be realized through the integration between physical processes and computation in a closed manner (**figure 3**).

[22] Concluded that the application of Industrial Revolution 4.0 provides high benefits for business processes and also the perpetrators, the benefits in are:

- Development in product is faster, flexible and efficient in utilizing resources;
- Improved productivity and increased income, increased demand for labor skilled and increased investment;
- Data utilization from each of existing production process;
- Fulfill the individual needs, more dynamic processes and optimal decision making; and
- Manifest the efficient and smart production process according to consumer needs at a reasonable cost.

Industry 4.0 focuses on making intelligent products, procedures and processes. Producers in industry 4.0 are described as 'Smart Factory' who are able to manage complexity, resistant to interference and able to produce goods more efficiently. Activities that occur in smart factories involve humans, machines, and resources that communicate with each other naturally as in social networks [12].

Similar with the Industrial Revolution 4.0, the term of Agriculture 4.0 is also known in the agricultural sector, namely the agricultural revolution in the future and must be green, approached with science and technology. Agriculture 4.0 in its application consider the demand and supply chain, uses technology not only for the sake of innovation but also to improve and handle actual consumer needs. The future will use advanced technologies such as robots, temperature and humidity sensors, aerial images, and GPS. This progress enables businesses to be better, more efficient, safer, and environmentally friendly [21].

Furthermore [21], describes the use of new technology and new solutions in Agriculture 4.0 could giving hope to the problem of food scarcity. Through the application of Agriculture 4.0, it is expected to produce different products using new technology, bringing food production that has been produced to consumers and increasing efficiency in the food chain by utilizing cross-industry technologies and applications. The world challenge in producing food will be greater, the world must produce 70% more food by 2050, use less energy, fertilizer, and pesticides. On the other hand, it also important to reduce the level of greenhouse gas emissions and overcome climate change effects. Old technology must be maximized, and new ones must be produced.

5. Discussion and solutions

The study of the Indonesian garlic farming system in this paper has revealed several obstacles to future development, such as the availability of land which is ready for planting, the readiness of local varieties suitable for both highland and lowland, efficient cultivation technology, even the marketing aspects of garlic products that must be ready to compete with imported products, are some of the tasks of many obstacles that must be found a solution. Especially if it is related to the Agriculture 4.0 concept as described previously, the effort to achieve garlic self-sufficiency must be well designed.

There are two questions related to the development of garlic, i.e. how to increase garlic production so that it can help the achievement of self-sufficiency in 2021?, and how to answer the challenge of the garlic industry facing the Agriculture 4.0 era so that production can compete with imported garlic? The second question is not an easy job in its implementation because careful planning is needed. Therefore, through this paper the concept of "Analysis of Farming Efficiency and Development of Smart Garlic Agriculture" was presented. The concept approach is both presented in **figure 3**.

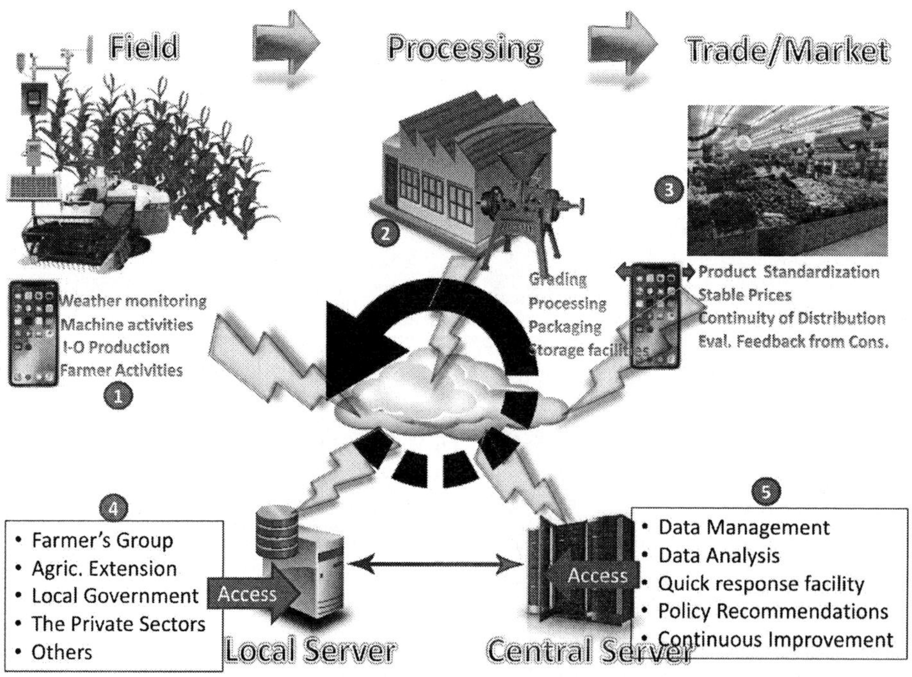

Figure 3. The Concept of Smart Farming on Garlic Farming

There are five design elements in the concept of smart farming on garlic farming and each other is interrelated. The function of the five elements can be explained as follows:

- Element 1 - Field Activities
  In the field activities, there are many activities that must be carried out, starting from land preparation, preparing the cultivation model, monitoring the use of production inputs, observing climate and monitoring the activities of agricultural machinery used during the production process. In this field activity, farmers are required to work together in one or more groups. Each group is guided by field experts. Activities carried out by farmers are monitored every day using manual worksheets. Farmers are asked to fill their daily activities during the cultivation process of garlic. All data is inputted into the computer system and stored on the local server. The data collected in this field activity is analyzed periodically to see the level of efficiency of garlic farming from each farmer using an econometric approach referring to equation (1) to equation (5). The indicators that are monitored include Technical Efficiency (TE), Economic Efficiency (EE), and in the long run followed by Environmental Efficiency analysis (EEnv). In addition, another analysis is carried out such as aspects of farming inefficiency or often called the analysis of managerial abilities of each farmer. The results of the analysis that have been tested, then made a model with computer that can be used as a monitoring activity periodically.
- Element 2 - Processing Activities
  Processing is the second stage activity after the farmer completes field activities. Activities in this phase include grading, processing, packaging and storage facilities. In Agriculture 4.0 implementation, this stage is a determining factor that bridges products from farmers to the market. Farmer products are usually diverse, so that with the construction of processing elements can be obtained a homogeneous and well-classified product. In conventional market systems, processing activities usually involve many stakeholders such as collectors, sub-district traders, wholesalers and so on. A breakthrough is also made through this stage so that farmers' products can be managed properly. For example, if production is abundant it must be carried out by storage to maintain the price effects that usually fall when the product is abundant. Storage facilities function is important to be held. All activities on the processing element are made by computer applications that are integrated in the server, so that it can be traced back to the products sent by farmers. In the long run, the grading, processing and packaging process is carried out by farmers through coordination in the respective farmer groups according to agreed quality. The processing system detects special codes from the products produced by each farmer. All farmers' products must meet agreed standards, then managed to be distributed according to market needs.
- Element 3 - Trade/Market Activities
  The marketing of garlic products from farmers is done by e-commerce approach, meaning that customers are given information, promotion and education to be able to access products produced by farmers online. However, before e-commerce is run, the preparation of standardization and product quality must be well developed, as well as the stock management must have been managed properly. In addition to marketing through the e-commerce system, products produced by farmers are also marketed through conventional mechanisms by selectively marketing consumers. At this stage, an application was also built that contained information on the origin of the product, when it was produced, product standards, and the application of complaints from consumers to the products purchased through application installed on the handphone. Thus, it is expected that all complaints or entries are recorded in the computer system and become evaluation material for improvement.
- Element 4 – Data Management in the Development Area (Local Server)
  In each development area, an operational sub-system is built where all activities are recorded on a local server. Not only data on production process activities are recorded on the local server, but all

other supporting activities such as climate data, observations of plant pests and diseases, farmer group activities, and other supporting data. All data is managed properly and correctly and is integrated with the central server.

- Element 5 – Data Management in the Center (Central Server)
  At the central level where the main analysts, planners and policy makers have activities supported by adequate facilities, including human resources who have various skills, real time data stored in the central server. Through these facilities various activities such as overall activity management, data analysis, complaint handling, policy recommendations and continuous improvement are formulated comprehensively. The server at the center is also designed to function as a data warehouse for other activities in the future.

The design and development of an efficient garlic farming system is still a concept and must be prepared comprehensively. Especially if it is associated with the Agriculture 4.0 phenomenon, it is necessary to take various follow-up steps in the development plan area of Parigi Moutong Regency, Central Sulawesi, including:

- Feasibility study
- Market survey
- Trial
- Socialization to farmers and interested farmers
- Development research in collaboration with academics / universities
- Study of product marketing

Agriculture 4.0 which is used as a reference in this paper is expected to provide benefits for the welfare of society in the future. On the other hand, it is also necessary to think about other impacts such as the emergence of resistance to changes in demography and social aspects, uncertainty in political conditions, limited resources, the risk of natural disasters and the demands for the application which have environmentally friendly technologies. But the real event hasn't happened yet and is still in the form of ideas [14].

6. Conclusion

Based on the results of reviews from various literature, the concept of garlic farming efficiency and management has been obtained comprehensively towards smart farming based on Agriculture 4.0 approach. Furthermore, the concept can be tested in areas that have been selected with various improvements. Thus, it is expected that improvements in concepts that have been tested to be replicated in other regions are expected to support the achievement of garlic self-sufficiency in 2021.

Agriculture 4.0, which has been echoed at the international level, has not yet occurred. Besides giving positive benefits, it is estimated that it can also have other social, political and natural disaster risks. So that scientific studies need to be carried out by collaborating with various parties including academics from universities.

References

[1] Farrel M J 1957 The measurement of productive efficiency *Journal of the Royal Statistical Society* **120(3)** 253-281

[2] Lau L J and Yotopulos P A 1971 A Test for relative efficiency and application to Indian agriculture *The American Economic Review* **61(1)** 94-109

[3] Battese G E and Coelli T J 1995 A model for technical inefficiency effects in a stochastic frontier production for panel data *Empirical Economics* **20** 325-332

[4] Coelli T J 1996 A guide to FRONTIER version 4.1: a computer program for stochastic frontier production and cost function estimation *CEPA Working Papers* Department of Econometrics University of New England (Australia: Armidale NSW 2351)

[5] Byantoro A, Wahyudi and Suyanto I 2004 Mapping of groundwater aquifers byonding resistivity method of Petapa, Pelawa and Binangga villages, Parigi Subdistrict, Parigi Moutong District, Central Sulawesi Province *Proceedings of the 29th Indonesian Geophysics Association* 582-591

[6] Ogundari K and Ojo S O 2006 An examination of technical, economic and allocative efficiency of small farmers: the case study of cassava farmers in Osun State of Nigeria *Journal of Central European Agriculture* **7(3)** 423-432

[7] Sarwadana S M and Gunadi I M A 2007 Potential development of garlic (*Alliun sativum* L.) lowland local varieties Sanur *Agritop* **26(1)** 19-23

[8] Kurniawan A Y 2008 *Analysis of Economic Efficiency and Competitiveness of Maize Farming on Dry Land in Tanah Laut District, South Kalimantan* [Thesis] (Indonesia: Bogor Agricultural University)

[9] Widyananto C S 2010 *Analysis of the Efficiency of the use of Production Factors on Garlic Farming (Case Study in Sopuran Sub-district, Wonosobo District*) (Indonesia: Faculty of Economics, Diponegoro University)

[10] Mkhabella T S 2011 *An Econometric Analysis of The Economic and Environmental Efficiency of Dairy Farms in the Kwa Zulu-Natal Midlands* [Dissertation] (South Africa: Stellenbosch University)

[11] Samijan, Prastiti T R and Pramono J 2011 Intensification of garlic cultivation *Minutes of the results of the study of "Horticultural Innovation in Central Java" BPTP Jawa Tengah* 62-81

[12] Kagermann H, Wahlster W and Heibig J 2013 Recommendations for implementing the strategic initiative industry 4.0 *Final report of the Industrie 4.0 working group* https://www.din.de/blob/76902/e8cac883f42bf28536e7e8165993f1fd/recommendations-for-implementing-industry-4-0-data.pdf [Acessed September 9, 2018]

[13] Kaisar I 2014 *Growth of Garlic (Allium sativum L.) Explants on Some Sucrose Concentration S And Activated Charcoal* [Thesis] (Bengkulu: Faculty of Agriculture, University of Bengkulu)

[14] Drath R and Horch A 2014 Industrie 4.0: hit or yype? *IEEE industrial electronics magazine* **8**(2) 56-58

[15] Miraj N and Ali S 2014 Estimation of technical efficiency of garlic farms in district Peshwar, Pakistan: A stochasticfrontier analysis *IJISR* **9(1)** 140-149

[16] Suriaatmaja M E, Muhaimin A W and Anindita R 2014 *Analysis of the Technical Efficiency of Garlic Farming (Case Study in Giripurno Village, Bumiaji Sub-District, Batu City*) http://agrise.ub.ac.id/index.php/agrise/article/view/119 [Acessed September 7, 2018]

[17] Hariwibowo P A, Anindita R and Suhartini 2015 The evaluation of Indonesia import policies of garlic *Green journal of business and management* **5(1)** 16-30

[18] Putri A H, Haryanto E T and Purnomo D 2015 Optimization of garlic tissue culture with various concentrations of yeast extract *Journal of sustainable agriculture* **30(1)** 30-32

[19] Abd El-Hady M and Eldasdiry E I 2016 Effect of different soil conditioners application on some soil characteristics and plant growthV-Effect of bentonite rates on the some soil chemical properties *International Journal of ChemTech Research* **9(10)** 38-44

[20] Hardiyanto, Devy N F and Supriyanto A 2017 Exploration, characterization and evaluation of several local garlic clones *J.Hort.* **17(4)** 307-313

[21] De Clercq M, Vats A and Biel A 2018 Agriculture 4.0: the future of farming technology *World Government Summit* https://www.worldgovernmentsummit.org/api/publications/document?id=95df8ac4-e97c-6578-b2f8-ff0000a7ddb6 [Acessed September 9, 2018]

[22] Prasetyo H and Sutopo W 2018 Industry 4.0: review of the classification of aspects and directions of research development *Jurnal Teknik Industri* **13(1)** 17-26

[23] [Pusdatin] Pusat Data dan Sistem Informasi Pertanian 2018 *Agricultural Statistics 2018* (Jakarta: Draft Pusat Data dan Sistem Informasi Pertanian, Kementerian Pertanian)

[24] Reily M 2018 *Ministry of Agriculture is targeting garlic self-sufficiency in 2021* https://katadata.co.id/berita/2018/04/12/lahan-tersedia-kementan-targetkan-swasembada-bawang-putih-2021 [Acessed August 22, 2018]

[25] Yun W 2018 2018 will be the most difficult year for garlic export: Chinese garlic at risk https://www.tridge.com/stories/2018-will-be-the-most-difficult-year-garlic-export-chinese-garlic-at-risk [Acessed September 6, 2018]

Artificial neural network based particle swarm optimization in predictions mortality rate of broiler chicken

D Hasman[1,3], D Gustian[2], R B Saputra[2], H Rahmasari[2], Lorentina[2] and S A Ratnasari[2]

[1] Department of Information Engineering, Nusaputra University, Indonesia e-mail: d_hasman@yahoo.co.id
[2] Department of Information System, Nusaputra University, Indonesia

Abstract. Success in broiler farms can be assessed from the mortality rate of chickens and also a solution to reduce the mortality rate of chickens. But in the process there are obstacles that is the death of hickens that tend to fluctuate so that of course resulted in financial losses for farmers. To prevent it required a method that can predict and control the mortality rate of broiler chickens. In this research, data mining method used is Artificial Neural Network (ANN) based on Particle Swarm Optimization (PSO), to produce accurate prediction with excellent iteration and also small error rate. The results of analysis in predicting mortality rate of broiler chickens by artificial neural network method in combination with Particle swarm optimization get better RMSE result (0.135) than Artificial Neural Network have not been optimized (0.381) with january mortality data and result of application quisioner made value 83,125 which can be categorized well and enough help the farmer in controlling prediction of chicken mortality rate broiler.

Keywords: artificial neural network, mortality rate, particle swarm optimization

1. Introduction

PT. Jafpa Comfeed Indonesia, Tbk Kalapa Nunggal 2 farm unit is one of the branches of PT. Jafpa Comfeed Indonesia Tbk is involved in breeding chickens, broilers, laying hens with a population of approximately 50.000 chickens. The large number of chicken populations, on the other hand, resulted in significant death. So that it can cause some losses such as in terms of financial, time, human resources etc. The company is trying to keep the death rate to a minimum. This is because the condition of each farmer has difficulty in handling conditions that are fast and efficient. The company wants the death rate to be kept to a minimum of around 10% of the total deaths, so as to reduce losses in financial terms.

In figure 1 can be seen that there is a problem that the number of uncertain mortality even in the second quarter of 2015 decreased chicken mortality, but a few months after the death of chicken increased significantly.

The method to be used is Particle Swarm Optimization (PSO) Neural Network (ANN), to produce accurate prediction value with excellent iteration and also small error rate. Prediction on back propagation neural network in the training and validation process to produce a percentage of 100% with one hidden layer architecture, the optimal parameters MSE = 0.0001, learning rate = 0 , 9, momentum = 0.4. As for the prediction of learners' achievements in the fuzzy inference system mamdani method by using S curve and bell curve (PI curve) to produce a percentage of 83.8%. [1]. While the method for optimization used the PSO method, because it is better than the method of genetic algorithm (GA). Classification accuracy of PSO based SOFM is 93.8 % and classification accuracy of GA based SOFM is 87.7 %.a comparative analysis is performed between the PSO and the GA optimization technique for abnormal retinal image classification. A 5-10% increase in classification accuracy is obtained for the PSO based technique over the GA based technique. The computational time period is reduced by 50% for the PSO based technique when compared with the GA based technique. [2]. It is expected that the PSO-based ANN method can be more accurate with a small RMSE value in predicting mortality rate of broiler chickens.

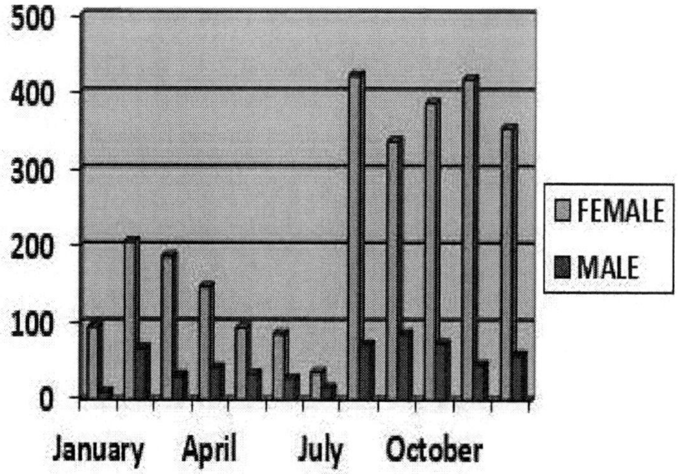

Figure 1. Broiler chicken mortality data 2015.

Therefore, with this method may provide a solution to control the chicken deaths are likely to increase, especially hen, As well as cause exists can be avoided. In addition it can control the mortality of chickens that tend to fluctuate and lead to more dead chickens compared with live chickens. So this condition is certainly very detrimental, especially the management.

This study can contribute to the management in predicting mortality rate of chicken and the factors causing death, so it is expected to take preventive measures early and can avoid the cause of death of chickens. Beside that, it can significantly increase profits for the company in the future because of the reduced death of the chicken.

2. Literature Review

Mohd Khalid Awang, Fadzilah Siraj (2013) conducted a study entitled "Utilization of Artificial Neural Networks in Prediction of Heart Disease". The problem under study is In addition to developing quality of life, increasing the economic burden of individual cardiovascular, family and community, and reducing the productivity of the State. Collection of data used as variables that most influence predictions, costs, and risks. Premature, smoking, cholesterol, hypertension, diabetes mellitus, hypercholesterolemia. The method used for Artificial Neural Networks that produces the best network model produces a predictive value of 88.89 percent. In addition, computer networks in medical diagnosis, this study also shows that brain tissue can be incorporated into the hospital information system as a predictor [3].

Abhijit Suresha, K.V Harisha, N. Radhikaa (2015) conducted a study entitled "Particle swarm optimization over back propagation neural network for the duration of prediction". Collection of data used in this study Input data obtained from the Index, the Main, and the Charlson Index. Special skills include categories such as Surgery, Internal and Emergency. Where this data is to predict the mortality of patients with co-morbid conditions. They compared the Artificial Neural Network with Partic Optimization Optimization Method which is a part that uses Particle Swarm Optimization (PSO) techniques compared to training for training. The algorithm evaluated based on the convergence of errors, sensitivity, specificity, value of precision and accuracy and the corresponding results are presented. Where there is an increase in the accuracy value of the ANN results optimized by PSO [4].

Reza Gharoie Ahangar, Mahmood Yahyazadehfar, Hassan Pournaghshband (2010) in their journal entitled "The Comparison of Methods Artificial Neural Network with Linear Regression Using Specific Variables for Prediction Stock Price in Tehran Stock Exchange" suggests that the Artificial Neural Network method is better and faster than the Linear Regression method because the Artificial Neural Network method can run parallel calculations and tolerate more errors and also this network can create rules without an implicit formula [5].

Seyed Ali Jafaria, Sepideh Jahandidehb, Mina Jahandidehc & Ebrahim Barzegari Asadabadid (2015). Conduct research with the title "Prediction of the level of road traffic death using neural networks optimized by genetic algorithms". The problem in this study is that traffic injury (RTI) is realized as a major cause of global, regional and national health problems. The dataset used as many as 178 countries with different Human Development Index (HDI) values (very high, high, medium and low). They use neural network methods optimized by genetic algorithms. Allows communication that can optimize artificial neural networks in predicting death with high performance with the lowest RMSE 0.0808 [6].

KienEe Lee, Izzatdin bin Abdul Aziz, Jafreezal bin Jaafar (2017) in their journal entitled "Adaptive Multilayered Particle Swarm Optimized Neural Network (AMPSONN) for Pipeline Corrosion Prediction" in their research suggests thatthe results have also revealed that the AMPSONN method shows a better performance even when the topologies of all methods are optimized, and when the size of training set is reduced. The adaptive multilayered PSO approach has proven itself to be a promising solution to target various individual damage mechanisms and it is also possible to use more datasets in the testing of the AMPSONN method to prove its effectiveness in other application areas [7].

SomayehYavari,Mohammad Javad Valadan Zoej, Mehdi Mokhtarzade, Ali Mohammadzadeh (2012) in their journal entitled "Comparison of Particle Swarm Optimization and Genetic Algorithm in rational function model Optimization" suggests that the results proved that both GA and PSO are able to determine the optimal terms of RFM to achieve rather the same accuracy. However, PSO shows to be more effective from computational time part of view. The other important achievement is that the algorithms are able to solve the RFM using less GCPs with higher accuracy in comparison to conventional RFM [8].

2.1. Framework of thinking

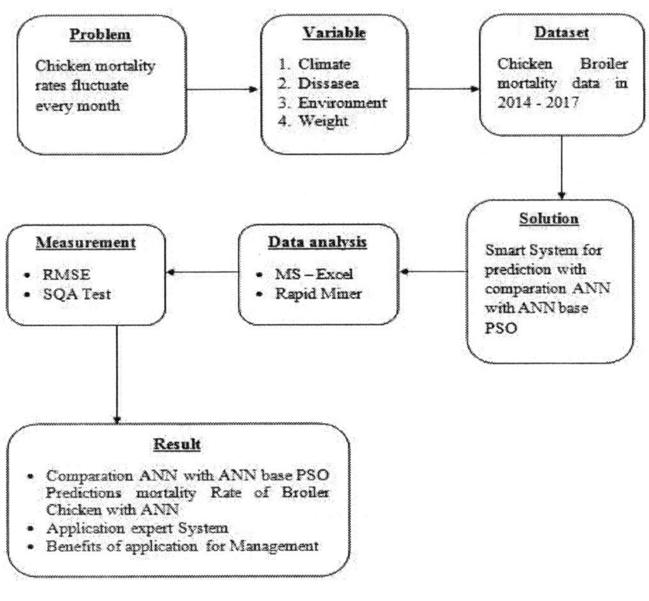

IOP Conf. Series: Earth and Environmental Science **335** (2019) 012007 doi:10.1088/1755-1315/335/1/012007

Figure 2. Framework of thinking.

3. Methodology

In this research used two methods that support each other is Particle Swarm Optimization Neural Network.

3.1. Step one with Artifical Neural Network [9]

For each input unit (x_i, i=1,2,3,...n) receive signal x_i and pass the signal on the layer above it (hidden layer). Every hidden layer (z_j, j=1,2,3,...,p) adds weighted input signals:

$$z\_in_j = v0_j + \sum_{i=1}^{n} xi\, v_{ij} \tag{1}$$

Use the activation function to calculate the output signal:z_j = f (z_in_j)and send the signal to all units in the top layer (output units). For each output units (y_k, k=1,2,3,...,m) calculate weighted input signals"

$$y\_in_k = w0_k + \sum_{j=1}^{p} z_j w_{jk} \tag{2}$$

Use the activation function to calculate the output signal.y_k= f (y_in_k)and send the signal to all units in the top layer (output units). For each output unit (yk, k=1,2,3,...,m) receiving target patterns connected with the learning input pattern. Then calculate the error information.

$$\sigma k = (tk - yk)\, f'(y\_ink) \tag{3}$$

then calculate the weight correction (which will be used to fix w_{jk}value) :

$$\Delta wjk = \alpha\, \sigma k\, zij. \tag{4}$$

also calculate the bias correction (which will be used to fix w_{0k}value) :

$$\Delta wjk = \alpha\, \sigma k. \tag{5}$$

send σ_k to the units in the bottom layer. Each hidden units (z_j, j=1,2,3,...,p) calculate the delta units (from units in the top layer):

$$\sigma\_in_j = \sum_{k=1}^{m} \sigma k\, w_{jk} \tag{6}$$

multiply this value withderivativefromits activation function to calculate information of error:

$$\sigma j = \sigma\_inj\, f'(z\_inj) \tag{7}$$

also calculate the bias correction (which will be used to fix v_{0j}value):

$$\Delta vjk = \alpha\, \sigma j\, xi \tag{8}$$

also calculate the bias correction (which will be used to fix v_{0j}value):

$$\Delta v0j = \alpha\, \sigma j. \tag{9}$$

Each output unit (y_k, k=1,2,3,...,m) improve bias and weight (j=0,1,2,3,...,p). w_{jk} (new) = w_{jk}(old) + Δw_{jk} for each hidden layer (z_j, j=1,2,3,...,p) improve bias and weight (i=0,1,2,3,...,n)

IOP Conf. Series: Earth and Environmental Science **335** (2019) 012007 doi:10.1088/1755-1315/335/1/012007

$$\mathrm{vij(new)} = \mathrm{vij(old)} + \Delta \mathrm{vij} \tag{10}$$

Test condition stopped [10].

3.2. Step two with particle swarm optimization

Where: X = particle position, V= particle velocity, w= weight of inertia, c1, c2 = acceleration coefficient, P= number of particles in swarm. The values of r1 and r2 are arranged so that the random value is intended to provide stochastic properties in the cognitive component and social components This stochastic nature causes each particle to move in a semi random way, strongly influenced in the direction of the best solution of the particles and the best globalsolution of swarm [11].

3.3. Step three software testing with SQA

There are three points of understanding of software quality, including as follows: (1) software requirements are the foundation from whichquality is measured; (2) specific standards that define development criteria thatguide the manufacture of a software; (3) there are implicit needs often overlooked (eg, desire for the best maintenance); and (4) scores of responder = <audibility score> * 0.10 + <scoreaccuracy> * 0.10 + completeness score> * 0.15 + <errortolerance score> * 0.10 + <score execution efficiency> *0.10 + <operability score> * 0.15 + < Simplicity score> *0.15 + <learning score> * 0.15 Evaluation based on criteria /average score of respondents [12].

4. Results and discussions

4.1. ANN Method

4.1.1. Dataset. The dataset used for this study is data on the death of broiler chickens PT. Jafpa Comfeed Indonesia Tbk unit kalapanunggal 2 from 2014 to 2017. The analysis process that predicts the mortality of broiler chickens for 2018 using the average data from 2014 to 2017 is reduced by 10% and the prediction for 2019 using 2018 data minus 10% in accordance with requests from management.

4.1.2. ANN Model. ANN model used is a backpropagation model which means a method that minimizes errors in output generated by the network. In Figure 3 below is the input unit setting, hidden layer and output on backpropagation. The hidden layer is used as much as 2 hidden layers. Input data is monthly data from chicken mortality data for the period 2014 to 2017 while the output unit is prediction data.

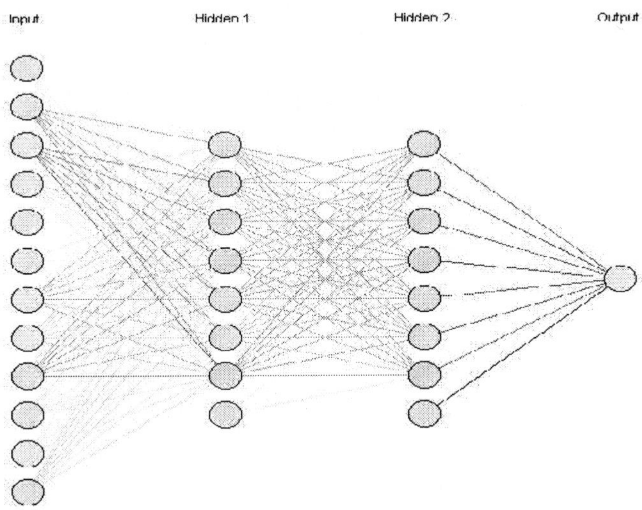

Figure 3. Configuration ANN formed.

In this study the settings used for artificial neural networks are 500 for training cycles, 0.1 for 0.1 for learning rates, and 0.5 for momentum. For configuration settings used for artificial neural networks as shown in the figure 4 below.

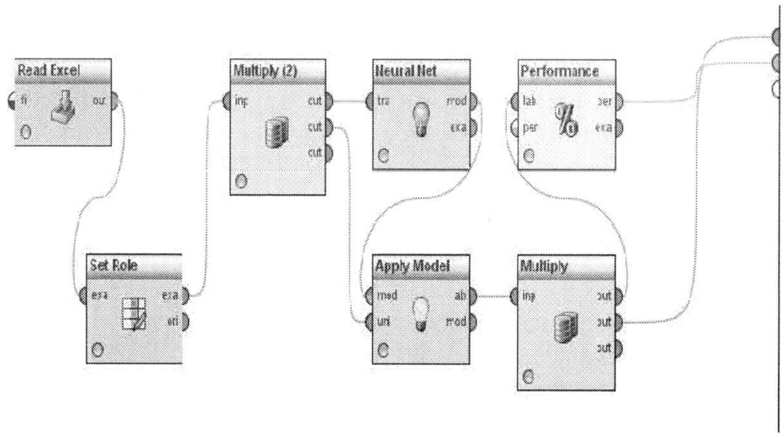

Figure 4. Configuration ANN created.

4.1.3. PSO-based neural networks. The configuration for the Artificial Neural Network method is based on Particle swarm optimization (PSO) which can be seen from the picture below:

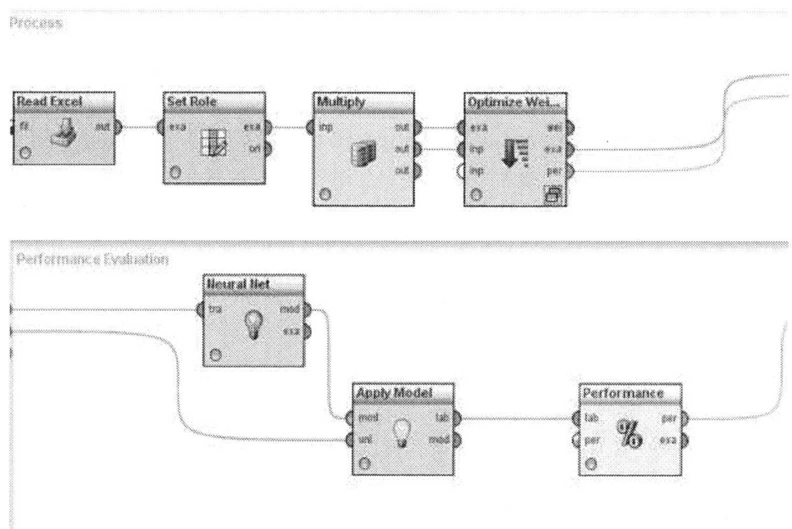

Figure 5. Particle swarm optimized neural network configuration.

4.1.4. Comparison between ANN and PSO-based ANN. The RMSE results obtained from this study using artificial neural networks and artificial neural networks with PSO can be seen in figure 6 below:

	JST	JST + PSO
Januari	9,710	7,199
Februari	8,767	3,549
Maret	0,785	0,589
April	13,141	2,870
Mei	2,583	0,493
Juni	0,730	0,223
Juli	1,028	0,227
Agustus	8,017	1,467
September	14,115	2,296
Oktober	5,737	1,115
November	4,198	0,612
Desember	5,854	2,032

Figure 6. Comparison of RMSE.

If seen from table 1 above it can be concluded that the results of ANN + PSO RMSE obtained less than the results obtained from the RMSE ANN shows that the RMSE from ANN + PSO can optimize the RMSE value obtained from ANN. For more details, see figure 7 below:

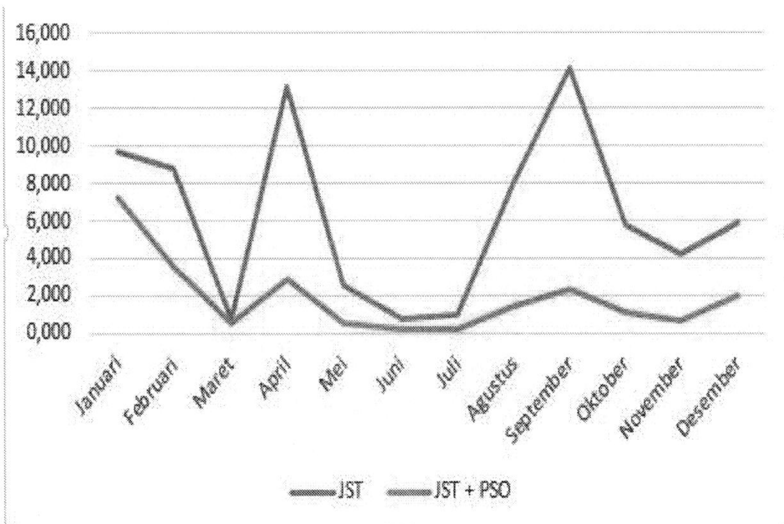

Figure 7. Comparison graph of RMSE results.

4.1.5. Prediction result data.

IOP Conf. Series: Earth and Environmental Science **335** (2019) 012007 doi:10.1088/1755-1315/335/1/012007

	Target 2018	Prediction 2018	target 2018	Prediction 2018
Januari	205	220	198	206
Februari	306	319	287	295
Maret	301	299	269	266
April	343	364	328	342
May	194	198	178	179
June	150	151	136	135
July	122	123	111	110
August	505	517	465	468
September	253	275	248	260
Oktober	297	306	275	285
November	344	336	302	314
Desember	322	332	299	301

Figure 8. Predicted data.

In this study, the analysis process that predicted broiler mortality for 2018 using the average data from 2014 to 2017 was reduced by 10% and the prediction for 2019 using 2018 data was reduced by 10% in accordance with the request from the management.

4.1.6. Implementation of the system. There are two diagrams presented use case diagram and class diagram.

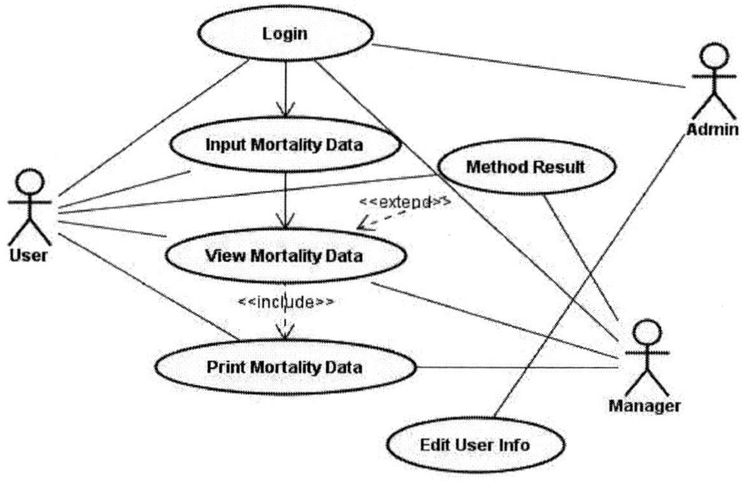

Figure 9. Use case designed.

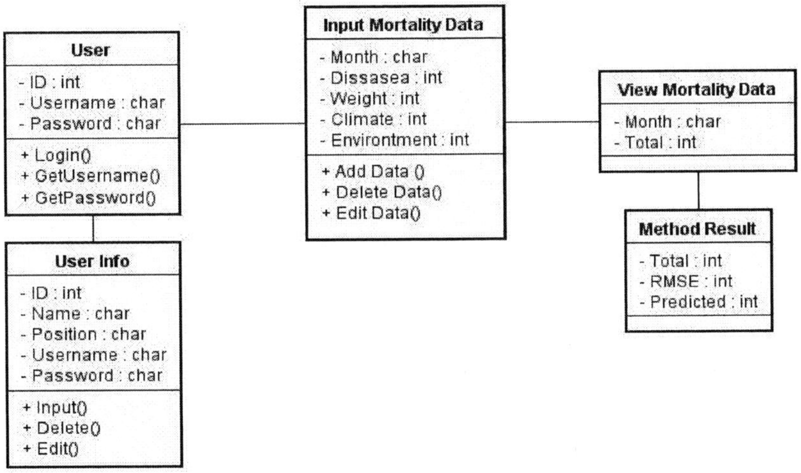

Figure 10. Class diagram created.

4.1.7. Interface created.

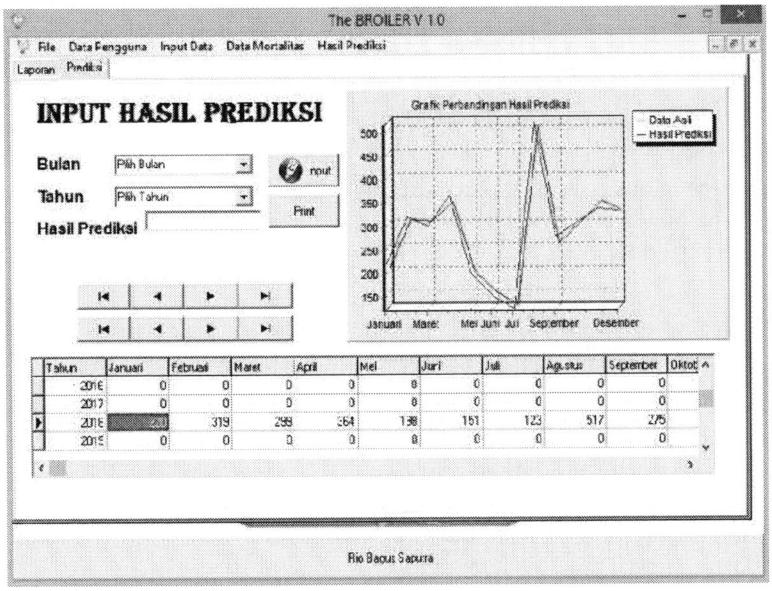

Figure 11. Applications created.

4.1.8. Interface created. To ensure that the software is created by SQA (Software Quality Assurance) method.Score = <85> * 0.125 + <82.6> * 0.125 + <83.4> * 0.125 + <82 > * 0.125 + <81.8> * 0.125 + <85.6> * 0.125 + <87.4> * 0.125 + <77.2> * 0.125. The resulting average score is 81,625, while the optimalvalue for a software that meets the quality standards based on theSQA test is 83.125.

4.2. Discussions

In this study, the analysis process that predicted broiler mortality for 2018 using the average data from 2014 to 2017 was reduced by 10% and the prediction for 2019 using 2018 data was reduced by 10% in accordance with the request from the management. While the application used in the calculation method is Rapidminer 5 with the settings used for Artificial Neural Networks, there are input units, output units and 2 hidden layers, 500 for training cycles, 0.1 for learning rates, and 0.5 for momentum and at optimize with the particle swarm optimization method. The RMSE results obtained from the method of artificial neural networks with backpropagation model based on particle swarm optimization are 5,854 (December), this is better than the results obtained from the artificial neural network method with backpropagation model before optimization is 2.032 (december). It can be seen from Figure 4.3 that the results of RMSE ANN + PSO are better than ANN before optimization. In addition to the results from RMSE, it was also predicted that broiler chickens would die for the coming year, which is the result of predictions from December that reached 332 broiler chickens mortality for 2018, 301 mortality for 2019. After getting the RMSE results and prediction results, then input into an application implemented with Delphi 7 which is designed to predict the death of broiler chickens to facilitate users in operation. With Admin to process application user data, User or Application user (Employee) who is tasked to input chicken death data, prediction data, RMSE results data obtained from calculations using Rapidminer 5 and Manager who can see and control all systems then there was an application test with the SQA method by distributing the quisioner to the relevant staff.

5. Conclusions

For accuracy and error resulting from the Neural Network method in predicting the mortality rate of chicken boiler, it was 5,854 (december). While the average RMSE results obtained from the method of artificial neural networks which is 6,222 each month. Existing data cannot affect the RMSE results obtained as evidenced by the results of the RMSE obtained in September is the highest result that is equal to 14.115 while the lowest results can be seen in June that is equal to 0.730. While the RMSE results obtained from the method of artificial neural networks with backpropagation models are enhanced by particle swarm optimization which is 2,032 (december). For the average RMSE obtained is equal to 1.889 better by 4.3% of the neural network method with the backpropagation model before being optimized. With the highest RMSE of 7.119 and the lowest RMSE of 0.223. And by using an application that can run the Neural Network method based on Particle Swarm Optimization and with the results of predictions with a good level of error and accuracy and the results of the application test from the management to the audience with the SQA results obtained that is equal to 83.125 this shows that the management is helped with this application and also take the steps needed to prevent mortality of broiler chickens from the prediction of ANN and PSO methods.

References

[1] Siti Helmiyah, Shofwatul 'Uyun, 2016. Comparison of Performance of Artificial Neural Networks and Fuzzy Inference System for Prediction of Students Journal of Informatics Engineering and Information Systems, ISSN: 2407-4322.

[2] J. Anitha, C.Kezi Selva Vijila D.Jude Hemanth. 2009. Comparative Analysis of Genetic Algorithm & Particle Swarm Optimization Techniques for SOFM Based Abnormal Retinal Image Classification", International Journal of Recent Trends in Engineering, Vol. 2, No. 3.

[3] Mohd Khalid Awang and Fadzilah Siraj, 2013. Utilization of an Artificial Neural Network in the Prediction of Heart Disease", International Journal of Bio-Science and Bio-Technology, Vol. 5, No. 4

[4] Abhijit Suresha, K.V Harish, N. Radhika, 2015. Particle Swarm Optimization Over Back Propagation Neural Network for length of stay prediction", International Conference on Information and Communication Technologies (ICICT 2014), Procedia Computer Science 46 (2015) 268 – 275

[5] Reza Gharoie Ahangar, Mahmood Yahyazadehfar, Hassan Pournaghshband, 2010. The Comparison of Methods Artificial Neural Networks with Linear Regression Using Specific Variables for Prediction Stock Prices in the Tehran Stock Exchange" (IJCSIS) International Journal of Computer Science and Information Security, Vol. 7, No. 2.

[6] Seyed Ali Jafari, Sepideh Jahandideh, Mina Jahandideh, Ebrahim Barzegari Asadabadi, 2013. Prediction of road traffic death rate using neural networks optimised by genetic algorithm", International Journal of Injury Control and Safety Promotion.

[7] Kien Ee Lee, Izzatdin bin Abdul Aziz, Jafreezal bin Jaafar, 2017. Adaptive Multilayered Particle

Swarm Optimized Neural Network (AMPSONN) for Pipeline Corrosion Prediction", International Journal of Advanced Computer Science and Applications (IJACSA), Universiti Teknologi PETRONAS, Vol. . 8, No. 11.
[8] Somayeh Yavari, Mohammad Javad Valadan Zoej, Mehdi Mokhtarzade, Ali Mohammad zadeh, 2012. Comparison of Particle Swarm Optimization and Genetic Algorithms in Rational Function Model Optimization" International Archives of the Photogrammetry, Remote Sensing and Spatial Information Sciences, Toosi University of Technology.
[9] Hermawan, A, 2006. Artificial Neural Networks Theory and Applications. Yogyakarta: Publisher Andi.
[10] Puspaningrum, D, 2006. Introduction to Artificial Neural Networks. Yogyakarta: Andi Offset.
[11] Blondin J, 2009. Particle swarm optimization tutorial.
[12] Handayani R.K, 2012. Review of Use of Software Supervision in the Process of Recurring Decision Making: A Case Study of Determination of High School Teachers.

Acknowledgements
Thank Dudih Gustian, Rio Bagus, Hela, etc. who have helped and collaborated with the authors so that they can complete this joint research. I also thank you to the IPB Lecturers who have supported and provided guidance to me. Hopefully this research can provide benefits to the readers and can be applied to the welfare of the community.

Environmental engineering and sustainability for smart agriculture: The application of UAV-based remote sensing to detect biodiversity in oil palm plantations

A Moreno[1], A Bhattacharyya[1], L Jansen[1], Y Arkeman[2], R Hartanto[1] and M Kleinke[1]

[1]Rhine-Waal University of Applied Sciences, Kleve, Germany.
[2]IPB Bogor Agricultural University, Bogor, Indonesia

Abstract. The United Nations strongly support the progress towards sustainable agriculture, food security, sustainable management of life on land in a simultaneous approach. Oil production, though on increasing demand, is causing severe loss of biodiversity in tropical areas, Indonesia being the country with the most environmental damage due to this crop. Geographical Information Systems (GIS) and Unmanned Aerial Vehicles (UAVs) for base-remote sensing are promising tools to help with environmental conservation in oil palm plantations and to improve production efficiency and quality of crops. While different UAVs exist in the market, multi-rotor UAVs have already been used for oil palm plantation monitoring. There are more than 100 different Vegetation Indices (VIs); the Normalized Difference Vegetation Index (NDVI) is often used in smart agriculture. Still, there is a huge need for algorithms that assess floristic diversity. To maintain some elements of the Indonesian biodiversity within oil palm plantations, experimentation with algorithms were carried out. The imagery shows great potential for future research.

Keywords: UAV; remote sensing; vegetation indices; biodiversity; oil palm plantations.

1. Introduction

The United Nations agreement in 2015 aim to generate progress towards sustainable agriculture, food security, climate action, and sustainable management life on land in a simultaneous approach (UN General Assembly, 2015) For that reason, the development of drone-based technologies for oil palm plantations is part of the efforts to make the industry more environmentally sustainable. During the last ten years, the scientific community has given increasing attention to oil palm expansion and its negative consequences for ecosystems (Dislich et al., 2017). The majority of taxonomic groups including fungi, plants, bees, beetles, ants, amphibians, birds, and mammals, show decreased richness and overall abundance in plantations (Foster et al., 2011). Palm oil has become the most consumed vegetable oil in the world, and the global demand for this fat is expected to increase (FAO, 2016). Besides, palm oil has the highest production among other oil crops, such as soybean, rapeseed and sunflower (Chong et al., 2017). The oil palm (*Elaeis guineensis*) is a palm species planted extensively in South-East Asia, especially in Indonesia, Malaysia, and Thailand (Petrenko, Paltseva and Searle, 2016; Chong et al., 2017). In 2017 the global oil palm production rose to 66,855 million tonnes, and Indonesia alone produced 34 Mt (Iskandar et al., 2017). Presently, Indonesia is the world's largest and most rapidly growing producer (FAO, 2017). Indonesia intends to realize and maintain a 7% annual GDP growth rate, and become one of the world's tenth largest economies by 2025 (Anderson et al., 2015). Thought the future demand for edible oil is estimated to amount to around 240,000 Mt in 2050, requiring an additional 12 million ha of palms, if average yields continue to rise as in the past years (Corley, 2009). Moreover, the production of palm oil is highly cost- and area-effective compared to other oil crops, this trend is projected to continue in Southeast Asia and other tropical regions (Fitzherbert *et al.,* 2008). To promote better oil palm management practices is necessary to monitoring oil palm plantations. However, it can be time-consuming and expensive (Chong et al., 2017). In the attempt to use time wisely, we should consider updated measurement techniques, instruments, tools, and technologies to describe vegetation in an area at different scales (Grebner et. 2012). Unmanned Aerial Vehicles (UAVs) have emerged as important and useful platforms for acquiring remotely sensed data of vegetation, offering the ability to collect imagery with high spatial and temporal resolution (Heaphy et al., 2017). The development of a biodiverse and properly functioning oil palm landscape is a vital conservation priority of the modern era. The **objectives** of this project research are to review the UAV technology applications within and oil palm plantations and to present an approach to measure plant diversity by detecting background vegetation. Furthermore, this research could be useful to improve the ecosystem services in oil palm plantations. Considering the role of oil palm for the local economy, food security and environmental conservation, a sustainable management is required (Camara et al., 2018).

2. Smart agriculture

2.1. Precision Agriculture

Diverse definitions for Precision Agriculture (PA) exist and many people have different ideas of what PA should encompass. Precision farming or precision agriculture is a farming or agro-industry management concept based on observing and responding to intra-field variations (Whelan and James, 2010). Smart agriculture can promote variable management practices autonomously within the agricultural fields, according to the site conditions (Norton and Swinton, 2018). It relies on technologies like Global Positioning System (GPS) and remote sensing imagery. This technology is capable to (1) optimize production efficiency; (2) optimize quality; (3) improve environmental site conditions and (4) minimize risk (Whelan and James, 2010).

3. Remote sensing

Remote sensing has become an essential tool in fields such as ecology, agriculture, forestry, geography as image capture from the air provides information about the earth's surface (Coops and Tooke, 2017). Remote sensing is a valuable method to monitor the status and progress of oil palm development, to assist in decision-making for an efficient plantation management and to investigate the effects of oil palm plantations on the environment (Zolkos, Goetz, and Dubayah, 2013; Chong et al., 2017). Thus, remote sensing devices detect a portion of the electromagnetic radiation (table 1), examples are gamma rays, x-rays, ultraviolet, visible light, infrared light, microwaves and radio waves; differences exist concerning their wavelengths (Padua et al., 2017). The sensors used in remote sensing can be categorized as either active or passive (Coops and Tooke, 2017).

According to Shamshiri et al., 2018, remote sensing operations in oil palm plantations can involve several activities, for example 1) Set up and operate a multi-rotor UAV remote sensing system for oil palm plantations; 2) To produce 2D and 3D visual maps of the fields under study; 3) To produce vegetation indices (VI) and Geographical Information System (GIS) maps for health and growth assessment and 4) to record diverse dendrometrics such as canopy size, crown diameters, palms distance and calculate yield per palm and yield per ha (Shamshiri et al., 2018). Despite the increasing importance of new survey tools such as unmanned aerial vehicles (UAVs) and processing software, the implications of how their spatial deployment may interact with species detection have not yet been assessed (Baxter and Hamilton, 2018).

Table 1. Electromagnetic spectrum of remote sensing.

	Ultra Violet	visible	NIR	Water absorption	SWIR	Water absorption	MWIR	LWIR	Micro waves	Radio waves
Wavelenght (nanometers)	10-400	390 – 750	750-1400	1450	1500-2800	2900	3000 - 8000	8000-14000		1mm-1m

IOP Conf. Series: Earth and Environmental Science **335** (2019) 012008 doi:10.1088/1755-1315/335/1/012008

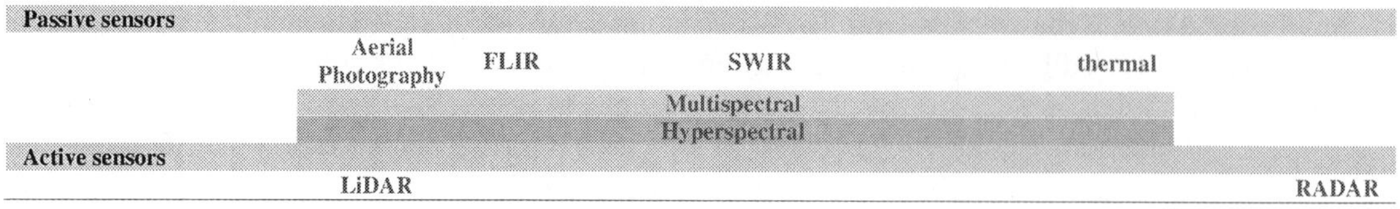

3.1. Sensors and vegetation indices
The essential component for carrying out remote-sensing activities is the imaging or sensing payload, which defines the capabilities and the usability of the UAV (Siebert and Teizer 2014; Padua et al., 2017). From low-cost commercial digital single lens reflex cameras to expensive equipment, such as LiDAR sensors, that are specially designed for drones, various options are available (Heaphy et al., 2017; Klemas 2015). The most common sensors (Figure 1) used for forestry and agriculture are, 1) Red-Green-Blue (RGB); 2) Near Infra-Red (NIR); 3) Multispectral; and 4) light detection and ranging (LiDAR). Examples of the imagery that each sensor generates are shown in Figure 2.

Figure 1. The Most common cameras in the market. **RGB:** (a) GoPro (b) Canon G9X (c) Panasonic DMC (d) Sony Alpha 7; **Near-infrared NIR:** (a) Canon S110 (b) Panasonic Lumix 7 and (c) Fujifilm X-M1; **Multispectral sensors:** (a) Parrot Sequoia; (b) multiSPEC 4C; (c) Tetracam ADC; and (d) MicaSense RedEdge; **LiDAR:** (a) the Routescene lidar Pod; (b) the Yellowscan Mapper; and (c) the Velodyne PUCK (Padua et al., 2017).

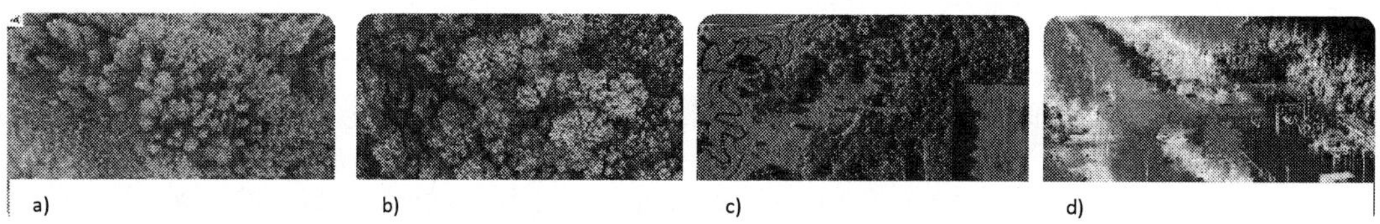

Figure 2. Example of sensor imagery: a) RGB (Goodbody et al., 2017); b) Near-infrared (geovantage.com); c) Multispectral: (skylab.global); d) LiDAR: (flight-evolved.com).

The Red, Green and Blue (RGB) imagery is often used to generate Digital Terrain Models in agriculture and forestry, this sensor is used for tasks where no special spectrum is required, for example; forest canopy gaps inspection (Getzin, Wiegand, and Schöning, 2012). The Near Infra-Red sensors, are used to detect plant health and plant identification (Torresan et al., 2017). LiDAR sensors can penetrate the tree cover, being able to generate accurate measurements of biomass and understory plants, but this type of sensor is more expensive and requires specialized software for imagery analysis (Padua et al., 2017). According to Shamshiri et al., 2018, Multispectral sensors can be used to 1) analyse oil palm stress factors, soil types, the need of fertilizers, or insecticides; 2) identify and differentiate plant species or recognize other plants such as weeds; 3) monitor soil and chemical conditions that are, in each case, able to be identified by their unique spectral signature and 4) graphically illustrate vegetation indices such as the Normalized Difference Vegetation Index (NDVI). A Vegetation Index (VI) refers to the ratio of reflectance values at different wavelengths and is commonly used to understand plant photosynthetic activity and structural variations of plants (Huete et al., 2000; Putra and Soni, 2018). Many VIs have been developed and tested for estimation of biophysical parameters of vegetation (Onyia, Balzter, and Berrio, 2018). Information about plant health, water content, environmental stress are often described as vegetation indices (VIs); there are more than 100 existing VIs, which are grouped into seven

IOP Conf. Series: Earth and Environmental Science **335** (2019) 012008 doi:10.1088/1755-1315/335/1/012008

categories (Xue, and Su, 2017). For example, the category of Broadband Greenness is used to describe the vigor and health of green vegetation (Ozdemir, and Sumer, 2013). Figure 3 shows the most common VIs belonging to the Broadband Greenness category.

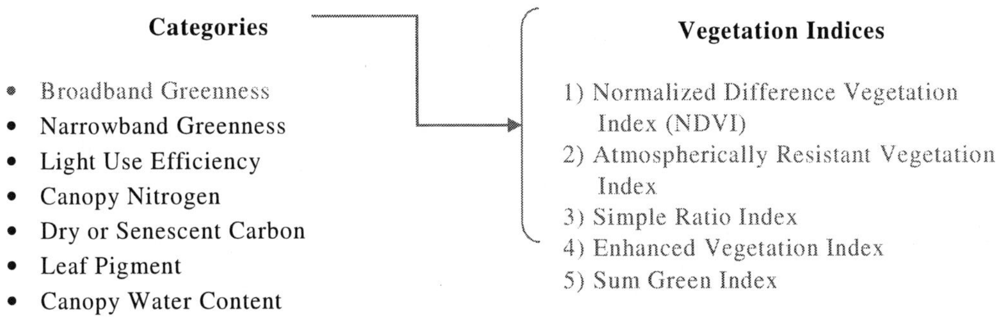

Figure 3. The seven different categories of Vegetation Indices for multispectral sensors. Broadband contains the most common indices used nowadays to describe changes in vegetations by remote sensing.

One of the important tasks of remote sensing operations is identification of native an exotic plant species by using the vegetation indices (Shouse, Liang and Fei 2013); results can propose a better and more adequate plant management within plantations. The determination of vegetation composition by remote sensing is based on the principle, e.g. thaevery tree species has its own spectral signature (Lisein et al., 2015). However, spectral reflectance patterns from plants are very similar even across different species, making differentiation between invasive and native species a very challenging task using remote sensing (Shouse, Liang, and Fei, 2013). For the identification and quantification of plant species, UAV high spatial resolution (at centimeter level) is therefore very useful as spectral differences among species can be detected (Ivosevic, Han, and Kwon, 2017). Just a few studies specifically focused on the identification of background vegetation using remote sensing (Lass et al., 2005, Asner et al., 2008; Ustin and Santos, 2010 and Wäldchen et al., 2018). Given the technological advances in remote sensing and the increasing demand for automated identification of plants among crop systems, this area of research can generate strong contributions to sustainable land management.

3.2. Unmanned Aerial Vehicles (UAVs)

The potential of UAV in the oil palm industry is enormous as it provides monitoring efficiency, high resolution imagery and it is especially useful for tropical countries where clouds are a serious obstacle for satellite image acquisition (Chong et al., 2017). There are two main types of small UAVs: fixed-wing and multi-rotor (Figure 4). Each type has its own advantages depending on the environment and required task (Padula et al., 2017). Advantages of drones are cost-effectiveness for small projects, very high resolution, positional accuracy; while disadvantages are: sensitivity to bad weather and subjection to governmental regulations (Padua et al., 2017).

The **fixed-wing** drone can travel several kilometers from the launch point, being suitable for mapping with applications in land surveying, agriculture, and forestry. This type of UAV can achieve image resolution up to centimetre level. For take-off is possible by hand or small launch ramp, the flight time is usually up to 1 h, and the payload capacity is small (Lisein et al., 2014). The **multi-rotor** drone relies on a set of propellers arranged around its core. Multi-rotor drones have an image resolution at millimeter level. They have vertical take-off and their flight time usually is low (approximately 30 min). The payload capacity depends on the number of rotors (Padua et al., 2017). Because of the low-cost and high efficiency, multi-rotor UAV has been used extensively for forestry, spatial ecology, and agriculture surveys, for example; Rokhmana, 2015; Yilmaz et al., 2017 and Goodbody et al., 2017.

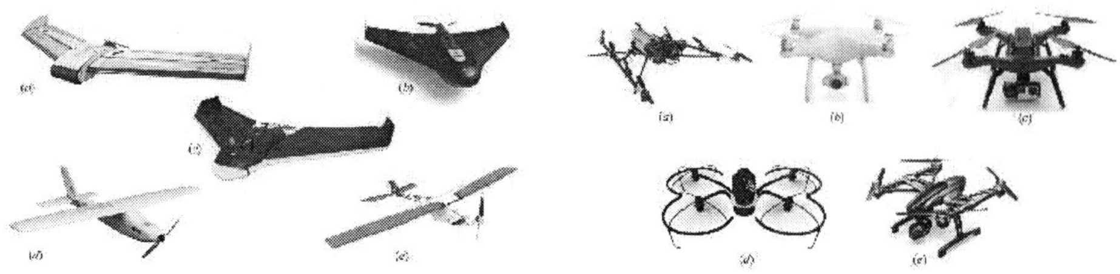

Figure 4. A diverse variety of fixed-wing and multi-rotor drones are offered in the market, being multi-rotor drones widely used for diverse surveys in agriculture and forestry (Padula et al., 2017).

4. Oil palm plantations

4.1. Plantation development of oil palm

Most of the world's species diversity is concentrated in the humid tropical forest, the ideal habitat for oil palm production (Hoffmann et al., 2010). The expansion of oil palm is, therefore, most likely to directly impact tropical biodiversity (Chong et al., 2017). Oil palm plantations without a proper sustainable management plan can affect severely biodiversity and promote land degradation (Schrier-Uijl et al., 2013). According to Dislich et al., (2017) after the productive time of oil palm plantations, in many cases is more feasible to establish a new plantation. The establishment of an oil palm plantation requires to clearing the land, either mechanically or with fire, remove lianas, epiphytes, shrubs and other plants. Between 2007 and 2010, 350,000 ha of oil palm plantations were established annually (Obidzinski et al., 2012). Nowadays, Indonesia is the world's largest and most rapidly growing producer of palm oil (FAO, 2017). About 10% of Indonesia's palm oil production comes from government plantations, 40% from smallholders, and 50% of the plantations are private (IPOC, 2006). Oil palms are now grown throughout 43 countries (18.1 million ha), with Indonesia (7.1 million ha) and Malaysia (4.6 million ha) together accounting for about 85% of global crude palm oil production FAO (2017). Indonesia's biodiversity has been severely damaged by the oil palm plantations (Curran et al., 2004; Mantel et al., 2006). However, according to Foster et al., 2011 and Anderson et al., 2015, investigations related to biodiversity in oil palm plantations are very scarce. Approximately 1100 about oil palms were published; only 4 percent of these focused on biodiversity and conservation, and more than half of those manuscripts have been focused on policy and meta-analysis, rather than the provision of field-based data (Foster et al., 2011).

4.2. Oil palm structure

Oil palm plantations have a simpler structure than forests: the upper canopy is dominated by only one species with a uniform tree age structure, lower canopy, sparse undergrowth, less stable microclimate, and greater human disturbance (Figure 5) (Danielsen et al., 2009; Foster et al., 2011). The understorey of oil palm plantations is hotter, drier, and receives more light than the forest understorey (Drescher et al., 2016). As a result, the high levels of disturbance and propagate pressure, oil palm plantations contain weedier and more unusual, successional plants species than forests (Foster et al., 2011). Oil palm starts to yield fruit 3 – 4 years after planting; Bunches ripen 5 – 6 months after pollination. Seeds normally require temperatures about 35 Celsius to germinate (Duke, 1983). Besides, oil palm requires fertile and well-drained soils (Danielsen et al., 2009). A planting density of 130–140 palms per ha is usually practiced but it varies according to the planting conditions and the type of oil palm breeds (Corley and Tinker 2008). The ability of oil palm plantations to provide habitat depends on plantation age and management intensity (Teuscher *et al.,* 2015).

Palm Structure	Young	Old	
Canopy Top (m)	4.31	13.38	***
Trunk Height (m)	1.31	9.63	***
Canopy Bottom (m)	2.21	8.17	***
Leaves Overhead	2.20	2.76	
Trunk DBH (m)	0.77	0.61	***
Trunk Surface Area (m^2)	3.16	18.49	***
Frond Stubs (per m^2)	30.82	19.42	***
Ground Structure			
Height Live Veg (cm)	21.77	11.46	**
Leaf Litter (cm)	9.84	20.61	**
Bare Ground Coverage	22%	22%	
Live Veg Coverage	46%	45%	
Leaf Litter Coverage	47%	47%	

Figure 5. Vegetation structure of oil palms, ground vegetation and leaf litter in plantations. Young plantations are 8 years since planting, old plantations are 22 years since planting. Statistical significance between old and young plantations indicated by *** for $p < 0.01$, ** for $p < 0.05$, and * for $p < 0.10$ (Luskin and Potts 2011).

4.3. Oil palm biodiversity understory

Because of the oil palm's light requirements, plantation development generally requires that all other vegetation is removed (Danielsen et al., 2009). However, many oil palm understory vegetation can still grow and play an important role in conserving some elements of the Indonesian biodiversity (Koh and Wilcove 2008). Basri et al., 1995 reports that companies which adopt integrated pest management approach without pesticides, favor the establishment of beneficial plants to attract the insect predators and parasitoids of oil palm pests such as the wasp *Dolichogenidea metesae*. According to Teuscher et al., 2016 a total of 92 plant species were found in oil palm plantations in Indonesia. Of the 92 plant morphospecies, 64 could be identified of which 25 were alien species. The three most frequent species (Figure 6) were *Clidemia hirta* (Melastomataceae), *Asystasia gangetica* (Acanthaceae) and *Paspalum conjugatum* (Poaceae), which are non-native species.

Figure 6. a) *Clidemia hirta* (keyserver.lucidcentral.org); b) *Asystasia gangética* (butterflycircle.blogspot.com); c) *Paspalum conjugatum* (singapore.biodiversity.online).

The conservation of forest species requires the preservation of large reserves of intact forest, but we must not lose sight of the importance of conserving biodiversity and ecosystem processes within the oil palm habitat itself (Foster et al., 2011). Besides, ecosystem functioning and the provision of ecosystem services within oil palm landscapes are potentially influenced by the relationship between understory vegetation and arthropods, e.g. biocontrol, pollination, decomposition and soil fertility (Foster et al., 2011).

5. Experiments and Results

5.1. Experiment Data

One goal of the project that is to be introduced is to determine the number and distribution of different species from the aerial imagery taken from a drone. As previously mentioned above, there are various different kind of algorithms and parameters that can be used for this purpose. Moreover, the height where the image is taken could influence the algorithms and the finding. Taking image above the canopy could help finding the number of the palm trees but not to find the underlying plants. Therefore, the drone should also take image from lower height.

Some algorithms can work for both kind of images regardless of the height were those images are taken. In order to test and validate the usefulness of the algorithms, some images were taken from the surrounding plantation in the university Rhine-Waal green house in Kleve, Germany. Figure 7 to 9 show examples of the images.

In the first phase, two algorithms have been evaluated namely, colour separation algorithm and pixel conversion algorithm. These algorithms are typically used to extract certain features before applying more complex ones, such machine learning approaches. Both algorithms are implemented in MATLAB using image processing toolbox. The images are stored in the RGB format where each pixel stores 24 bits value with 8 bits for each channel.

5.2. Colour Separation Algorithm

The colour separation algorithm is used for separating and analysing the image on each different colour channel. The algorithm converts the image into three different greyscale images using three different colour channels. The intensity of the grayscale images depends on the applied colour channel. This algorithm is quite fast and helpful in separating objects which have different colour properties, e.g. separating tomatoes from their leaves. The images were scaled into the desired size before applying this algorithm.

The grayscale images can be easily evaluated, as the darker the image the lesser the amount of the colour in the corresponding channel and vice versa. Figure 7 and Figure 8 shows two images with corresponding grayscale images after applying the colour separation algorithm. A scale ranging from 0 to 255 shows the transition from black to white respectively as reference.

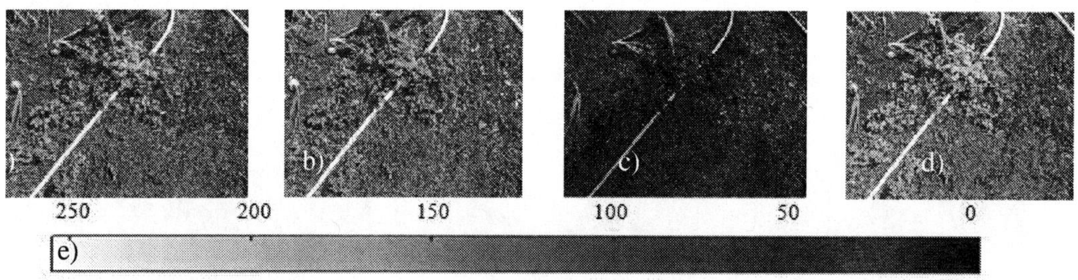

Figure 7. a) Original RGB image of onion plants and a hose pipe. b) Red channel separated. c) Blue channel separated. d) Green channel separated. e) Scale for comparison.

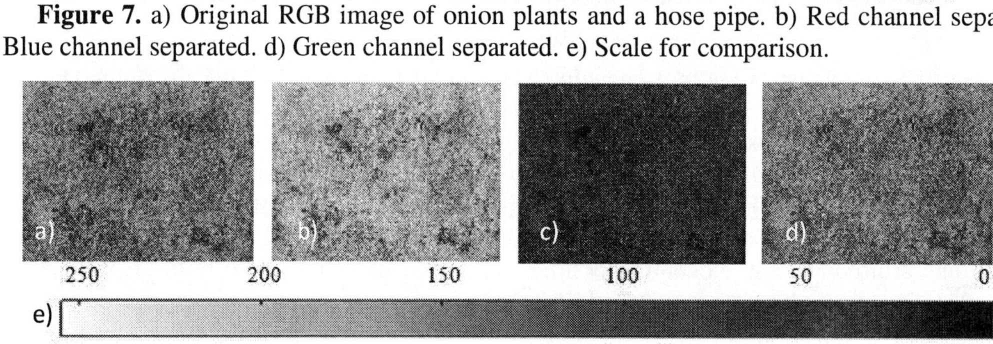

Figure 8. a) Original RGB image of dead grass with some green plants. b) Red channel separated. c) Blue channel separated. d) Green channel separated. e) Scale for comparison.

The grayscale images from each channel can be processed further on to find certain plants or objects. Applying simple threshold value on a channel can immediately remove the background image and leaving the object of interest in the foreground for further process. Two different channels can be subtracted from one to another for finding certain colour scheme of the objects. Other operation can also be applied in order to find different objects.

5.3. Pixel Conversion Algorithm

The pixel conversion algorithm is a further development of the colour separation algorithm. This algorithm can be used to differentiate whether the target object presents in a particular image or not. Depending on the main colour channel of the sought object, the original image and the three greyscale images can be compared in order to discover some region which might contain hidden object with the same features which difficult to distinguish on the original image with all colour displayed.

An example of the application of this algorithm is to determine the occurrence of certain species on the image. In the example in Figure 7, the red channel greyscale image from the colour separation algorithm is used to distinguish between grass and leaves with other species or objects. The pixel which denotes the species can be retained on the image resulting an annotated greyscale image. The result of this process is shown in Figure 9, where the grass and leaves are still visible in their original colour while other objects or species in greyscale. This algorithm might be too trivial for further processing with computer but it provides great advantages for human operator. This is needed in the later process where machine learning will be used for further processing the images, where human expert can work on the annotated images and mark the species in the training data set.

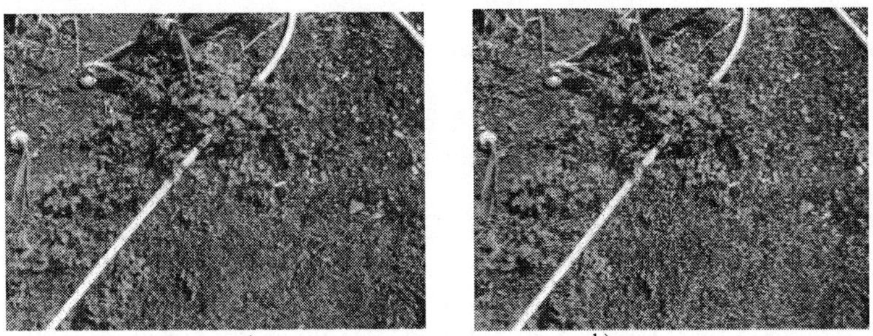

a) b)

Figure 9. a) Red channel grey scale image; b) Annotated image to find leaves and grass.

6. Conclusion and Outlook

According to Chong et al., 2017, some relevant applications of remote sensing in oil palm plantations are land cover classification, change detection, tree counting, age estimation, pest and disease detection and yield estimation. The type of sensor should be considered, because technical limitations may affect data analysis. The UAVs available in the market can be grouped into fixed-wind and multi-rotor. For small projects, multi-rotor UAVs are cost-effective while also obtaining hight resolution data. The vegetation indexes are an essential part of smart agriculture in terms of identifying plant spots that may be related to the proper function of ecosystem services among the fields. As oil palm expansion among the tropics tends to keep on increasing, studies are needed to understand how species diversity varies in different oil palm contexts (Meijaard and Sheil, 2013).

In planted areas, biodiversity management considers the diversity and structural complexity of vegetation and promotes diversity by increasing the height and coverage and by letting plants that cover the ground thrive (Koh et al., 2009). A successful and efficient conduct of planting operations requires a good layout design where remote sensing could contribute to providing the necessary information prior to and during the planting process. The investigation of Teuscher et al., 2016, reveals possibilities of integrating botanical information from ground plants into an algorithm. Plant diversity must be conserved within the remainder of the landscape to provide potential support to a variety of other ecosystem functions such as pollination, biological control, litter and dung decomposition, maintenance of water quality and environmental awareness (Foster et al., 2011).

The results obtained using two simple algorithms show great potential for finding the amount and distribution of species on the image automatically. In this preliminary experiment, the algorithms can create annotated image with the target species in focus which allow human expert to count the species objectively and then label the image with the corresponding value. This step is the first step for creating training data set for machine learning which will be applied on the aerial imagery taken form the UAV. In addition to the machine learning algorithm, further object classification and recognition algorithms will be investigated further. Evaluation of various different algorithms will also be conducted to evaluate the performance of those algorithms. The performance metrics will later on be used to justify whether there is one powerful algorithm which can perform the task alone or various different kind of algorithms are needed to come with reasonable results. With the help of this kind of algorithms, a method shall be developed to determine the biodiversity of palm oil plantations. Indicators can then be derived from this in order to improve the management of the plantations.

References

[1] Anderson Z R, Kusters K, Obidzinski K and McCarthy J 2015 May. Growing the Economy: Oil palm and green growth in East Kalimantan, Indonesia. In *Conference Paper* (No. 20).

[2] Asner G P, Jones M O, Martin R E, Knapp D E, Hughes R F 2008 Remote sensing of native and invasive species in Hawaiian forests. *Remote Sensing of Environment* 112 p1912–1926.

[3] Basri M W, Norman K and Hamdan A B 1995 Natural enemies of the bagworm, Metisa plana Walker (Lepidoptera: Psychidae) and their impact on host population regulation. *Crop Protection* 14 (8): 637-644.

[4] Baxter P W J, and G. Hamilton 2018 Learning to fly: integrating spatial ecology with unmanned aerial vehicle surveys. *Ecosphere* 9(4):e02194. 10.1002/ecs2.2194

[5] Camara B, Gosme M, Ngom D, Gomis Z D, Badji M Sanogo D and Dupraz C 2018 Ecological characterization and evolution of *Elaeis guineensis* Jacq. traditional parklands in Lower Casamance (Senegal). *Agroforestry Systems* pp.1-10.

[6] Corley R H V and P B H Tinker 2008 *The Oil Palm* John Wiley & Sons.

[7] Coops N C, Tooke T R 2017 Introduction to Remote Sensing. In: Gergel S., Turner M. (eds) Learning Landscape Ecology. Springer New York NY

[8] Curran L M, Trigg S N, McDonald A K, Astiani D, Hardiono Y M, Siregar P 2004 Lowland forest loss in protected areas of Indonesian Borneo. Science 303 1000–1003.

[9] Chong K L, Kanniah K D, Pohl C and Tan K P 2017 A review of remote sensing applications for oil palm studies. *Geo-spatial Information Science*, *20*(2) p184-200.

[10] Danielsen F, Beukema H, Burgess N D, Parish F, Brühl C A, Donald P F, Murdiyarso D, Phalan B E N, Reijnders L, Struebig M and Fitzherbert E B 2009 Biofuel plantations on forested lands: double jeopardy for biodiversity and climate. *Conservation Biology 23*(2) pp.348-358.

[11] Dislich C, Keyel A C, Salecker J, Kisel Y, Meyer K M, Auliya M, Barnes A D, Corre M D, Darras, K, Faust H and Hess B 2017 A review of the ecosystem functions in oil palm plantations, using forests as a reference system. *Biological Reviews 92*(3) p1539-1569.

[12] Duke A J. Handbook of Energy Crops. p. 1-5 1983. Available at: https://www.hort.purdue.edu/newcrop/duke_energy/Elaeis _guineensis.html (accessed on 06.08.18).

[13] Drescher J, Rembold K, Allen K, Beckschäfer P, Buchori D, Clough Y, Faust H, Fauzi A M, Gunawan D, Hertel D and Irawan B 2016 Ecological and socio-economic functions across tropical land use systems after rainforest conversion. *Phil. Trans. R. Soc. B 371*(1694) p.20150275.

[14] FAO (2015). Production/Crops: oil, palm fruit, statistics division. Available at http://faostat3.fao.org/browse/Q/QC/E. Accessed: August 2018.

[15] FAO (2016). State of the World's Forests 2016. Forests and agriculture: land-use challenges and opportunities. Rome. Accessed: August 2018.

[16] FAO (2017). Production/Crops: oil, palm fruit, statistics division. Available at

http://faostat3.fao.org. Accessed: August 2018.
[17] Fitzherbert, E. B., Struebig, M. J., Morel, A., Danielsen, F., Bruehl, C. A., Donald, P. F. & Phalan, B. (2008). How will oil palm expansion affect biodiversity? *Trends in Ecology & Evolution* 23, 538–545.
[18] Foster, W.A., Snaddon, J.L., Turner, E.C., Fayle, T.M., Cockerill, T.D., Ellwood, M.F., Broad, G.R., Chung, A.Y., Eggleton, P., Khen, C.V. and Yusah, K.M., 2011. Establishing the evidence base for maintaining biodiversity and ecosystem function in the oil palm landscapes of South East Asia. *Phil. Trans. R. Soc. B*, *366*(1582), pp.3277-3291.
[19] Getzin, S., Wiegand, K. and Schöning, I., 2012. Assessing biodiversity in forests using very high-resolution images and unmanned aerial vehicles. *Methods in Ecology and Evolution*, *3*(2), pp.397-404.
[20] Goodbody, T.R., Coops, N.C., Marshall, P.L., Tompalski, P. and Crawford, P., 2017. Unmanned aerial systems for precision forest inventory purposes: A review and case study. *The Forestry Chronicle*, *93*(1), pp.71-81.
[21] Heaphy, M., Watt, M.S., Dash, J.P. and Pearse, G.D., 2017. UAVs for data collection-plugging the gap. *New Zealand Journal of Forestry*, *62*(1), pp.23-30.
[22] Huete, A.; Didan K., Miura T., Rodriquez E. P., Gao X. & Ferreria L. G. (2000). "Overview of the radiometric and biophysical performance of the MODIS vegetation indices". *Remote Sensing of Environment.* 83: 195–213.
[23] Hoffmann, M., Hilton-Taylor, C., Angulo, A., Böhm, M., Brooks, T.M., Butchart, S.H., Carpenter, K.E., Chanson, J., Collen, B., Cox, N.A. and Darwall, W.R., 2010. The impact of conservation on the status of the world's vertebrates. *science*, p.1194442.
[24] IPOC (Indonesian Palm Oil Commission), (2006) Statistik Kelapa Sawit Indonesia 2005. Jakarta, Indonesia: Department of Agriculture.
[25] Iskandar, M.J., Baharum, A., Anuar, F.H. and Othaman, R., 2017. Palm oil industry in South East Asia and the effluent treatment technology—A review. *Environmental Technology & Innovation.*
[26] Ivosevic, B., Han, Y.G. and Kwon, O., 2017. Monitoring butterflies with an unmanned aerial vehicle: current possibilities and future potentials. *Journal of Ecology and Environment*, *41*(1), p.12.
[27] Kattenborn, T., Sperlich, M., Bataua, K. and Koch, B., 2014. Automatic single tree detection in plantations using UAV-based photogrammetric point clouds. *The International Archives of Photogrammetry, Remote Sensing, and Spatial Information Sciences*, *40*(3), p.139.
[28] Klemas, V.V., 2015. Coastal and environmental remote sensing from unmanned aerial vehicles: An overview. *Journal of Coastal Research*, *31*(5), pp.1260-1267.
[29] Koh, L.P. and Wilcove, D.S., 2008. Is oil palm agriculture really destroying tropical biodiversity?. *Conservation Letters*, *1*(2), pp.60-64.
[30] Koh, L. P., Levang, P. & Ghazoul, J. (2009). Designer landscapes for sustainable biofuels. *Trends in Ecology & Evolution* 24, 431–438.
[31] Lass, L.W., Prather, T.S., Glenn, N.F., Weber, K.T., Mundt, J.T., Pettingill, J., 2005. A review of remote sensing of invasive weeds and example of the early detection of spotted knapweed (Centaurea maculosa) and babysbreath (Gypsophila paniculata) with a hyperspectral sensor. Weed Science 53, 242–251.
[32] Lisein, J., Pierrot-Deseilligny, M., Bonnet, S. and Lejeune, P., 2013. A photogrammetric workflow for the creation of a forest canopy height model from small unmanned aerial system imagery. *Forests*, *4*(4), pp.922-944.
[33] Luskin, M.S., and Potts, M.D., 2011. Microclimate and habitat heterogeneity through the oil palm lifecycle. *Basic and Applied Ecology*, *12*(6), pp.540-551.
[34] Norton, G.W., and Swinton, S.M., 2018, April. Precision agriculture: global prospects and environmental implications. In Tomorrow's Agriculture: Incentives, Institutions, Infrastructure, and Innovations-Proceedings of the Twenty-fouth International Conference of Agricultural Economists: Incentives, Institutions, Infrastructure and Innovations-Proceedings of the Twenty-fouth International Conference of Agricultural Economists (p. 269).

IOP Conf. Series: Earth and Environmental Science **335** (2019) 012008 doi:10.1088/1755-1315/335/1/012008

[35] Mantel, S., Wösten, H. and Verhagen, J., 2007. Biophysical land suitability for oil palm in Kalimantan, Indonesia. *Report-ISRIC World Soil Information*, (2007/01).
[36] Meijaard, E., and Sheil, D., 2013. Oil-palm plantations in the context of biodiversity conservation. In *Encyclopedia of biodiversity* second edition, Volume 5, pp. 600-612. Elsevier Science Publishers, Netherlands.
[37] Obidzinski, K., Andriani, R., Komarudin, H. and Andrianto, A., 2012. Environmental and social impacts of oil palm plantations and their implications for biofuel production in Indonesia. *Ecology and Society*, *17*(1).
[38] Onyia, N.N., Balzter, H. and Berrio, J.C., 2018. Normalized Difference Vegetation Vigour Index: A New Remote Sensing Approach to Biodiversity Monitoring in Oil Polluted Regions.
[39] Padua, L., Vanko, J., Hruška, J., Adão, T., Sousa, J.J., Peres, E. and Morais, R., 2017. UAS, sensors, and data processing in agroforestry: a review towards practical applications. *International Journal of Remote Sensing*, *38*(8-10), pp.2349-2391.
[40] Petrenko, C., Paltseva, J. and Searle, S., 2016. Ecological impacts of palm oil expansion in Indonesia. Washington (US): International Council on Clean Transportation.
[41] Putra, B.T.W. and Soni, P., 2018. Enhanced broadband greenness in assessing Chlorophyll a and b, Carotenoid, and Nitrogen in Robusta coffee plantations using a digital camera. *Precision Agriculture*, *19*(2), pp.238-256.
[42] Rokhmana, C.A., 2015. The potential of UAV-based remote sensing for supporting precision agriculture in Indonesia. *Procedia Environmental Sciences*, *24*, pp.245-253.
[43] Siebert, S. and Teizer, J., 2014. Mobile 3D mapping for surveying earthwork projects using an Unmanned Aerial Vehicle (UAV) system. *Automation in Construction*, *41*, pp.1-14.
[44] Shamshiri, R.R., Weltzien, C., Hameed, I.A., Yule, I.J., Grift, T.E., Balasundram, S.K., Pitonakova, L., Ahmad, D. and Chowdhary, G., 2018. Research and development in agricultural robotics: A perspective of digital farming. *International Journal of Agricultural and Biological Engineering*, *11*(4), pp.1-14.
[45] Shouse, M., Liang, L. and Fei, S., 2013. Identification of understory invasive exotic plants with remote sensing in urban forests. *International Journal of Applied Earth Observation and Geoinformation*, *21*, pp.525-534.
[46] Schrier-Uijl, A.P., Silvius, M., Parish, F., Lim, K.H., Rosediana, S. and Anshari, G., 2013. Environmental and social impacts of oil palm cultivation on tropical peat: a scientific review. *Roundtable of Sustainable Palm Oil, Kuala Lumpur, Malaysia*, pp.131-168.
[47] Teuscher, M., Gérard, A., Brose, U., Buchori, D., Clough, Y., Ehbrecht, M., Hölscher, D., Irawan, B., Sundawati, L., Wollni, M. and Kreft, H., 2016. Experimental biodiversity enrichment in oil-palm-dominated landscapes in Indonesia. *Frontiers in plant science*, *7*, p.1538.
[48] Torresan, C., Berton, A., Carotenuto, F., Di Gennaro, S.F., Gioli, B., Matese, A., Miglietta, F., Vagnoli, C., Zaldei, A. and Wallace, L., 2017. Forestry applications of UAVs in Europe: A review. *International Journal of Remote Sensing*, *38*(8-10), pp.2427-2447.
[49] UN General Assembly, *Transforming our world : the 2030 Agenda for Sustainable Development*, 21 October 2015, A/RES/70/1, available at:
http://www.refworld.org/docid/57b6e3e44.html (accessed: 30 August 2018).
[50] USDA, N., 2011. The PLANTS Database. National Plant Data Team, Greensboro, North Carolina.
[51] Ustin, S.L. and Santos, M.J., 2010. Spectral identification of native and non-native plant species. In *Proceedings of ASD and IEEEGRS; Art, Science and Applications of Reflectance Spectroscopy Symposium* (Vol. 2).
[52] Wäldchen, J., Rzanny, M., Seeland, M. and Mäder, P., 2018. Automated plant species identification—Trends and future directions. *PLoS computational biology*, *14*(4), p.e1005993.
[53] Whelan, B. and James T., *An introduction to Precision Agriculture for Australian grains*, Australian Centre for Precision Agriculture, 2010. Change for Harvard citation.
[54] Xue, J. and Su, B., 2017. Significant remote sensing vegetation indices: A review of developments and applications. *Journal of Sensors*, *2017*.
[55] Yilmaz, V., Serifoglu Yilmaz, C., Tasci, L. and Gungor, O., 2017. Determination of Tree Crown

IOP Conf. Series: Earth and Environmental Science **335** (2019) 012008 doi:10.1088/1755-1315/335/1/012008

Diameters with Segmentation of a UAS-Based Canopy Height Model. *Ipsi bgd transactions on internet research, 13*(2).

[56] Zolkos, S.G., Goetz, S.J., and Dubayah, R., 2013. A meta-analysis of terrestrial aboveground biomass estimation using lidar remote sensing. *Remote Sensing of Environment, 128*, pp.289-298.

Spatial model for predicting sugarcane crop productivity using support vector regression

A W M Gaffar[1] and I S Sitanggang[1]

[1] Department of Computer Science, Faculty of Natural Science and Mathematics. Bogor Agricultural University (IPB), Jl. Meranti Wing 20 Level V, Kampus IPB Dramaga, Bogor, Indonesia 16680, e-mail: imas.sitanggang@ipb.ac.id

Abstract. Sugarcane which is the main ingredient of sugar production has a very high production demand so it must have a balance with the level of productivity. Sugarcane productivity in Indonesia can only reach 5.8 tons/ha which is still relatively low. In recent years sugarcane farmers have felt the decline in sugarcane production due to climate change which has affected the success of sugarcane planting so that sugar production has not been in line with national needs. This study aims to build a prediction model for the productivity of sugarcane crops in each area of Java Island based on climatic factors. The climate data used are collected from Indonesian Agency for Meteorology, Climatology, and Geophysics from 2006 to 2015. The method used is Support Vector Regression (SVR) using a radial base kernel function with a generalized neighbourhood matrix as a spatial model. The prediction model has RMSE of 0.1203954 and correlation of 0.9459743. The results of visualization of sugarcane productivity mapping show that areas in East Java have a higher value of productivity increase than areas in other provinces. Based on the results obtained, the model used has a quite good performance in predicting the productivity of sugarcane.

Keywords: climate, prediction model, sugarcane, support vector regression

1. Introduction

Sugarcane (*saccharum sp.*) is one of the most important plantation products that is used as the main ingredient in sugar production so it have high economic value [1]. In 2013 there was an increase about 3% in sugar consumption due to high industrial demand, which initially the consumption was only 2.6 million tons increased to 2.7 tons. The high demand for sugar production should be balanced by increasing the productivity of sugarcane [2].

According to the Central Statistics Agency (2014), the productivity of sugarcane plantations in Indonesia reached 56% of the area of sugarcane plantations in Indonesia which reached 194.9 hectares owned by the company and 247.8 hectares of smallholder plantations. Although every year there is a development of sugarcane harvest area in Indonesia, it affects the increase in sugarcane production but the amount of sugarcane productivity in Indonesia can only reach 5.8 tons/ha which is still relatively low.

East Java, Central Java, Lampung, West Java and Yogyakarta have a significant production contribution of sugarcane plants at 98.9% of the total amount of sugar production. East Java has a contribution of 69.75 %, it is the highest position as the most significant contributor to sugar production in Indonesia [3].

IOP Conf. Series: Earth and Environmental Science **335** (2019) 012009 doi:10.1088/1755-1315/335/1/012009

Climate and weather are factors that influence the success of sugarcane planting. The last few years sugarcane farmers feel the decline in sugarcane production due to weather changes so that sugar production does not meet national needs. In 2014, sugarcane production was targeted at 2.8 million tons but only 89.9% was achieved [4].

A study on crop productivity predictions has been carried out by Rahul Tripathi (2014) which predict productivity and yield of rice in Odisha and India by processing data on wide of area, production and productivity using the Autoregressive Integrated Moving Average (ARIMA) model. The result is an analysis of productivity trends along with the results of forecasting and validation [5]. Another study by Reza Septiawan (2012) developed a predictive model for the productivity of red chili and cayenne pepper in West Java and East Java based on climate data derived from weather stations. The prediction uses the Self-Organizing (SOM) clustering method from Kohonen [6]. The prediction of plant productivity in both studies still uses nonspatial data. The spatially related study was carried out by A. Podznoukhov (2011) to build spatial data on estimated avalanches in the Scottish region based on historical data, meteorological conditions and snowpack observations in the field using Support Vector Machine [7].

This study aims to develop a spatial model to predict the productivity of sugarcane crops in each area in Java Island based on climate influence using Support Vector Regression (SVR) technique. This study processes the spatial data and the results of sugarcane plantations with climate data in each area. The output of prediction model is the productivity of sugarcane crops with the aim to monitor the development of sugarcane productivity. The model can provide early warning to sugarcane farmers in case of decreasing productivity which is predicted to reduce the risk due to climate change.

2. Material and method

2.1. Data

The data used in this research are productivity, area, and yield of sugarcane crops in each area in the island of Java. Those annually data are collected from the website of the data center and Indonesian Plantation Statistics report (Sugarcane) published by the Ministry of Agriculture in 2006 to 2015. Climate data in the form of daily minimum and maximum temperature, humidity, rainfall and duration of sunshine are taken from Indonesian Agency for Meteorology, Climatology, and Geophysics at several UPT weather stations spread across Java. The attributes of data can be seen in Table 1.

2.2. Spatial prediction model using Support Vector Regression

Spatial data mining is the extraction of specific spatial information from spatial databases. Spatial data focus and refer to a dataset that has location or space aspects. Spatial data mining is usually used to detect image databases, remote sensing, navigation and route controllers, geographic information systems and various other sectors that require spatial information.

The characteristics of spatial data include the data has a very large amount of information so that an efficient algorithm is needed, have attributes that indicate spatial data such as latitude and longitude, there is ambiguity in spatial relationships or uncertainty of spatial location, and have a relationship with neighboring objects so that the data are more complicated [8].

Support Vector Machine is one of the methods developed in machine learning for classification and regression problems. Regression-based SVM (SVR) is a type of optimization model and can also be used to model nonlinear processes, thus the model can minimize error predictions and complexity model [9]. This method has an output in the form of continuous or real numbers so that it can overcome overfitting and performance that is better than the SVM algorithm. The SVM algorithm in the case of regression aims to build a hyperplane that is close to as much data as possible. Therefore, the optimal hyperplane is the one that matches all input data with the smallest possible error by mapping the input vector into a higher dimension [10]. In this study, SVR is applied to predict the productivity of plantation crops, especially sugarcane.

Spatial prediction is carried out using a generalized neighborhood matrix w model. Regression coefficients are constant in some of the observations with appropriate parameter estimates. The neighborhood matrix is related to the dependence on the number of neighbors. The productivity prediction of sugarcane crops was built by prediction models using SVR algorithm by considering the most optimal hyperplane search [11]. The steps of Regression Vector Support algorithm [10]:

2.2.1. An optimal hyperplane. Function of the regression line as optimal hyperplane:

$$f(x) = wx + b \quad w \in X, b \in \mathcal{R} \tag{1}$$

2.2.2. Quadratic Programming. Looking for the best hyperplane using Quadratic Programming optimization method by minimizing:

$$\frac{1}{2} w^T w \tag{2}$$

So, it will produce optimized functions:

$$\min\{\frac{1}{2} ||w||^2\} \tag{3}$$

The optimized function will make the function as small as possible so that it can control the function capacity. Assumed that all points are in the range $f(x) \pm \varepsilon\ (feasible)$, in terms of infiesibility, where there are some points that may come out of the range $f(x) \pm \varepsilon\ (feasible)$ the slack variable ξi is added to overcome the not feasible, so that optimization is carried out formulated as follows:

$$\min\{\frac{1}{2} \parallel w \parallel^2\} + C \sum_{i=1}^{l} (\xi i + \xi^* i)\} \tag{4}$$

2.2.3. Lagrangian function. Build the following multiplier Lagrangian function:

$$L = \frac{1}{2} \parallel w \parallel^2 + C \sum_{i=1}^{l} (\xi i + \xi i^*) - \sum_{i=1}^{l} \lambda i(\varepsilon i + \xi i - yi + wxi + b) - \sum_{i=1}^{l} \lambda i^* (yi + \varepsilon i + \xi i^* - yi + wxi + b) - \sum_{i=1}^{l} (\eta i \xi i + \eta i^* \xi i^*), \quad and\ \lambda i, \lambda i^*, \eta i, \eta i^* \geq 0 \tag{5}$$

2.2.4. Substitution of derivative calculations. To get the optimal solution, substitution calculation is substituted, as for the optimal w equation:

$$w^* = \sum_{i=1}^{l} (\lambda i - \lambda i^*) xi \tag{6}$$

Then the optimal hyperplane is written below:

$$f(x) = \sum_{i=1}^{l} (\lambda i - \lambda i^*) xi\, x + b^* \tag{7}$$

2.2.5. Kernel function. The learning process in finding vector support points requires kernel function in solving the formulation of feature transformation into new spaces that are of a higher dimension efficiently [12].

3. Result and discussion

3.1. Pre-processing data

Preprocessing data begins by identifying the missing values in the dataset, using the summary (dataset) function. For climate datasets, since the weather station is not in all areas, the climate data in each area are determined based on the location of the closest weather station. The climate dataset which is the daily data must be averaged annually. The year data is ignored because in this study temporal prediction models were not carried out.

The spatial model in this study is shown by the neighbor matrix, where the productivity value of a district is influenced by the productivity value of the neighboring districts. For example in the Figure 1,

IOP Conf. Series: Earth and Environmental Science **335** (2019) 012009 doi:10.1088/1755-1315/335/1/012009

Bojonegor has neighbors, namely Tuban, Blora, Ngawi, Madiun, Nganjuk, Jombang and Lamongan. Then the productivity value is normalized so that it has a value in the range from 0 to 1.

3.2. Build model prediction using Support Vector Regression

The dataset is divided into two, namely training data and testing data. The spatial prediction model is built using training data which is 80% of the entire dataset. Model building was done using the Support Vector Regression (SVR) technique with the radial base kernel function (RBF) with a default gamma of 0.125 which is processed using R.

The model that has been built was then tested using data testing to produce a predictive value of productivity. Then the comparison of the actual productivity value of the dataset and the value of productivity predictions was measured by the RMSE. The resulting RMSE is 0.1335017, and the predictive model correlation is 0.9371245. The graph in Figure 2 shows the comparison of the actual value symbolized by the red line and the prediction value symbolized by the blue line.

3.3. Evaluating prediction model

The model determination is a procedure to measure the average error in predicting the model using Root Mean Squared Error (RMSE) [13] which is formulated as follows:

$$RMSE = (\frac{1}{n}\sum_{i=1}^{n}(\hat{y}i - yi)^2 \quad (8)$$

The multiple correlation coefficient (R^2) is a relative measure of the relationship between one variable and another variable, so it can be known how strong the relationship between variables X and variable Y. With the formulation as follows [14] :

$$R^2 = 1 - \frac{\Sigma(\hat{Y}_i - \bar{Y})^2}{\Sigma(Y_i - \bar{Y})^2} \quad (9)$$

After building a spatial prediction model with the default gamma parameter of 0.125, then tuning parameters is performed to see the potential for performance improvement on the model. Tuning parameters are presented on gamma parameters with a range of values between 0.100 and 10 so that several models are formed which are then evaluated using the RMSE and the correlation coefficient which can be seen in Table 2.

3.4. Visualization predicting model

Visualization is a method to present data or problem into a graphic or image format so that it is easier to understand. The model that has been built previously using SVR produces prediction of the productivity of sugarcane crops in each area in Java Island. The results of the prediction are then visualized in the form of mapping in Java Island using QuantumGIS, wherein each area is symbolized by a color that means the range of values of sugarcane productivity in the area.

In the mapping as shown on Figure 3, four-color symbols have a range of values for the productivity of sugarcane. The exact white color in the area has no productivity value or does not have the sugarcane planting area. Yellow color indicates productivity value below 3000 Kg/Ha, light green has a range between 3000 to 4000 Kg/Ha, dark green means a range of 4000 to 5000 Kg/Ha. Whereas the blue color has a range between 5000 to 7000 Kg/Ha and the dark blue color has a productivity value above 7000 Kg/Ha.

4. Conclusion

This study results a spatial model to predict the productivity of sugarcane with a quite good performance, where the prediction results depend on neighboring matrix. The prediction model uses SVR techniques that process data using a radial base kernel function (RBF). RMSE of the model generated with the default gamma is 0.1335017, and the correlation is 0.9371245. Then tuning parameter was performed on gamma parameter value. The result is the model with lower RMSE, but its correlation is getting higher. Based on the experiment, the best gamma value is 0.100 with the RMSE of 0.1203954 and the correlation coefficient of 0.9459743. Based on the results visualization, this study concludes that the sugarcane productivity in area in East Java has a higher value than those in other area in the province of Java.

5. Appendix

Table 1 Attribute of the data.

Attribute Name	Type	Explanation
Year	Numeric	Year of calculating the productivity of sugarcane
Area	Character	Area where sugarcane is produced
Province	Character	Province of Area
Minimum temperature	Numeric	The minimum temperature in area is expressed in centigrade (%)
Maximum temperature	Numeric	The maximum temperature in area is expressed in centigrade (%)
Average humidity	Numeric	The average humidity in an area
Rainfall	Numeric	Rainfall in an area stated in mm
Duration of sunshine	Numeric	The duration of sunshine in an area within hours
Wide area	Numeric	Area planted with sugarcane stated in Ha
Production	Numeric	Production of sugarcane which is expressed in tons
Productivity	Numeric	The productivity value of sugarcane plants which is stated in Kg/Ha

Table 2 The results of tuning parameter.

Parameter (*Gamma*)	RMSE	Correlation
0.100	0.1203954	0.9459743
0.125	0.1335017	0.9371245
0.50	0.1749498	0.7997868
1	0.2170132	0.7034049
2	0.3119699	0.5895436
5	0.3708898	0.5001312
10	0.3929847	0.4671223

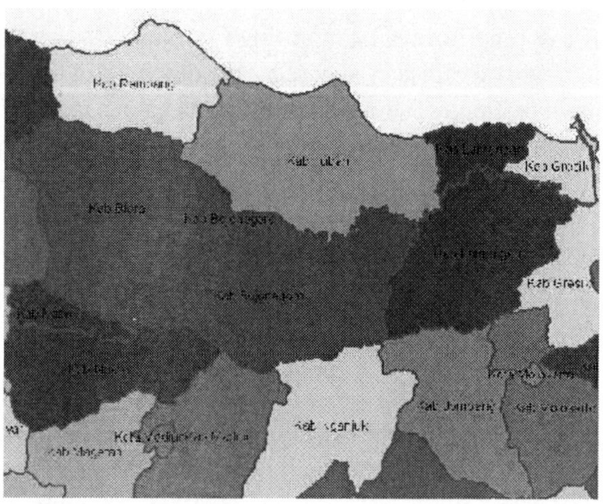

Figure 1. Example of determining matrix neighbors.

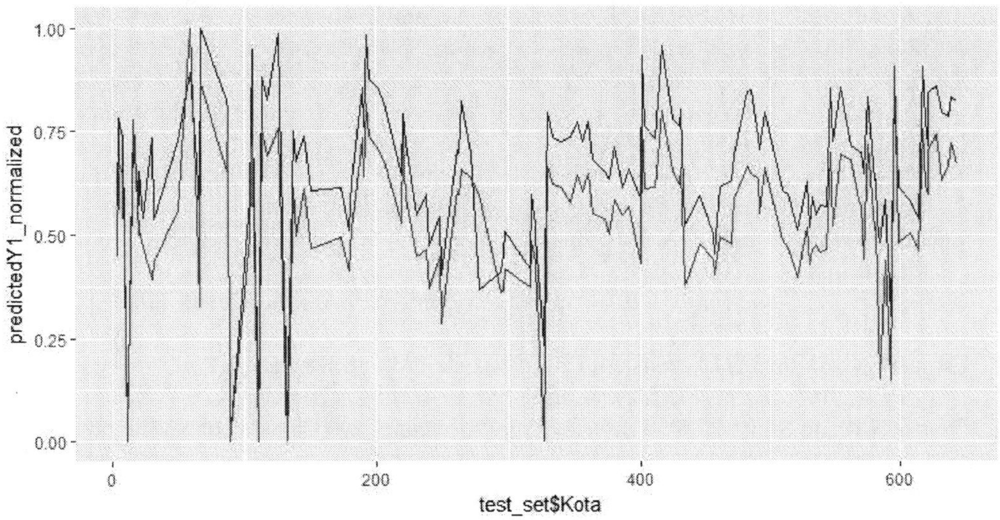

Figure 2. Graph comparison of actual values and predictive values.

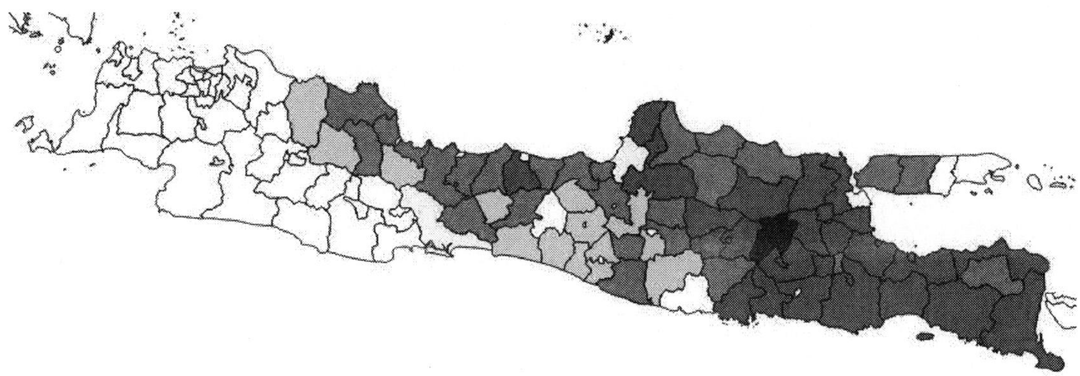

Figure 3. Visualization of sugarcane productivity in Java.

References

[1] Mariska I and Rahayu S 2011 J Litbang Pertanian *Pengadaan bibit tebu melalui kultur jaringan* 15

[2] Diana N E 2017 J Ilmu Pertanian Indonesia *Pertumbuhan, Produktivitas, dan Rendemen Pertanaman Tebu Pertama (Plant Cane) Pada Berbagai Paket Pemupukan* **21** 3

[3] Apriawan D C, Irham I and Mulyo J H 2014 Agro Ekonomi *Analisis Produksi Tebu dan Gula di PT. Perkebunan Nusantara VII (Persero)* **26** 159-67

[4] Mustadjab M M, Fahriyah F and Hana I 2012 Agric Socio-Economics *Dampak Perubahan Iklim Terhadap Pendapatan Petani Tebu (Studi Kasus Pada Kptr (Koperasi Petani Tebu Rakyat)"Tani Mulya" Di Kabupaten Sidoarjo, Jawa Timur)* **12** 1

[5] Tripathi R, Nayak A K, Raja R, Shahid M, Kumar A, Mohanty S, Panda B, Lal B and Gautan P 2014 Advances in Agriculture *Forecasting rice productivity and production of Odisha, India, using autoregressive integrated moving average model* **2014** 9

[6] Septiawan R, Komaruddin A, Sulistya B, Alfi N and Shanmuganathan S 2012 Prediction model for chilli productivity based on climate and productivity data (Bandung: Geise Parahyangan Catholic University)

[7] Pozdnoukhov A, Matasci G, Kanevski M and Purves R S 2011 Nat Hazards Earth Syst Sci. *Spatio-temporal avalanche forecasting with support vector machines* **11** 2

[8] Bist A and Faridi M A 2017 Int. J. of Eng. Trends and Tech. *Survey: On Spatial Data Mining* **46** 327
[9] Sánchez A S, Nieto G, Fernández R, Díaz D and Rodríguez I 2011 Math Comput Model *Application of an SVM-based regression model to the air quality study at local scale in the Avilés urban area (Spain)* **54** 5–6
[10] Trafalis T B and Ince H 2000 Support vector machine for regression and applications to financial forecasting (Italy:IEEE) 6
[11] Cheng T, Wang J and Li X 2007 In: Proc Geocomputation *The support vector machine for nonlinear spatio-temporal regression*
[12] Maji S, Berg A C and Malik J 2013 IEEE Trans Pattern Anal Mach Intell. *Efficient classification for additive kernel SVMs* **35** 1
[13] Esmaeili M, Salimi A, Drebenstedt C, Abbaszadeh M and Bazzazi A Arab J Geosci *Application of PCA, SVR, and ANFIS for modeling of rock fragmentation* **8** 9
[14] Draper N R and Smith H 2014 *Applied regression analysis* (Canada:John Wiley & Sons) 326

Adaptive Neuro-Fuzzy Inference System implementation for farmer's term of trade forecasting in West Sumatra

Y Aufar[1] and I S Sitanggang[1]

[1] Department of Computer Science, Faculty of Natural Science and Mathematics, Bogor Agricultural University (IPB), Jl. Meranti Wing 20 Level V, Kampus IPB Darmaga, Bogor, Indonesia 16680, e-mail: imas.sitanggang@ipb.ac.id

Abstract. The farmer's term of trade (NTP) is used by the Central Bureau of Statistics (BPS) as one of the indicators for measuring the level of welfare or purchasing power of farmers. To prepare preventive measures when the NTP index falls from the previous period, the relevant parties need to predict NTP for the coming period. The purpose of this study is to measure the performance of Adaptive Neuro-Fuzzy Inference System (ANFIS) in predicting NTP of West Sumatra Province in the coming month. The data used are monthly NTP data of West Sumatra Province obtained from the BPS website www.sumbar.bps.go.id from 2013 to 2016 for the network training process. Model evaluation was done by comparing the test results with actual data in 2017. Forecasting systems are divided into two types, namely time series model and the multivariate model. The test results show that the time series model has the smallest RMSE of 0.3430 for the training set and the RMSE value is 1.4570 for the testing set.

Keywords: ANFIS, farmer's term of trade, forecasting.

1. Introduction

Agriculture is still one of important sectors in the economy of West Sumatra, where each year has a percentage of more than 20 percent in the Gross Regional Domestic Product (PDRB) of West Sumatra Province [1]. This sector is expected to increase farmers' income and can help alleviate poverty. Therefore, the role of the government and related parties in carrying out strategic policies to improve the welfare of farmers is required. The Central Statistics Agency (BPS) has published data on the farmer's term of trade (NTP) as an indicator that can measure the level of welfare of farmers. NTP describes the ratio between the price index received by farmers (It) and the price index paid by farmers (Ib) in percentage. The higher the NTP, the more prosperous the level of life of the farmer. The right method is needed to find out or predict how many NTP indices will be in the coming period so that action can be taken if the NTP index has decreased from the previous period.

Some methods that are often used in time series data forecasting are neural networks [2] and fuzzy logic [3]. Neural networks have the advantage of computing by simulating structures and functions such as performance in brain nerve tissue. The advantages of the neural network can be used to deal with conditions with increasing uncertainty [4]. Whereas in fuzzy logic, it has advantages in dealing with various uncertain or unclear conditions [5]. Besides having advantages, of course, neural networks and fuzzy logic also have disadvantages. Neural networks have deficiencies in complex processing stages so that they are not effective on networks which are large enough and do not have reasoning capabilities

IOP Conf. Series: Earth and Environmental Science **335** (2019) 012010 doi:10.1088/1755-1315/335/1/012010

like those in fuzzy logic. Whereas in fuzzy logic, it has the disadvantage of not having the ability to learn and adapt like a neural network. Therefore, the Adaptive Neuro-Fuzzy Inference System (ANFIS) method is developed which combines the two methods [6].

In this study, the ANFIS method is used to predict NTP in West Sumatra Province. The use of the ANFIS method in the prediction process is expected to produce a more accurate predictive value. ANFIS analysis in this study uses the order-1 Sugeno model. Order-1 is chosen with consideration of simplicity and ease of calculation. The learning algorithm used is a hybrid optimization method. The software used is MATLAB.

2. Related research

NTP-related research in Indonesia was made by Hendayana using regression to determine the relationship between factors influencing the NTP in North Sumatra. Hendayana said that NTP is a relationship between agricultural products sold with other goods and services purchased by farmers [7]. Another research on NTP for time series datasets was conducted by Khusniyah and Sutikno; they studied the predictions of NTP in East Java Province using backpropagation neural network algorithms. The smallest percentage error or the highest accuracy is obtained when the number of hidden layer nodes is 7, and the learning rate value is 0.1. But neural networks are not effective on large enough networks because they require long and complicated processing times [8].

Related research on ANFIS method has been widely applied to time series data prediction problems. Putra et al. applying the ANFIS method to predict rainfall in Kalimantan [9]. Faulina et al. applied the ANFIS method and compare it with the Autoregressive Integrated Moving Average (ARIMA) method for predicting monthly rainfall data at certain area in East Java Province, Indonesia, namely Pujon and Wagir area. Based on the Root Mean Squared Error (RMSE) in the test dataset, the results showed that the ANFIS method produced an estimate of the monthly data of rainfall in the Pujon area more accurately while the ARIMA model produced a good estimate of monthly rainfall data in the Wagir area [10]. Wei applied the ANFIS method in the case of TAIEX stock forecasting [11]. Dewi and Himawati used the ANFIS method to predict the unemployment rate [12]. Forecasting the level of atmospheric turbidity by the ANFIS method was carried out by Nou et al. [13]. The studies carried out show that the ANFIS method approach is well known and widely used as a time series data prediction algorithm. Therefore, this study uses ANFIS as a method to predict NTP in the Province of West Sumatra with the results of the right and accurate NTP prediction.

3. Methodology

There are several stages of the process to predict NTP in West Sumatra Province by using ANFIS method. The process stages are illustrated in figure 1.

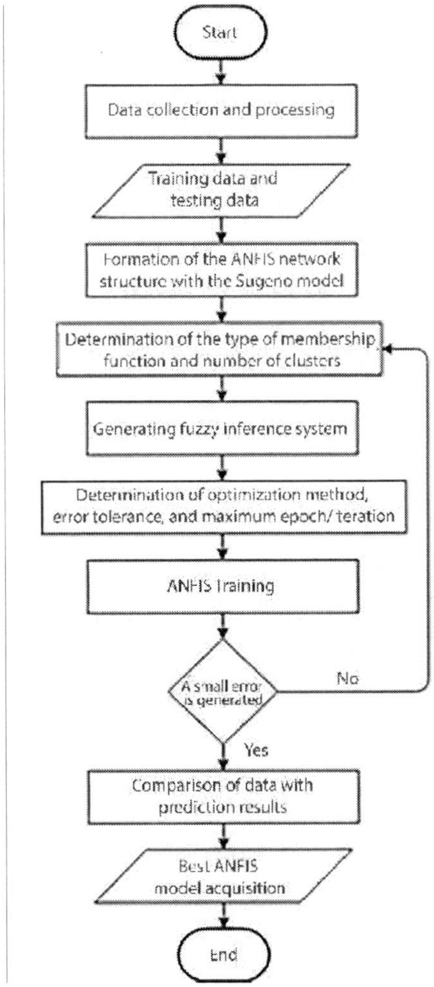

Figure 1. Research Methodology.

3.1. Data Collection and Processing

The data used in this study are secondary data obtained indirectly through intermediate media. This study uses NTP data in West Sumatra from 2013 to 2017 obtained from the website of the Central Bureau of Statistics (BPS) www.sumbar.bps.go.id and accessed on April 14, 2018. The data obtained are divided into two parts, first as network training data, and the second as test data. Training data contain the West Sumatra NTP data in the period of 2013, 2014, 2015, and 2016. Testing data contain the West Sumatra NTP data in 2017.

3.2. Formation of ANFIS network

In this study, in general, the monthly NTP forecasting system is made to predict future NTP. Based on the data, this NTP forecasting system uses the ANFIS method with two input and output models, namely the time series model where all input and output variables are the same variables and multivariate models with different input variables with output variables [14]. ANFIS time series model, for example, to predict NTP, the input used is current NTP and NTP data in one month ago, while the output is NTP data in one month ahead. The ANFIS model is a double variable, the type of input data variable is different from the type of output variable, for example, to predict the NTP in one month to come, the input used is the current data of the Farmer Accepted Price Index (It) and the Price Index Paid by Farmers

(Ib). There are several input-output combinations that are formed to get the best predictor of performance. The following is the input-output arrangement that was made as an NTP predictor, as given in Table 1 and Table 2.

Table 1. ANFIS predictors for monthly time series data.

Input 1	Input 2	Output
Farmer Exchange Rate this month (t)	Farmer Exchange Rate one month ago (t-1)	Farmer Exchange Rate next month (t + 1)

Table 2. ANFIS predictors of multi-variable monthly data.

Input 1	Input 2	Output
Price Index Received by Farmers this month (t)	Price Index Paid by Farmers this month (t)	Farmer Exchange Rate next month (t + 1)

4. Determination of Membership Function and Number of Clusters

A neuro-fuzzy program implementation uses the ANFIS method. ANFIS trains FIS with FIS initialization, which sets the initial price of the membership function parameters in the FIS. FIS initiation includes selection of membership functions, selection of membership function types, and selection of training iterations (epoch).

4.1. ANFIS Training

In the program development phase, training was carried out with ANFIS to find parameters in the IF section, namely parameters a and c (premise parameters), and parameters in the THEN section (consequent parameters). After these parameters are known with a small error, these parameters are used in the next stage. At this stage, we use Fuzzy C-Means (FCM) to do clustering to determine data assigned to a cluster. The next step is calculating the mean and standard deviation for each cluster, the mean value, and this standard deviation will then be the value of a and c of the first parameter value. Furthermore, the output in the first layer is the degree of membership of each data, in the second layer is the degree of membership multiplication, in the third layer is normalized, the fourth layer looks for consequent parameters, in the fifth layer is the sum of the fourth layer. Recursive least-squares estimation is used to correct consequent parameters until we get a small error value in the forward step, while in the reverse step we use error propagation to fix the premise parameters.

4.1.1. Fuzzyfication layer. After getting the results of fuzzy clustering we then calculate the first layer by calculating the mean and standard deviation. The Generalized Bell membership function is done with the following formula:

$$\mu = \frac{1}{1+\left|\frac{A-c}{a}\right|^{2b}} \tag{1}$$

where A as an input, A $\{A_{1,t}, A_{2,t}\}$ and $\{a, b, \text{and } c\}$ are parameters, usually $b = 1$. If the values of these parameters change, then the shape of the curve will change too. These parameters are usually called the premise parameters [15].

4.1.2. Second Layer. This layer is the result of the degree of membership from the first layer, as follows:

$$w_i = \mu_{Ai} \cdot \mu_{Bi} \tag{2}$$

The results of this calculation are called firing strength of a rule. Each neuron represents the rule to-i [16].

4.1.3. Third Layer. Each neuron in this layer is a fixed neuron (given the symbol N) which is the result of calculating the ratio of i (w_i) firing strength to the total number of firing strengths in the second layer, as follows:

$$\overline{w} = \frac{w_i}{w_1+w_2}, i = 1,2 \tag{3}$$

This result is also called normalized firing strength [16].

4.1.4. Fourth Layer. This layer is a neuron which is an adaptive neuron to an output, as follows:

$$\overline{w}_i\ f_i = \overline{w}_i(p_i Z_{1,t} + q_i Z_{2,t} + r_i) \tag{4}$$

$\overline{w}_i$ is normalized firing strength in the third layer and p_i , q_i , and r_i are the parameters on the neuron. These parameters are usually called consequent parameters [16].

4.1.5. Fifth Layer. This layer is a single neuron (symbol $\sum$) which is the sum of all outputs from the fourth layer, as follows [17]:

$$\sum \overline{w}_i\ f_i = \frac{\sum_i w_i f_i}{\sum_i w_i} \tag{5}$$

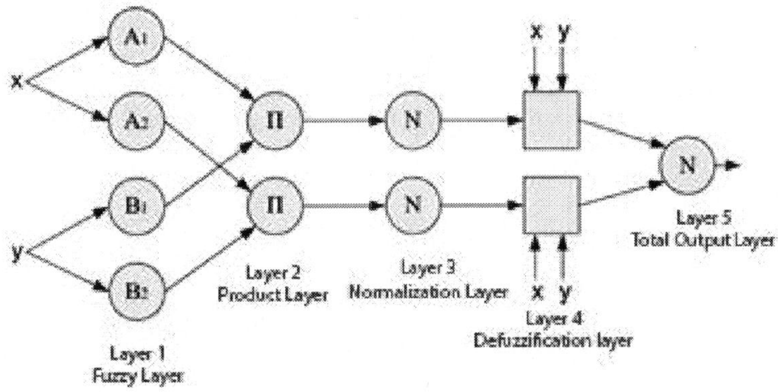

Figure 2. ANFIS Structure [18].

4.1.6. Recursive LSE. If the m element is owned by the vector Z_t (Z_t measuring m × 1) and n parameters θ (θ measuring n × 1), with the row i, in the matrix [A:Z_t] is denoted as [a_i^T :Z_t], Least-squares estimator is written as follows:

$$A^T A\hat{\theta} = A^T Z_t \tag{6}$$

If $A^T A$ is nonsingular and $\hat{\theta}$ unique, it can be given:

$$\hat{\theta} = (A^T A)^{-1} A^T Z_t \tag{7}$$

or by removing ^ and assumed the number of lines of A and Z_t pair is k then it is obtained:

$$\theta_k = (A^T A)^{-1} A^T Z_t \tag{8}$$

there is a recursive LSE added a data pair [a_i^T :Z_t], so there are as many as m+1 pairs of data. Then LSE θ_{k+1} is calculated with the help of θ_k. Because there are as many parameters as n then with the inversion method, as follows:

$$P_n = (A_n^T A_n)^{-1} \tag{9}$$

and

$$\theta_n = P_n A_n^T Z_{t(n)} \tag{10}$$

Then the iteration starts from data (n+1), with P_0 and θ_0 is calculated by the equation P_n and θ_n, the value of P_{k+1} and θ_{k+1} can be calculated as follows:

$$P_{k+1} = P_k - \frac{(P_k a_{k-1} a_{k+1}^T P_k)}{1 + a_{k+1}^T P_k a_{k+1}} \tag{11}$$

$$\theta_{k+1} = \theta_k + P_{k+1} a_{k+1} (Z_{t(k+1)} - a_{k+1}^T \theta_k) \tag{12}$$

ANFIS trains FIS by modifying membership function parameters until a minimum error between FIS output and output training data is obtained [17].

4.2. Comparative Data with Prediction Results

The model validation is the FIS testing process that has been trained by ANFIS but uses input/output data that has not been trained by FIS.

5. Results and Discussion

5.1. ANFIS analysis on monthly time series data

5.1.1. Cluster Selection. By using the gbellmf (Generalized Bell) membership function, which is available in MATLAB, in each ANFIS training with a different number of clusters, a summary of the results is given in Table 3.

Table 3. ANFIS training on monthly time series data based on the number of clusters

ANFIS Input	Number of Clusters	RMSE	
		Training Data (in-sample)	Data Testing (out-sample)
NTP (t) and NTP (t-1)	[2 2]	0.6856	0.8605
	[3 3]	0.6399	1.1279
	[4 4]	0.5018	1.3868
	[5 5]	0.3875	1.4022
	[6 6]	0.3418	1.4768
	[7 7]	**0.3039**	**1.5436**
	[8 8]	0.1943	3.8771

According to Table 3, the best number of clusters is determined, namely those that produce RMSE on the smallest training and testing data. The results of testing the number of clusters show that up to the cluster [8 8], the RMSE value has decreased in training data. But there was a high increase in testing data. This happens because the data with the number of clusters [7 7] has exceeded the optimal state. Thus, in this experiment that the best input is NTP (t) and NTP (t-1) with the number of clusters [7 7].

5.1.2. Selection of Membership Functions. Based on the previous analysis, the best input is NTP (t) and NTP (t-1) with the number of clusters [7 7]. Then ANFIS training was carried out on NTP (t) and NTP (t-1) input with the number of clusters [7 7] on several different membership functions. The following results are obtained, as shown in Table 4.

Table 4. ANFIS training on monthly time series data based on membership functions

Membership Function	RMSE	
	Training Data (in-sample)	Testing Data (out-sample)
trimf	0.4468	1.5177
trapmf	0.5963	1.5055
gbellmf	0.3039	1.5436
gaussmf	**0.3430**	**1.4570**

According to Table 4 the best membership function is determined, which produces the smallest RMSE on training and testing data. Although gbellmf (Generalized Bell) gives the smaller RMSE on training data than RMSE from the gaussmf (Gaussian), but for testing data, Gaussian is much better than Generalized Bell in resulting small RMSE. It is concluded that the best membership function is Gaussian.

5.1.3. Results Analysis. From the training conducted on the data, it is concluded that the best ANFIS models are NTP (t) and NTP (t-1) inputs with number of clusters [7 7] and the Gaussian membership function. RMSE on training data is 0.3430 and RMSE on testing data is 1.4570.

5.2. ANFIS analysis on multi-variable monthly data

5.2.1. Cluster Selection. By using the gbellmf (Generalized Bell) membership function, which is available in MATLAB, in each ANFIS training with a different number of clusters, a summary of the results is provided in Table 5.

Table 5. ANFIS training on monthly multi-variable data based on the number of clusters

ANFIS Input	Number of Clusters	RMSE	
		Training Data (in-sample)	Data Testing (out-sample)
It (t) and Ib (t)	[2 2]	0.9238	35.8984
	[3 3]	0.6320	34.2002
	[4 4]	0.5864	32.9146
	[5 5]	0.4961	31.6605
	[6 6]	0.4865	28.5457
	[7 7]	0.4798	32.0965
	[8 8]	**0.4241**	**27.0349**

According to Table 5, the best number of clusters is determined, namely those that produce the smallest RMSE in training and testing data. It is concluded that the best inputs are It (t) and Ib (t), with the best number of clusters [8 8].

5.2.2. Selection of Membership Functions. Based on the previous analysis, the best inputs are It (t) and Ib (t) with the number of clusters [8 8]. Then the input of ANFIS training are It (t) and Ib (t) with the number of clusters [8 8] on several different membership functions. The following results are obtained, as given in Table 6.

Table 6. ANFIS training on multi-variable monthly data based on function membership

Membership Function	RMSE	
	Training Data (in-sample)	Data Testing (out-sample)
trimf	0.5180	49.8184
trapmf	0.6200	39.7706
gbellmf	**0.4241**	**27.0349**
gaussmf	0.4874	26.3147

According to Table 6, the best membership function is determined, which produces the smallest RMSE on training and testing data. Although the RMSE from Gaussian is smaller for data testing than the RMSE from gbellmf (Generalized Bell), but for Generalized Bell, RMSE on training data is much smaller than those of Gaussian. It is concluded that the best membership function is Generalized Bell.

5.2.3. Results Analysis. From the training conducted on the data, it is concluded that the best ANFIS model has the input It (t) and Ib (t) with the number of clusters [8 8] and the Generalized Bell membership function. RMSE on training data is 0.4241 and RMSE on testing data is 27.0349.

6. Conclusions and Suggestions

This study has successfully applied the fuzzy neuro model, namely the ANFIS on NTP data from January 2013-December 2017. Based on experimental results we conclude that the best model for inflation prediction data according to the smallest RMSE in data training and data testing is obtained on the time series monthly data with input NTP (t) and NTP (t-1), number of clusters [7 7], and the Gaussian membership function. On training data prediction of NTP using the model the ANFIS has RMSE of 0.3430 whereas on the testing data has RMSE of 1.4570. These results are still better than prediction of NTP for multi-variable monthly data with input It (t) and Ib (t), the number of clusters [8 8] and the Generalized Bell membership function which produce RMSE of 0.4241 on the training data and RMSE of 27.0349 on the testing data. RMSE in the testing data is very large, it can be concluded that the prediction of NTP using the ANFIS model is not suitable for multi-variable monthly data, but it is suitable for monthly time series data.

Based on the results of this study a suggestion can be considered in the further study. Further work is expected to be able to use other methods so that they can be compared in term of their performance.

References

[1] V. Wardi 1943 Nilai Tukar Petani Provinsi Sumatera Barat 2013-2017 T. Sugiyarto, Ichsan, Y. Buchari, M. H. Bahar, and Deswaty, Eds. Badan Pusat Statistik Provinsi Sumatera Barat

[2] W. S. Mcculloch and W. Pitts 1943 A logical calculus of ideas immanent in nervous activity Bulletin of Mathematical Biophysics

[3] L. Zadeh 1965 Fuzzy sets Information and Control 8 pp. 338–353

[4] S. Hardwinarto, M. Aipassa et al. 2015 Rainfall monthly prediction based on artificial neural network: A case study in tenggarong station, east kalimantan-indonesia *Procedia Computer Science* 59 pp. 142–151

[5] M. Hasan, X. F. Shi, T. Tsegaye, N. U. Ahmed, and S. M. M. Khan 2013 Rainfall prediction model improvement by fuzzy set theory *J. Water Resource and Protection* 5 no. 01 p. 1

[6] J.-S. Jang 1993 Anfis: adaptive-network-based fuzzy inference system IEEE transactions on systems, man, and cybernetics 23 no. 3 pp. 665–685

[7] R. Hendayana 2001 Analisis faktor-faktor yang mempengaruhi nilai tukar petani Pusat Penelitian dan Pengembangan Sosial Ekonomi Pertanian, Badan Penelitian dan Pengembangan Pertanian Bogor

[8] T. W. Khusniyah and S. Sutikno 2016 Prediksi nilai tukar petani menggunakan jaringan syaraf tiruan backpropagation *Scientific Journal of Informatics* 3 no. 1 pp. 11–18

[9] B. S. Putra, R. S. Wahono et al. 2011 Simulasi penerapan anfis pada sistem lampu lalu lintas enam ruas Kursor 6 no. 2

[10] R. Faulina, D. A. Lusia, B. W. Otok, H. Kuswanto et al. 2012 Ensemble method based on anfis-arima for rainfall prediction Statistics in Science, Business, and Engineering (ICSSBE) International Conference on. IEEE pp. 1–4.

[11] L.-Y. Wei 2011 An expanded adaptive neuro-fuzzy inference system (anfis) model based on ar and causality of multi-nation stock market volatility for taiex forecasting 5 no. 15, p. 6377

[12] C. Dewi and W. W. Himawati 2015 Prediksi tingkat pengangguran menggunakan adaptif neuro fuzzy inference system (anfis) *Proceedings Konferensi Nasional Sistem dan Informatika (KNS&I)*

[13] J. Nou, R. Chauvin, A. Traor´e, S. Thil, and S. Grieu 2014 Atmospheric turbidity forecasting using side-by-side anfis *Energy Procedia* 49 pp. 2387–2397

[14] N. Chasanah 2015 Perancangan sistem prediktor cuaca dengan metode anfis untuk menentukan produktivitas panen sayuran kubis putih (brassica oleracea var. capitata) di karangploso kabupaten malang Ph.D. dissertation Institut Teknologi Sepuluh Nopember,

[15] Tjahjono, E. Martiana, and T. H. Ardhinata 2011 Penerapan adaptive neuro fuzzy inference system (anfis) untuk sistem pengambilan keputusan distribusi obat pada sistem informasi terintegrasi puskesmas dan dinas kesehatan Electronic Engineering Polytechnic Institute of Surabaya (EEPIS) Indonesia 4 no. 1, pp. 338–344

[16] S. Sri Kusumadewi 2010 Neuro-fuzzy: Integrasi sistem fuzzy & jaringan syaraf Yogyakarta: Graha Ilmu

[17] S. Kusuma 2003 Artificial intelligence (teknik dan aplikasinya) edisi pertama Yogyakarta: Graha Ilmu

[18] J.-S. R. Jang, C.-T. Sun, and E. Mizutani 1997 Neuro-fuzzy and soft computing-a computational approach to learning and machine intelligence [book review] *IEEE Transactions on automatic control* 42 no. 10 pp. 1482–1484

IOP Conf. Series: Earth and Environmental Science **335** (2019) 012011 doi:10.1088/1755-1315/335/1/012011

Requirement analysis of blockchain systems on cocoa supply chain

D A Iswari[1], Y Arkeman[2], and Muslich[2]

[1] Postgraduates Student of Agroindustrial Technology Departement, Faculty of Agricultural Technology, Bogor Agricultural University e-mail: dianardifahiswari2@gmail.com
[2] Agroindustrial Technology Departement, Faculty of Agricultural Technology, Bogor Agricultural University

Abstract. The long cocoa supply chain in Indonesia makes cocoa difficult to trace, data in several countries in Indonesia related to national cocoa data. National cocoa production data is needed in creating policies that are right on target, so these things need to be done. The development of industry 4.0 makes the creation of blockchain technology that can overcome these problems. To create a blockchain system for the cocoa supply chain needs a requirement analysis for the time of making the application. The purpose of this research is to identify the needs and know the interactions that occur in the blockchain chain system in the cocoa supply chain. The method used is requirement analysis, use case diagram, and sequence diagram. The results of this study are two inputs: the structure of the cocoa supply chain and activities that occur in the cocoa supply chain, four stakeholders: farmers, collectors, agroindustry, and exporters, one output: the blockchain system from the cocoa supply chain. Interactions that occur in the system is log in to the system, add the activity, execution phase, ordering phase, and validation phase.

Keywords: blockchain systems, cocoa, supply chain

1. Introduction

Indonesia is the third largest cocoa producing country after Ivory Coast and Ghana. According to [1] "cocoa beans are seeds from cacao (Theobroma cacao L.) which come from noble cocoa beans or cocoa beans act after fermentation, washed or without washing, drying, and cleaning". Cocoa plays a role in the national economy as a contributor to the country's foreign exchange, the provider of raw materials for domestic industries, providers of employment, and environmental conservation [2]. In its circulation, Indonesian cocoa has a long supply chain. Based on research conducted by [3], the cocoa supply chain consists of farmers, collectors, agroindustry, exporters, and consumers. The length of the supply chain makes the information cut off in the middle of the supply chain. The farmers does not know how much cocoa that industries need, and industries does not know how much cocoa that farmers produce. It is become big problem when all the stakeholder do not know the transparency of the amount of cocoa.

The current situation about the amount of cocoa is the different data from the different institutions. There are institutions which say that domestic cocoa production in 2017 is 600 thousand to 700 thousand tons, but there are also other institutions that say the national cocoa production is only 300 thousand tons [4]. The difference in data on the amount of cocoa production occurs because each party has a different approach in calculating production capacity. Some parties calculate production practically by calculating the amount of export-import cocoa, and other parties estimate the production of cocoa from the area of the crop and the level of productivity. This different data is always a polemic, because, with the difference in the amount of cocoa data, the policy can be missed [5]. Therefore, it is necessary to resolve the problem of differences in national cocoa production data.

In the industrial 4.0 era, where all things will be digitized can certainly help in overcoming the problem of differences in national cocoa production data. An advanced digital technology, the blockchain, will be able to overcome these differences. The blockchain is a distributed ledger system that has several advantages namely the system will be decentralized, transparent, the data can not be changed, and each stakeholders will validate the data each other [6]. The blockchain will be record the data from the farmers until data from exporters. So, the amount of cocoa will be known and blockchain can help the problem the lack of the transparency about data.

In building a blockchain system for the cocoa supply chain, a system requirement analysis is needed to determine what components are required to build the blockchain system and also make it easier to continue the coding process. In this study two diagrams of Unified Modeling Languages (UML) are used, namely use case diagrams and sequence diagrams. UML diagram is used to analyze the system requirement because UML provide a language visual model that make the developer easier to make the blueprint of the system [7]. The purpose of this study is to identify the elements needed by the blockchain system in the cocoa supply chain and analyze the interaction of activities that occur between stakeholders.

The results of this paper is the blockchain need two inputs: the structure of the cocoa supply chain and activities that occur in the cocoa supply chain, four stakeholders: farmers, collectors, agroindustry, and exporters, one output: the blockchain system from the cocoa supply chain. Interactions that occur in the system is log in to the system, add the activity, execution phase, ordering phase, and validation phase.

2. Research Methodology

This research in knowing the system requirement described in the input-output diagram where this diagram will illustrate several components of the system including input, output, resources, stakeholders, role, mission, objective, threats, and control. UML diagram for interactions that occur between stakeholders in the system. The flow of this study can be seen in figure 1.

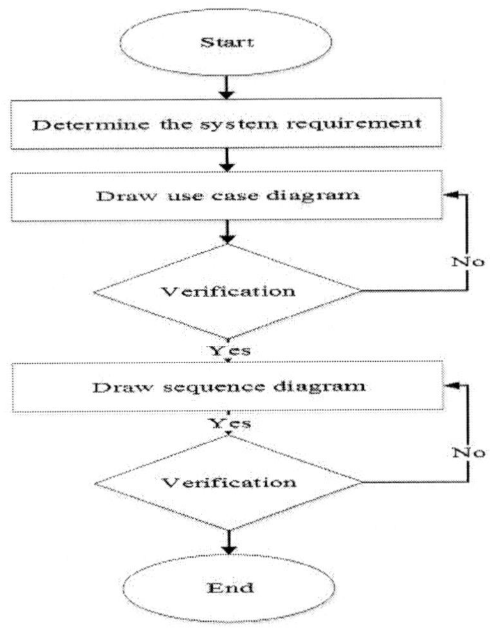

Figure 1. Flow Chart.

3. Result and Discussion

3.1 Requirement Analysis

Requirement analysis is to identify what components are needed by the system. In this study, what elements are required in building a blockchain system in the cocoa supply chain. Requirement analysis is identified using the input-output diagram. The component that described in the input-output diagram is stakeholders, objectives, threats, and resources. The input-output diagram can be seen in figure 2.

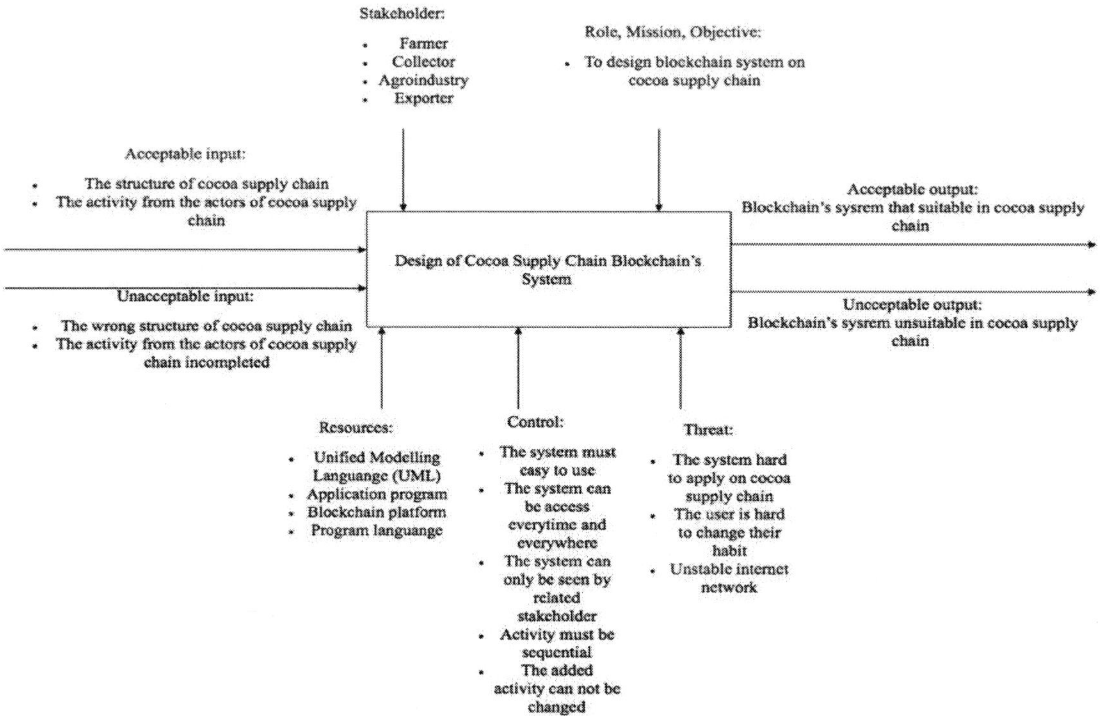

Figure 2. Input Output Diagram.

The input-output diagram, which can be seen in Figure 2, is a component needed to build a blockchain system in the cocoa supply chain. To get the output which is blockchain system that is suitable for the cocoa supply chain, two inputs are needed, namely the structure of the cocoa supply chain and the activities of each actor. The structure of the cocoa supply chain is used to determine how many stakeholders are involved in the system. This structure will determine how many ledgers will be formed on the blockchain system. In addition to knowing the structure of the chain, activities in the supply chain must also be identified to determine the design of the ledger display that will build. Existing activities must also be sequential and detailed because the blockchain system wants each activity in sequence when validating process. The closed blockchain is a system that executes order-execute architecture, so its execution is executed sequentially [10].

In the stakeholder component, there are four relevant stakeholders, namely farmers, collectors, agroindustry, and exporters. Farmers, collectors, agroindustry, and exporters are members of the cocoa supply chain. These stakeholder will play a role in the blockchain. They will interact each other, when one party record their data, the other parties will be verified and validate the data. They have an agreed contract and this contract is the basis for verifying and validating the data. The data that fit with the contract will be verified and validate and add into the block, but if the data does not fit with the contract the other stakeholders will not verified and validate the data.

The blockchain system must be able to be accessed by all relevant stakeholders because the purpose of the blockchain system in the cocoa supply chain is to utilize its traceability and transparent data, so that the system must be built with the aim of the system being able to be implemented. In the resources component, there is four resources including unified modelling languages (UML), application programs, blockchain platforms, and programming languages. UML is used to describing the blockchain system that will be built in the diagram form. UML will make it easier for programmers when coding runs. Application programs that used to draw the UML diagram is power designer, rational rose, star uml, and so on. The blockchain platform is used as a coding medium when want to create a program. There are several blockchain platforms like hyper ledger, multichain, and others. The programming language is the language used for making applications later. Programming languages that can be used are C language, R language, and so on.

In the control component there are several controls to make the blockchain system, they are the system must be easy to use, accessible at any time, can only be seen by the relevant stakeholders, the activity must be sequential, and the activity that has been added cannot be changed. Control is used to monitor the course of the system. The system must be easy to use and can be accessed at any time is a control to allows users who have high mobility. The system must be visible to the relevant stakeholders as a monitoring process, the added activity must be sequential because the validation process based on the previous procedure, and the added activity cannot be changed to support the system so the all data in the system are true.

In the threat component, there are three threats, namely the system is difficult to apply to the cocoa supply chain, it is difficult for stakeholders to change their habits, and internet network instability. The cocoa supply chain consists of different people which will certainly be difficult when applying an application that is used together. At this time stakeholders rarely record their activities, changing their habits is a difficult thing especially not everyone is open to change. Internet networks are needed in this application, while the internet network in Indonesia is still not spread evenly, this will certainly complicate the implementation of blockchain applications.

3.2 Activity Interaction

Activity interaction is an important thing to be analyzed to see how each stakeholder contributes to the making of the system. Activity interactions are described in two diagrams, use case diagrams and sequence diagrams. This diagram can be seen in Figure 3 and Figure 4.

Figure 3 show the interaction between stakeholder when blockchain applied on the supply chain of cocoa, while figure 4 shows the detailed activity based on use case diagram. Figure 4 show about the activity of farmer when they want to adding the activity into blockchain application. In the first activity is log in to the system, this activity did by all stakeholder. Every stakeholder that wanted to report their activity must log in to the system. When stakeholder want to log in to the system, they must write the user name and password.

The second activity adds the activity. This activity is done by all stakeholders too. After stakeholder log in to the system, they must add the activity into the system. The information that attaches to the system is the name of the activity, with whom the activity occurred, and the amount of cocoa when the interaction occurred. The third activity is the execution phase, the execution phase is a phase that occurs when a stakeholder wants to add their activity. In this phase the stakeholder sends the proposal that consists identity, transaction payload, and the identifier from chaincode [10]. For example in figure 4 farmer wants to insert their activity into the block, so farmer must send the proposal. Collector and agroindustry will execute the activity and give the endorsement to the proposal. The proposal that has enough endorsement will be continue to the ordering phase.

The fourth activity is the ordering phase, the ordering phase is a phase to spread the endorsement and determine the consensus for the activity. So, the activity will be grouped into the block and generate the sequence of hash chain from the block that consists of the transaction [10]. For example, the activity that sends from farmer spread into collector, agroindustry, and exporter. The last activity is validation phase, validation phase is phase to validate the transaction. In this phase, if validator does not satisfy with the endorsement, so the activity stated do not valid or will be ignored, while the valid activity will be add into block [10].

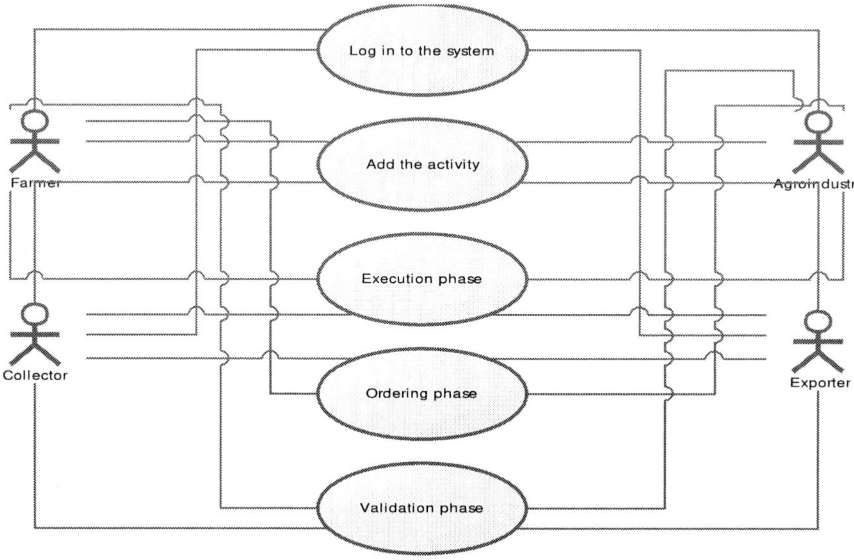

Figure 3. Use Case Diagram.

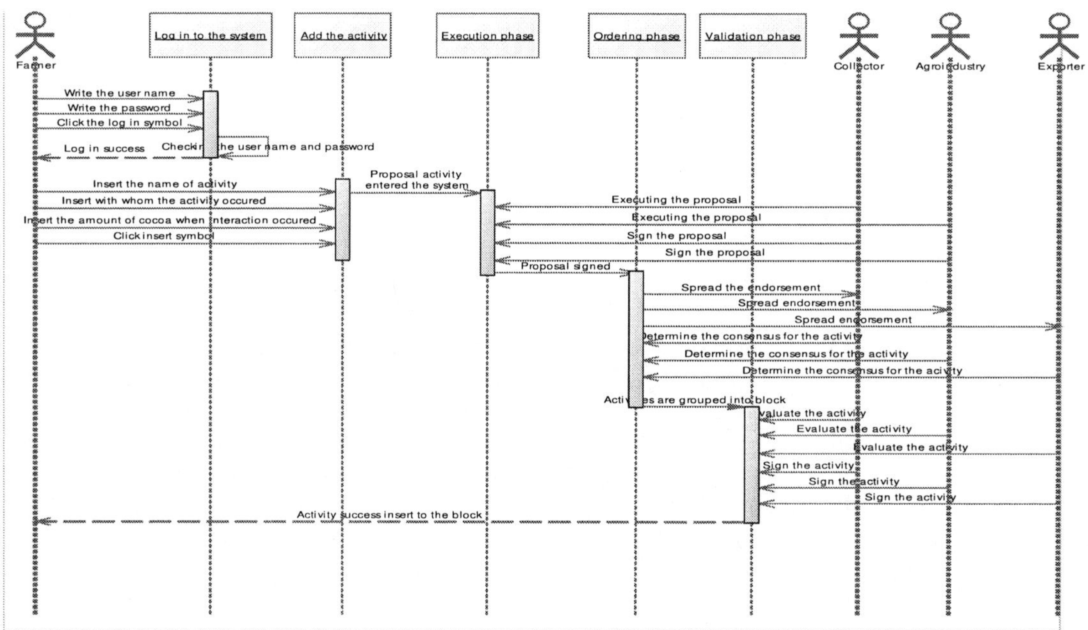

Figure 4. Sequence Diagram.

4. Conclusion

This study analyzes the needs of the blockchain system that will be applied to the cocoa supply chain and knows the interaction between stakeholders in the system. The result is that in producing blockchain system output in the cocoa supply chain, two inputs are needed, namely the cocoa supply chain structure and activities for each actor. Four stakeholders involved in the system, namely farmers, collectors, agroindustry, and exporters, the goal is all of information is transparent so the real amount of cocoa can be known, four resources namely UML, application programs, blockchain platforms, and direct programs. Five controls namely the system must be easy to use, accessible whenever and wherever, only visible to relevant stakeholders, activities must be sequential, and activities the added cannot be changed, and the threat is that the system is difficult to implement, the user has difficulty changing habits, and the instability of the internet network. The interaction between stakeholders in the system in designing blockchain applications for cocoa supply chains is log in to the system, add the activity, execution phase, ordering phase, and validation phase.

References

[1] Peraturan Menteri Pertanian Republik I Indonesia Nomor 67/Permentan/OT.140/2014 2014 Kakao
[2] Sudjarmoko B 2013 State of the art industrialisasi kakao Indonesia *SIRINOV* **1** 1
[3] Yusrina, Yandra A, Sapta R, and Purwiyanto H 2016 Analisis titik-titik kritis ketertelusuran pada rantai pasok kakao Aceh *J Teknologi Industri Pertanian* **26** 1
[4] Yuniartha L and Dessy R 2017 Askindo: *Data Produksi Kakao Harus Akurat* Available: https://industri.kontan.co.id/news/askindo-data-produksi-kakao-harus-akurat
[5] Laoli N 2016 *Data Produksi Kakao Jomplang, Kebijakan Meleset* Available: https://industri.kontan.co.id/news/data-produksi-kakao-jomplang-kebijakan-meleset
[6] Iansiti M and Karim RL 2017 *The Truth about Blockchain* (Cambridge: Harvard Business Review)
[7] Padeli, Hunderi, and Suyatno Membangun (E-procurement) pengadaan barang dan jasa dengan prinsip good corporate governance dengan visual UML *J Creative Communication and Innovative Technology* **2** 1
[8] Sulistiyorini P 2009 Pemodelan visual dengan menggunakan UML dan rational rose *J Teknologi Informasi DINAMIK* **14** 1
[9] Novianti A and Ami F 2009 *Sistem Informasi Sekolah Dasar Berbasis SMS* (Indonesia: Seminar Nasional Aplikasi Teknologi Informasi Yogyakarta)
[10] Androulaki E, Artem B, Vita B Christian C, Konstantinos, Christidis, Angelo D C, David E, Christopher F, Gennady L, Yacov M, Srinivasan M, Chet M, Binh N, Manish, Sethi, Gari S, Keith S, Alessandro S, Chrysoula S, Marko V, Sharo WC, and Jason Y 2018 *Hyperledger Fabric: A Distributed Operating System for Permissioned Blockchains (IBM)*

Acknowledgments
We would like to thank Irman Hermadi, the lecture of Computer Science, Bogor Agricultural University for giving knowledge about blockchain and giving suggestion to this research. We also would like to thank Karina Meidayanti for the discussion about this research and giving the support. And thank you for my parents, that support my study and my research until now.

Analysis and design of beef supply chain traceability system based on blockchain technology

K Meidayanti[1], Y Arkeman[2], and Sugiarto[2]

[1] Postgraduates Student of Agroindustrial Technology Departement, Faculty of Agricultural Technology, Bogor Agricultural University e-mail: karinameidayanti@gmail.com
[2] Agroindustrial Technology Departement, Faculty of Agricultural Technology, Bogor Agricultural University

Abstract. The need for beef supply chain traceability system is vital due to increase in counterfeiting and use of excessive preservatives and hazardous chemicals. To overcome these problems, it is needed a system which provides the option for the consumer to verify the origin of all input materials for food product based on one step forward and one step backward in supply chain mechanism. Blockchain technology is used as the solution to this problem. This paper aims to identify the requirement and designing the traceability system that could capture all relevant data. The design of beef supply chain traceability system will be described as a business process flow with BPMN (Business Process Model and Notation), in which BPMN will be described the stakeholders involved with the task and needs of the designed system. Based on requirement analysis, this system composed of 4 stakeholders, namely the farmer, feedlot, industry, and retailer. The result showed that the blockchain can create a transparent and efficient supply chain through sharing data between the stakeholders.

Keywords: beef, blockchain technology, supply chain, traceability system

1. Introduction

Beef one of the foods that have high nutritional value, especially the protein that makes it preferred by consumers to fulfill the daily nutritional needs, so the demand for beef continues to increase. According to the directorate general of food and animal health, the amount of demand for beef in 2016 was 675,220 tons, while beef production was 441,761 tons. Based on the data, it can be seen that beef production is not proportional to the level of consumption. The lack of fresh meat supply is influenced by several factors, including the low level of domestic meat production, the lack of transportation to distribute livestock from production centers to consumption centers, and the long chain of local beef trading systems.

Another challenge in a beef supply chain is food safety. This is related to the characteristic of beef which is easily damaged so that it requires extra handling in its distribution. A traceability system is needed to monitor product movements along the chain to ensure quality products to consumers. Traceability is a tool that can help an organization operate in the food or feed supply chain to achieve the results that have been identified in the management system [6]. The main purpose of the traceability system is to report and design material, process, and product distribution processes to consumers. By using a good traceability system, leakage or inefficiency in a supply chain can be avoided.

IOP Conf. Series: Earth and Environmental Science **335** (2019) 012012 doi:10.1088/1755-1315/335/1/012012

One of the technologies that can be used is Blockchain technology. The blockchain is basically a new database, where data is stored chronologically according to the time sequence. In general issues, blockchain as a public ledger or general ledger that records all transactions in general and can be seen by everyone involved. With blockchain technology, all beef data will be accessible to all group members including consumers.

The outline in this paper consists of an introduction, research methodology in section 2, the result in section 3, and section 4 is the conclusion of this paper.

2. Research Methodology

The research design of the beef supply chain traceability system based on blockchain is described as follow:

2.1. Identification System

The identification system is done to understand the initial problems in the system. We need to define the system by identifying the stakeholder, rule, role, mission, input, and output.

2.2. Analyze the interaction between stakeholder

This step is done by modeling using unified modeling language (UML) with use case diagram which then followed by modeling in BPMN 2.0 and simulated to answer system requirement. The use case diagram is composite attribute that describes the interaction between the actors involved in the system. Actors can consist of people, places, events, and real or virtual objects [11]. The use case diagram shows the attributes that exist in the system along with their functions.

2.3. System Analysis and Design

BPMN 2.0 is used to design the model. System analysis and design are started by making a flowchart. The flowchart describes the stakeholders, role, process, data, and information. System analysis is conducted to parse a system to be studied into components so it the interactions between components and their environment can be seen [3].

3. Result and Discussion

3.1 Identification System

Identification of the system was begun by making the input-output diagram. Then determined the system requirements and entities to build the system, such as stakeholders, input, output, rule, purpose, and role (fig. 1).

Input output diagram is a component needed to build a system in beef supply chain. Input output diagram describes the system requirements, role, mission, objective, threat, and opportunities. Based on fig. 1, we can see that beef supply chain traceability system based on blockchain need an input such as, an age of the cattle, weight, sex, delivery and receiving dates, health history, and barcode. This input was collected by stakeholder involved. There are four stakeholders, namely farmer, feedlot, slaughterhouse, and industry. Farmer (Breeder) are actors who will input the cattle data into the system, feedlot is actors who will input data related to health history of the cattle, the slaughterhouse is actors who will input the date of receiving the cattle, and input the history of cattle before being a slaughter. The industry is actors who will input the data of meat processing and packaging, and also make the barcode. This stakeholder will interact with each other in building the system.

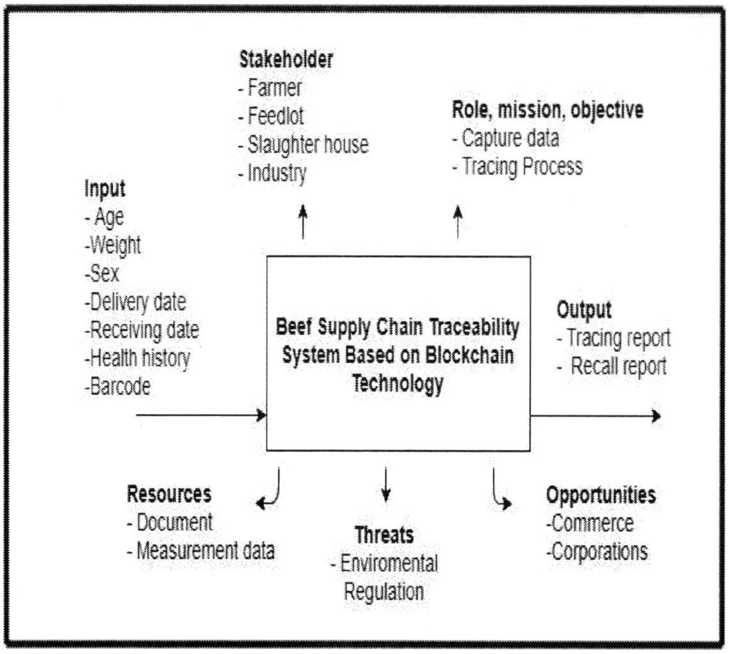

Figure 1. Input output diagram.

The beef supply chain traceability system based on blockchain there is an object that is to capturing data and tracing process. The resources of this system are document and measurement data. The environmental regulation became a threat for this system because if the adoption of blockchain is increased, the energy burden will become more considerable. There is likely to be a lot of opposition as far as blockchain technology large-scale implementation is concerned.

3.2 Analyze the interaction between stakeholder

Use case diagrams show interactions between actors or "species". In the beef supply chain traceability system, there are four actors involved, namely cattle farm (breeder), feedlot, and meat industry. Activities in the system are begins by initiating transactions by the cattle farm, then the cattle farm will broadcast the details of the transaction to be carried out. After the transaction is broadcast, all stakeholders must approve the activity and then the block will be formed. Each stakeholder inputs data related to each activity and the data is transparent so it can be seen all over the chain. After uploading the data, each actor is required to verify the activities that have been done in the supply chain processes. After all, transactions are verified, the block will be given a special code as a seal so that the data that has been arranged into blocks cannot be lost and cannot be changed. The Interaction between actors was described in the use case diagram (fig. 2).

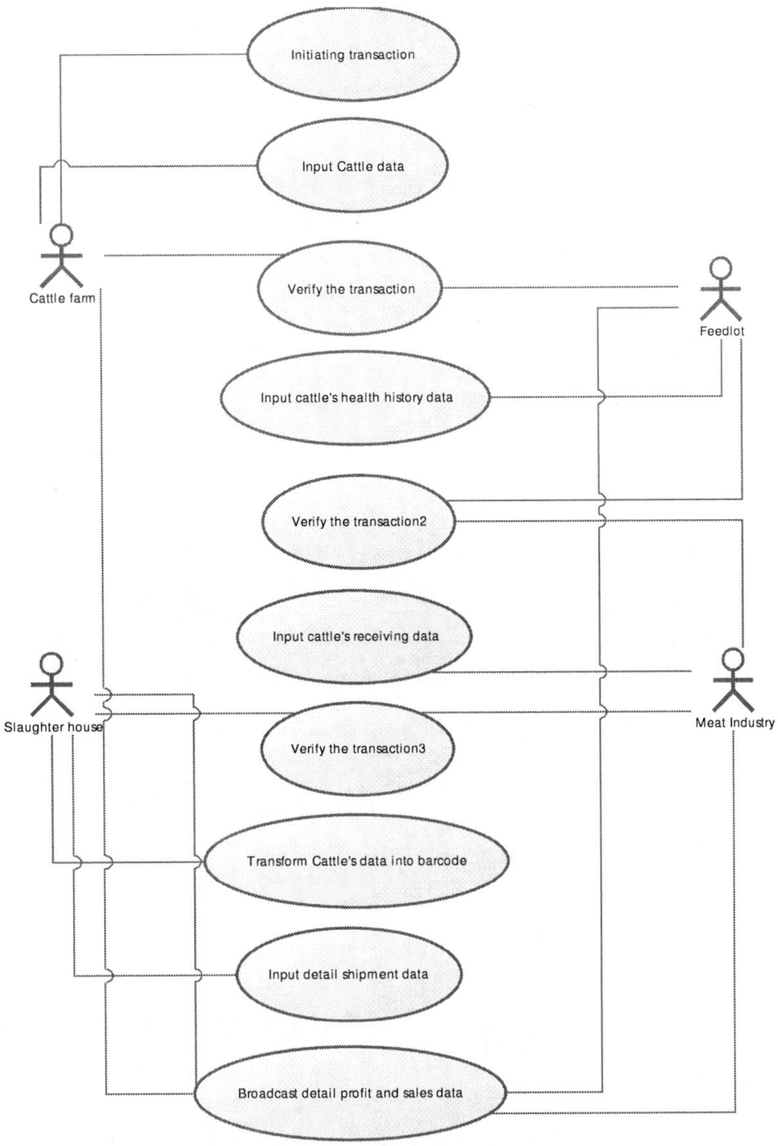

Figure 2. Use case diagram.

3.3. System Analysis and Design

System analysis and design are modeled in BPMN 2.0 and the software used is power designer, where the model is right if the output contains zero error (s) and zero warning (s). The business model is divided into a four-lane that represents the supply chain actor in a beef supply chain. The business process is linked procedures or activity which aims to achieve overall business goals, defines the functional roles and relationship in the context of an organizational structure [3]. The results of the business process were described in fig. 3,4,5 and 6.

IOP Conf. Series: Earth and Environmental Science **335** (2019) 012012 doi:10.1088/1755-1315/335/1/012012

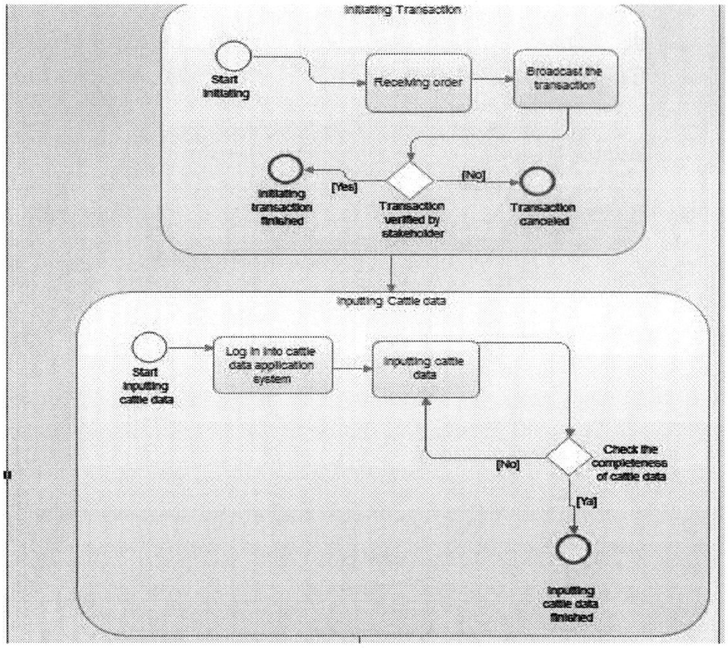

Figure 3. The BPMN of the cattle breeder.

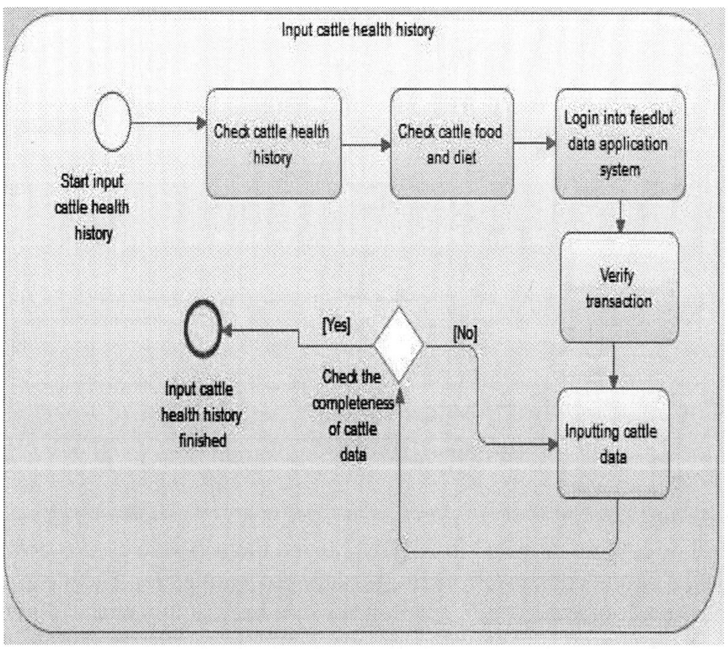

Figure 4. The BPMN of feedlot.

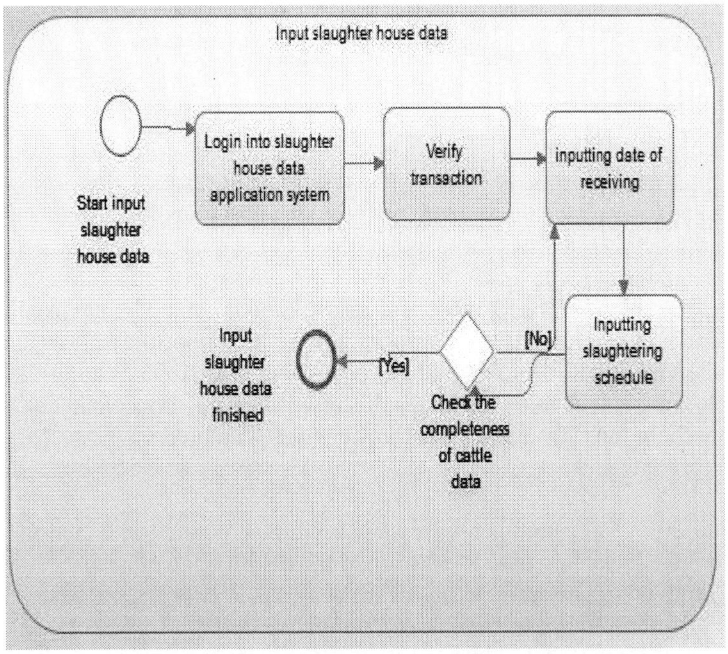

Figure 5. The BPMN of the slaughterhouse.

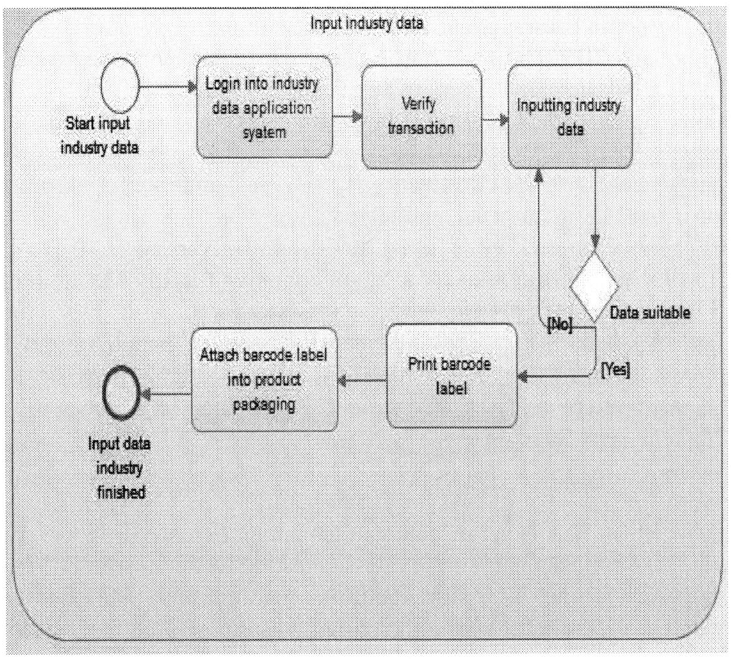

Figure 6. The BPMN of the meat industry.

Based on the results of BPMN that have been made, it can be seen that the system begins with the breeder receiving orders, and then broadcasting it to all stakeholders. If the transaction is approved by all actors, the transaction is considered valid and can proceed. If the transaction is disapproved, then the transaction will be canceled. After the transaction approved, the stakeholders will log in into the application to start entering data related to their activity. At the industry level, data that has been uploaded by previous stakeholders will be checked again, and then the industry will transform the data into a barcode that will be attached to each product package.

4. Conclusion

This study analyzes the needs of the blockchain system that will be applied to the beef supply chain and knows the interaction between stakeholders in the system. The result is that in producing blockchain system output in the beef supply chain, several inputs are needed, namely the beef supply chain structure and activities for each actor. Four stakeholders involved in the system, namely cattle breeder, feedlot, slaughterhouse, and industry. The goal is to build a beef supply chain traceability system using blockchain, two resources namely document and measurement data. The environment regulation is a threat for the system, while commerce and corporation are the opportunities for the system. Based on the analysis using BPMN 2.0, it is known that there was four stakeholder taking the role in traceability system based on blockchain technology. The result showed that the blockchain can create a transparent and efficient supply chain through sharing data between the stakeholders.

References

[1] Axfoundation 2017 SKL kommentus Swedish country council and regions Martin & servera Kairos future Blockchain use cases for food traceability and control
[2] Biswas K, Muthukkumarasamy V and Tan W L 2017 Blockchain based wine supply chain traceability system future technologies Conference Griffith University
[3] Djatna T and Luthfianti R 2016 An analysis and design of responsive agroindustrial supply chain for multi-products of tropical fruit [Thesis] (Bogor: Bogor Agricultural University)
[4] Ginantika A 2015 An analysis and design of frozen shrimp traceability system based on digital business ecosystem [Thesis] (Bogor: Bogor Agricultural University)
[5] Hua A V and Notland J S 2016 Blockchain enabled trust & transparency in supply chains [Thesis] Retrieved from Norwegian University Science and Technology
[6] ISO 22005 2007 Traceability in the feed and food chain Available https://www.iso.org/standard/36297.html
[7] Nian L P and Chuen D 2015 Introduction to bitcoin handbook of digital currency: bitcoin, innovation, financial instruments, and big data
[8] Sander F 2017 From Farm to Fork [Thesis] Maastricht University
[9] Swan M 2015 Blockchain Blueprint for a New Economy O'Reilly Media, Inc
[10] Tian F 2016 An agri-food supply chain traceability system for china based on rfid & blockchain technology Available https://www.semanticscholar.org/paper/An-agri-food-supply-chain-traceability-system-for-%26-Tian/24cdeb7d7421012c2fdd362b8e2816c105b7071f
[11] Wasson C S 2006 System analysis, design, and development concept, principles, and practices John Wiley & Sons New Jersey (CA)

Acknowledgments

We would like to thank Irman Hermadi, the lecture of Computer Science, Bogor Agricultural University for giving knowledge about blockchain and giving suggestion to this research. We also would like to thank Dian Ardifah Iswari for the discussion about this research and, thank Sea Galifiannisa, Anitsa Puri, Lintang Harwina, Utari Yolla for giving the unlimited support. And thank you to my parents, that support my study and my research.

Operational risk evaluation and mitigation for palm oil supply chain: a case study at x co.

Safriyana[1], Marimin[1], E Anggraeni[1] and I Sailah[1]

[1]Department of Agroindustrial Technology, Faculty of Agricultural Engineering and Technology, Kampus IPB Darmaga, PO Box 220 Bogor 16680, Indonesia e-mail : anayirfas@apps.ipb.ac.id

Abstract. Risk analysis in the palm oil suppy chain is essential to achieve the optimum efficiency and effectiveness of the supply chain flow. One of the most common types of risk found in the palm oil supply chain is operational risk. The objectives of the research were to model palm oil supply chain risk identification and to model its risk mitigation. Measurement of operational risk required in an effort to improve the performance of the palm oil supply chain, the method used in the assessment of this type of risk is Failure Mode and Effect Analysis-House of Risk (FMEA-HOR) through two stages of modelling, the analysis and evaluation stages used to know the risk event and risk agent known as HOR 1. Meanwhile the stages of risk mitigation measures applied to handle the potential risk types known as HOR 2. Based on the results of operational risk identification in the palm oil supply chain, 31 types of risk events and 20 risk agents were obtained. Furthermore, risk mitigation measures were based on the potential risk agents generated, converged into 8 risk proactive actions that are prioritized based on costs and resources effectiveness. The results of priorities proactive actions show that 1) maintenance's labor training, 2) committing routine preventive and predictive maintenance, and 3) preparing alternate stand by equipment and machine units were necessary mitigation actions to be done.

Keywords: FMEA-HOR, palm oil supply chain, risk agent, risk event

1. Introduction

Indonesia dominancy in palm oil production respectively can be seen from plantation area growth and palm oil agroindustry itself, considered of high profit margins while the palm oil production are simple and low in costs. Furthermore, the significant increase of palm oil global demand made the business interesting and open up the opportunities for palm oil stakeholders due to the important roles of this sector in affecting national economy and improving community welfare [1], [2]. However, along with the increasing of palm oil supply chain, there are unavoidable impacts and pressures in Indonesia's CPO exports which gave influences on the operational sustainability of palm oil agroindustry thats are directly related to the palm oil supply and demand stabilization [3], [4]. This things need to be considered due to its direct relation to the impacts that arise as potential risks to the palm oil supply chain because of economic and social changes from various palm oil supply chain stakeholders, also the black campaign or market sentiment towards palm oil [4]. As an example, in a palm oil agroindustry company, the impact can cause a loss or decrease in company profits and further lead to money loss and unemployement of employess as a social impact. Likewise for palm oil farmers, if there are incompatibility in the cultivation processing until harvesting, it will cause effects to the price and production continuity, therefore the palm oil farmers will receive lower incomes [5].

Hence, it is required to consider palm oil supply chain activities' diverse risks due to the issue of risk and uncertainty in agriculture and the decisions relating to risks are needed [6] and also needed to optimize supply chain [7]. According to internal and external occurence of risk in supply chain management factors [8], the frequently types of risks appear is the operational risk factor [9], [10]. These operational risk factors are namely faults planning, raw material shortages, capacity constraints, machine failure or down time, imperfect results, inefficiencies, process changes, property losses due to accidents/disasters, transportation risks (delays, damage during expedition), warehouse risks (imperfect costumer orders, insufficient inventory, and others), budget expenditures, technology disruptions' emergence, terms of agreement (minimum and maximum costumer requests), and interuptions of communication or information system [11].

Considering to the direct impacts of operational risks on palm oil supply chain performance, a measurable and systematic investigation is needed to mitigate these type of risks. Through to the supply chain risks identification, it can be found the overall risks data and information that occur in supply chain that can be optimally utilized to obtain the mitigation actions from palm oil supply chain problems to improve the supply chain performance. This background makes the study of supply chain agricultural risks done a lot and some studies has already proposed several method such as sustainable palm oil supply chain risk judgment based on performance [9], risk balancing model of agro-supply chain [12], agent based modelling for palm oil supply chain operational risk [13], supply chain risk in organic rice product using Fuzzy-FMEA method [14], poultry feed production's risk analysis using Fuzzy-FMEA [15], supply chain risk in rafinated sugar using HOR method [10], and fuzzy logic for risk dependency chain in dairy agroindustry [16]. Yet, the types of risks involved in palm oil supply chain are different and to our knowledge, there is no research has been investigate this operational risk in the palm oil supply chain. Thus, this paper will investigate palm oil supply chain operational risk and proposed risk mitigation actions.

Moreover, it is necessary to analyze the palm oil supply chain risk factors and its treatment that can be applied. The purposes of this research were to model supply chain risk identification, calculate and analyze the risk priority index value to the causes of risk in palm oil supply chain activities, and to conducts palm oil supply chain risk mitigation modelling based on potential risk index. In order to fulfill these objectives, this research uses House of Risk method to evaluate the types of potential risks and evaluate it, and mitigation model for potential risks obtained. House of Risk (HOR) modelling is a modification method to prioritize potential risk agent that are chosen to get effective actions in achieving objectives to reduce potential risks from risk event. From the statement above, it is known that HOR method is beneficial because its ability to integrate the risk identification till risk mitigation in one compreherence model, but the drawback of the method is time consuming and complexity data is needed to build the model.

2. Methodology

2.1. Research Framework

House of Risk (HOR) is a modification method of FMEA (Failure Mode and Effect Analysis) and House of Quality (HOQ) models to select and prioritize risk agent that will be given the action in purpose to decrease the risk agents' potencies [10]. Risk assessment using FMEA is well-known done through quantitative score to the risk priority number (RPN) as a mathematical model of the risks to identify the most critical failure mode, leading to corrective actions [15], [17]. Different from the FMEA model, where both the severity and occurence are correlated with the risk events, in FMEA-HOR the occurence is assigned to the risk agents and the severity is assigned to the risk events and measure the Aggregate Risk Potential (ARP) of a risk agent. The HOQ model is adapted to conclude which risk agents should prioritize for preventive actions due to each risk agent's rank [18].

The HOR model is depend on the nation that a preventive supply chain risk management should attempt to concern on proactive actions [18] that would prevent the risk events to appear. In determining HOR's risk analysis, it is based on the SCOR (*Supply Chain Operations Reference*) business process parameters model to represent, analyze, and configure the supply chains based on its basic pillars: plan, source, make, deliver, and return [8], [19] in order to describe business process along palm oil supply chain that centralize the risk event, risk agent, and preventive or mitigation actions of risk agents. From the supply chain, it will be carried out some problems mitigation that may arise out of the supply chains before the impacts realized spread [20].

House of Risk (HOR) method used in this paper divided to two deployments models, called HOR 1 and HOR 2. The HOR 1 is used to identify, analyze, and determine the risk agents and the risk events which are to be measured and prioritize the potential risk agents as the evaluate step. The HOR 2 is used to provide preventive actions considered effective but with acceptable costs and resources.

2.2. *Data Collection Procedure*

The data used in this research were obtained from primary dan secondary sources. The primary data used was resulted from field observation, questionnaires, interviews, and experts' discussion. The field observation was conducted in X Co., Cigudeg, West Java, Indonesia about supply chain operational risks and experts assessment for each risks measurement and risks mitigation. The secondary data was used from scientific study, previous research, and other relevant data.

2.3. *House of Risk 1 Data Analysis*

2.3.1. Risk Identification. The risk identification begins with a supply chain structures preliminary study and supply chain members relationships. It is conducted as an evaluation and efficiency control of the business process [20] and extensively able to control each dimension of the supply chain and detected the whole business process activities. Data collection is done through literature study, field observation, filling out questionnaires, and in-depth interviews with industry practitioners (professionals) involved in the palm oil supply chain. The supply chain risks is assessed due to severity of risk events and occurence of risk agents based on the FMEA approach and identified based on model and alternative adjustments according to company needs. Each of indicators determine directly using depth interview with the expert and interpreted in the HOR modelling and determined critical points of each activities.

2.3.2. Risk Analysis. The risk analysis is performed by determining risk events' severity (Si), risk agents' occurrence (Oj), and risk event and risk agent's correlation (Rij). The severity value using 1-10 scale (no failure-must occured failure), occurence value using 1-10 scale (almost never happened-often), and correlation with Rij (0, 1, 3, 9); 0 is no correlation and 1, 3, 9 show low, moderate, and high correlation, correlation relationship is expressed as Rij ε {0,1}, Rij=1, there is correlation between risk i with risk agent j and Rij=0, there isn't correlation between risk i and the risk agent j. The weighting of risk components (as a modified aggregation method from the previous HOR researches) is carried to provide its value, through the use of formula for experts combination [21]:

$$Si = \text{median } [Si_1, Si_2, ..., Sik] \quad \forall i \tag{1}$$

$$Oj = \text{median } [Oj_1, Oj_2, ..., Ojk] \quad \forall j \tag{2}$$

Si = Severity level of risk event i, j = 1, 2, ..., n
Oj = Occurence level of risk agent k = k-person measurement

The calculation of the risk aggregation value is obtained by calculating the ARPj value [18]:

[illegible] x Rij. (3)

2.3.3. isk Evalution. The risk evaluation aims to determine ARPj ranking to risk priority, potential impacts of risk event, and correlation among risk agents of HOR 1's roof. After the ARPj value (risk analysis' result) obtained, the rank is made using the formula as follows [22]:

P[illegible]j ∑[illegible] x (Rij x wij) ∀ j (4)

Pj = Risk priority index
Oj = Occurence level of risk agent
j = 1, 2, …, m; Rij [illegible],1}

Si = Severity level of risk event
Rij = The correlation between risk event i with risk agent j
wij = The weighted correlation between risk event i with risk agent j

Furthermore, from ARP value rank, the risk mitigation is conducted using Pareto diagram (80:20 rule) which gives sequencing classification data description to find the most important risk agents to resolve and mitigate immediately [22]. In general, the HOR 1 stages are carried out as follows:

1. Identification of weaknesses and risk events in each business process is done as the result of supply chain business process mapping.
2. The risk events' estimating impacts assess with the 1-10 scale value (shows the severity (Si)).
3. The determination of the potential caused/impact by each risk event, expressed as Ci, determined by experts who are experienced in their field.
4. The occurence's frequency of each risk agents determined with a set of 1-10 scale (Oj).
5. The matrices relationships development. Determine the relationship among risk agent and risk event, Rij (0, 1, 3, 9), 0 shows no correlation and 1, 3, 9 show low, medium, and high correlations.
6. The calculation of potential risk value (*Aggregate Risk Potentials of risk agent j* = ARPj) is determined as in the equation ARPj = Oj Σ Si Rij
7. The relationship among risk agents determine using Raij notation (0, θ, *, •); 0 shows no correlation and θ, *, • respectively show weak, moderate, and strong correlation [22].
8. The ranking determination of risk agents based on the ARPj value.

2.4. House of Risk 2 Data Analysis

House of Risk 2 modeling is applicated to decide the risk mitigation actions carried out by considering the resources and costs effectiveness in mitigate the risk agents so that it is not difficult to implement and effectively reduces the occurence of risk agents. The steps taken for the risk mitigation process are, identify the risk mitigation, evaluate the risk mitigation, determine the correlation among mitigation actions, and select the mitigation actions priority. The steps of HOR 2 model are explained below:

1. Choose a number of risk agents with high priroty rank based on ARPj's Pareto analysis, placed as to be treated risk agent (Aj).
2. Identify the relevant preventive or proactive action which simaltenously can decrease the possibility of more than one risk agents' occurence, the action called preventive action (PAk).
3. The correlation value (Ejk) weighting of risk agent and its mitigation action, using Ejk scale (0, 1, 3, 9). Ejk can be considered as the actions' effectiveness level in reducing the risk agents occurence.
4. The calculation of each total effectiveness actions: TEk = $ARPj\ Ejk\ \forall kj$
5. Estimated the preventive actions with actions difficulty degree, Dk (*difficulty*) using Likert scale (3, 4, 5) which reflects the costs and other resources needed to carry out these risk mitigation actions.
6. The calculation of total effectiveness with difficulty, ETDk = TEk/Dk. Determine each of actions' priority level (Rk), the first level indicates the action with the highest ETDk.
7. The relationship among mitigation actions determine using Rajk notation (0, θ, *, •); 0 shows no correlation and θ, *, • respectively show weak, moderate, and strong correlation [22].

3. Results and Discussions

3.1. Palm Oil Supply Chain's Operational Risk Analysis (House of Risk 1)

Palm oil supply chain performance which has not reached the maximum value as in [8] indicate the needs to evaluate and increase its performance level because it could be regarded as a loss caused by the supply chain problems that directly or indirectly disturb the palm oil supply chain, which is defined as a supply chain risk. Septiani [23] also mentions that the supply chain risks are interconnected and affect the supply chain performance. The risk identification is conducted on the palm oil business process, based on the SCOR dimension decomposed into each subprocess so that common or potential errors or risks occur throughout the supply chain can be known in detail [16]. Therefore, it is necessary to control the supply chain risk in order to avoid the ongoing and sustainable effects that occur at every point in supply chain network by carrying out the supply chain risk analysis [24]. Supply chain risk analysis from the operational side using HOR method is determine by doing risk event identification, risk agent identification, and its correlation.

3.1.1. Risk event identification. The risk event identification is carried out in plan, source, make, deliver, and return stages of the palm oil supply chain business process [8] which are break down into each sub-process, so the details of faults or risks that are common and potentially occur along the supply chain can be known [16]. The identification is based on the critical points of each sub-process activities related to the palm oil supply chain impacts. Based on the results of risk identification conducted, it determines 31 risk events as the critical point that linked to the impact of supply chain. The identification result of the risk events with the severity result depicted in appendix A.

3.1.2. Potential impact identification. Identification of risk event's possibility effect (potential impact) is shown in appendix B, the potential impact declaring the disturbance that might arise if a risk event occurs [21]. The potential impact of 31 risk events is the result of the competent experts' judgement.

3.1.3. Risk agent identification.. The risk agent identification is based on the results of previous identified risk events. Each of identified risk agents may present one/more risk events and vice versa, one risk event may be caused by one/more risk agents causing the disruption [18], the identification result of palm oil supply chain risk agents and its occurence is obtained as shown in table 1.

Table 1. The risk agents' occurence.

Code	Risk Agent	Oj
A1	Production planning miscalculation	1
A2	Sudden/unexpected demand from consumers	1
A3	Insufficient of raw material to production capacity	1
A4	Incompetent labors	1.5
A5	The wrong selection of raw material suppliers	1
A6	The supplier does not fulfill contract	1
A7	Technical disturbance/production stagnation	4.5
A8	The production target is relatively high	3
A9	The defect to machine production	5
A10	The lack of production machines' maintenance	1.5
A11	Negligence of labor	2
A12	Overlong stacking of raw material/product	2.5
A13	Raw materials defect during transportation	2
A14	Process efficiency factor	3
A15	Reliability factor of process machine and equipment	3
A16	The decline of product quality during transportation	2.5
A17	The lack of coordination in the shipping process	1.5
A18	The distance and communication factors between producers and consumers	1
A19	Environmental conditions	1
A20	Limited transportation facilities	2

3.1.4. The correlation between risk event and risk agent. The analysis between each risk event and risk agent correlation is shown in the 0, 1, 3 and 9 value scales based on each level of each other correlation. The correlation is defined if a risk agent occurs, due to the emergence of risk events with each level of correlation low, medium, and high based on expert judgment. For example, due to production planning miscalculation (A1), it results several risk events, such as incompatibility between capacity planning and processing realization (E1), the gap between planning and on-hand material (E2), supply chain incompatibility with cost budget (E3), production target is not achieved (E5), raw materials don't meet the capacity planning (E7), the changes in the raw materials quality (E10), non-conformity of material sent by supplier (E12), the production process is inefficient (E15), production activities aren't done/stopped (E21), the demand can't be fulfilled (E22), the lack of products in the distribution center (E25), and the stability of production is disrupted (E29) as shown in appendix C. It is known, that from one risk agent (A1) can results 12 risk events, so that if a risk agent can be handled or mitigated then the risk event can be reduced or even eliminated.

3.1.5. The correlation among risk agents. The correlation among risk agents in the House of Risk 1 is placed in the HOR's roof component which states the risk agents' correlation. The correlation consists of four categories, namely a strong correlation (•) if a risk agent will have a very significant influence with other risk agent, a moderate correlation (*) if a risk agent will provide enough influence with other risk agent, weak correlation (θ) if a risk agent will provide a weak relationship with other risk agent, and there is no correlation (0) between risk agents. From 20 risk agents identified, the relationship/correlation among risk agents as shown as the roof in appendix C.

3.1.6. Aggregate risk potentials analysis. Risk measurement preparation as a series of coherent, systemic, and explicit actions that accomodate the risk correlated to the supply chain activities [25]. The weight determination of each risks occured is based on the experienced experts assessment, resulting Aggregate Risk Potentials (ARP) value from the result of severity, occurence, and the correlation will be used to determine the risk priority to handle, shown in table 2.

Based on the results of ARP value priority's calculation, it shows that the order for 20 identified risk agents, it is known that the highest value of ARP is the defect to machine production (A9) while the lowest value is the decline of product quality during transportation (A16). From the overall of ARP value results, then it is conducted the analysis using Pareto diagram to prioritize mitigations will be given. Depend on the classification, there are 8 risk agents selected prioritized to mitigate. Sequentially, the results of ARP value calculation then classified according to Pareto diagram to select the priority of risk agents from the overall risk to be handled as an effort to minimize the risk occurrences. The risk agents classification is based on the Pareto 80:20 diagram, then classified into 3 classifications; i.e. classification A (high-level risk agents, 50% of overall risk agents), classification B (medium-level risk agents, 30% of overall risk agents), and classification C (low risk agents, 20% of overall risk agents) [9]. The classification results indicate that 50% of the risk sources in classification A (emerging rate of 56.29%), 30% risk sources in classification B (26.37% incidence rate), and 20% in classification C (17.34% incidence rate) as displayed in Table 2. The overall HOR 1 model as depicted in appendix C.

3.2. Risk Mitigation Analysis (House of Risk 2)

Risk mitigation actions are proposed based on selected potential risks for corrective action, conducted in the form of risk reduction, risk-transfer, risk-aversion, and risk-taking [15]. The risk mitigation model refers to every effort to identify and manage/prevent risks from happens and make sure this situation handled [7]. Risk handling stages are conducted by giving effective mitigation action to decrease the risk agents possibility's priority based on the House of Risk 2 model. Based on the risk agents ranking, the preventive measures identified are divided into 8, there are performing the preventive and predictive maintenance on a regular basis (PA1), preparation of stand by equipment and machine units as substitutes the engine failure (PA2), coordination management and related sections (PA3), improved production design based on the suitability of field conditions (PA4), improved cooperation and coordination with stakeholders related to raw materials (PA5), coordination with the transporter (PA6), optimization of production process in accordance with the provisions and work instructions (PA7), the labors' training on maintenance (PA8). Completely, HOR 2 modelling as depicted in figure 1.

Table 2. The ARP value priority for risk agent.

Rank	Code	Risk Agent	ARP Value	% Total Cumulative ARP	ABC Class
1	A9	The defect to machine production	1680	28.62	A (high level risks)
2	A7	Technical disturbance/production stagnation	1224	49.48	
3	A12	Overlong stacking of raw material/product	400	56.29	
4	A13	Raw materials defect during transportation	368	62.56	B (medium level risks)
5	A8	The production target is relatively high	360	68.69	
6	A10	The lack of production machines' maintenance	350	74.66	
7	A15	Reliability factor of process machine and equipment	243	78.8	
8	A1	Production planning miscalculation	227	82.66	
9	A14	Process efficiency factor	210	86.24	C (low level risks)
10	A11	Negligence of labor	156	88.9	
11	A3	Insufficient of raw material to production capacity	150	91.46	
12	A4	Incompetent labors	104	93.23	
13	A5	The wrong selection of raw material suppliers	85	94.68	
14	A19	Environmental conditions	67	95.82	
15	A20	Limited transportation facilities	54	96.74	
16	A2	Sudden/unexpected demand from consumers	48	97.56	
17	A18	The distance and communication factors between producers and consumers	44	98.3	
18	A6	The supplier does not fulfill contract	35	98.9	
19	A17	The lack of coordination in the shipping process	34.5	99.49	
20	A16	The decline of product quality during transportation	30	100	

3.2.1. The correlation between risk agent and mitigation action. The analysis between each risk agent and mitigation action correlation that will be handled with the planned mitigation action is shown in the 0, 1, 3, and 9 scale value based on each other level of correlation. The correlation is defined as the effectiveness of a mitigation action in reducing the risk agent. Performing the preventive and predictive maintenance on a regular basis (PA1) mitigation action shows 9 value (high correlation) with the defect to machine production (A9), technical disturbance/production stagnation (A7), the lack of production machines' maintenance (A10), and reliability factor of process machine and equipment (A15) risk agents. The other correlation between mitigation actions (PAk) and its risk agent as shown in figure 1.

3.2.2. The correlation among risk mitigation actions. The correlation among risk mitigation actions in the HOR 2 is place in the HOR's roof component that states each of mitigation action correlation. The correlation consists of four categories, namely a strong correlation (•) if a mitigation action will have a very significant influence with other mitigation action, a moderate correlation (*) if a mitigation action will provide enough influence with other mitigation action, weak correlation (θ) if a mitigation action will provide a weak relationship with other mitigation action, and there is no correlation (0) between mitigation actions. From 8 mitigation actions identified, the relationship/correlation among mitigation actions as shown above in figure 1.

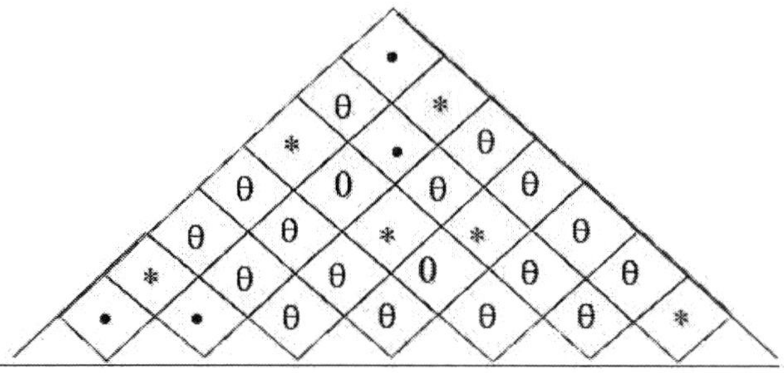

To be treated risk agent (Aj)		PA1	PA2	PA3	PA4	PA5	PA6	PA7	PA8
A9	1680	9	9						9
A7	1224	9	9						9
A12	400			9	9		9	3	
A13	368			1		3	1		
A8	360			1	3	1		9	
A10	350	9		1					9
A15	243	9	9					3	1
A1	227			9	1				
	TEk	31473	28323	6721	4907	1464	3968	5169	29529
	Dk	4	4.5	3	3.5	3	3	3	3.5
	ETD	7868.25	6294	2240.33	1402	488	1322.67	1723	8436.86
	Rank	2	3	4	6	8	7	5	1

Figure 1. House of Risk 2 model result.

3.2.3. Risk mitigation action priority. Risk mitigation actions assessment is performed by considering the costs and resources effectiveness, each actions' total effectiveness calculation, and difficulty level (Dk) of implementation; to conclude the effectiveness value to mitigation action as shown in table 3. The result of mitigation action as the solution or alternative for the risk agents, i.e. the PA8 action which has the highest rank will minimize the correlated risk agents and so on as described in *3.2.1* section.

Table 3. Mitigation actions' rank.

Rank	Code	Mitigation Action	TEk	Dk	ETD
1	PA8	The labors' training on maintenance	29529	3.5	8436.86
2	PA1	Performing the preventive and predictive maintenance on a regular basis	31473	4	7868.25
3	PA2	Preparation of stand by equipment and machine units as substitutes the engine failure	28323	4.5	6294
4	PA3	Coordination management and related sections	6721	3	2240.33
5	PA7	Optimization of production process in accordance with the provisions and work instructions	5169	3	1723
6	PA4	Improved production design based on the suitability of field conditions	4907	3.5	1402
7	PA6	Coordination with the transporter	3968	3	1322.67
8	PA5	Improved cooperation and coordination with stakeholders related to raw materials	1464	3	488

Based on the priority mitigation actions measures with consideration to the resources and costs effectiveness, the handling risk results are classified in the Pareto diagram to select the best and feasible action as shown in figure 2. The accumulated results of the effectiveness action level percentage indicate four priority actions which is included in 83.42% of the total value. The mitigations' priority are the labors' training on maintenance (PA8=28.34%), performing the preventive and predictive maintenance on a regular basis (PA1=26.43), preparation of stand by equipment and machine units as substitutes the engine failure (PA2=21.14%), and coordination management and related sections (PA3=7.52%).

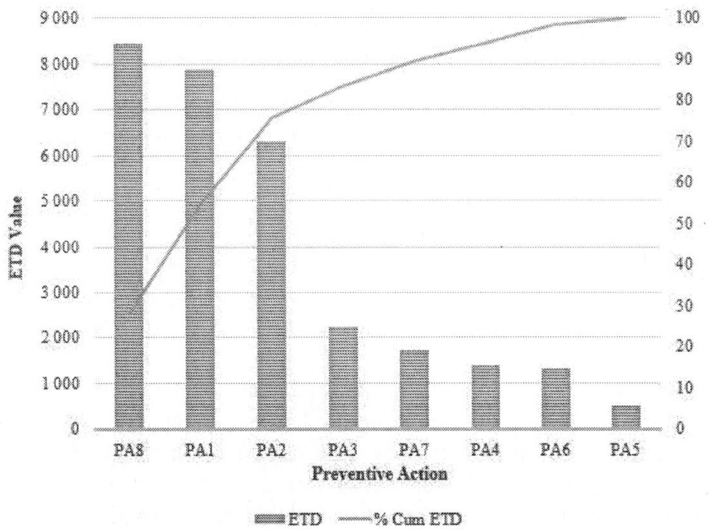

Figure 2. Pareto diagram of mitigation actions' ETD

4. Conclusions and Recommendations

The operational risk identification by using the House of Risk approach concluded 31 risk events and 20 risk agents that obtained in all palm oil supply chain business process phases. From the HOR 1 model, it was known that each of risk events had correlations to risk agents with various values and generates aggregate values of potential risk agents that form the framework for implementing mitigation actions. From the 20 risk agents identified, 8 risk agents were selected were chosen to get preventive action immediately in HOR 2. In the HOR 2 model, 8 mitigation actions were prioritized to be implemented based on resources and costs effectiveness. There were four priority mitigation actions, 1) the labors' training on maintenance, 2) performing the preventive and predictive maintenance on a regular basis, 3) preparation of stand by equipment and machine units as substitutes the engine failure, and 4) coordination management and related sections.

For further research, it is necessary to conduct a broader analysis substance and detail identification to other types of potential risks through overall information and each palm oil supply chain actors especially related crucial actors such as smallholder farmers as the main suppliers of palm oil who do have uncertainty positions due to its situational conditions. In addition, to facilitate work in this digitalization era, an integrated information system related to the palm oil supply chain is needed.

References

[1] [VSII] Van der Schaar Indonesia Investments 2016 Minyak kelapa sawit *http://www.indonesia-investments.com/id/bisnis/komoditas/minyak-sawit/item166?* Aug 22nd 2018

[2] Machmud M 2018 Pembentukan badan pengelola dana perkebunan kelapa sawit (bpdpks) untuk mendukung sawit indonesia yang berkelanjutan *Seminar Nasional Sawit Indonesia* (Jakarta, Aug 20th, 2018)

[3] Hadrjowiryono M 2018 Kerangka kelembagaan dan fleksibilitas bpdpks dalam penerapan pola pengelolaan keuangan blu *Seminar Nasional Sawit Indonesia* (Jakarta, Aug 20th, 2018)

[4] Halim S 2018 Harapan industri kelapa sawit terhadap peran bpdpks untuk mendukung peran sawit berkelanjutan *Seminar Nasional Sawit Indonesia* (Jakarta, Aug 20th, 2018)

[5] Papilo P, Marimin, Hambali E and Sitanggang I S Sustainability index assessment of palm-oil based bioenergy in Indonesia *J. of Cleaner Production* **196** pp 808-820
[6] Septiani W, Marimin, Hediyeni Y and Haditjaroko L 2017 Risk based milk pricing model at diary farmers level *Media Peternakan* doi:https://doi.org/10/5398/medpet.2107.40.3.218
[7] Teniwut W A, Betaubun K D, Marimin and Djatna T 2017 A conceptual mitigation model for asymmetric information of supply chain in seaweed cultivation *IOP Conf. Series: Earth and Environ. Sci.* **89** doi:10.1088/1755-1315/89/012022
[8] Marimin and Safriyana 2018 Evaluation of palm oil supply chain's performance, added value, and performance improvement: a case study at x co. *IOP Conf. Series: Earth and Environ. Sci.* **196** pp 1-12
[9] Hadiguna R A 2012 Model penilaian risiko berbasis kinerja untuk rantai pasok kelapa sawit berkelanjutan di Indonesia *J. Tek. Ind.* **14** pp 13-24
[10] Ulfah M, Maarif M S, Sukardi and Raharja S 2016 Analisis dan perbaikan manajemen risiko rantai pasok gula rafinasi dengan pendekatan house of risk. *J. Tek. Ind. Per.* **26** pp 87-103
[11] Deleris L A and Erhun F 2007 *Risk Management in A Supply Network: A Case Study Based on Engineering Risk Analysis Concepts, Handbook of Production Planning in Operation Research and Management Science* (Kluwer Academic Publishers)
[12] Suharjito and Marimin 2012 Risks balancing model of agri-supply chain using fuzzy risks utility regression *J. of Theoretical and Applied Information Tech.* **41** pp 134–144
[13] Hidayat S and Marimin 2014 Agent based modeling for investment and operational risk considerations in palm oil supply chain *Int. J. Sup. Chain. Mgt.* **3** pp 34-40
[14] Rohmah D U M, Dania W A P and Dewi I K 2015 Risk measurement of supply chian organic rice product using fuzzy failure mode effect analysis in mutos seloliman trawas mojokerto *Agriculture and Agriculturual Science Proc.* **3** pp 108-113
[15] Wessiani N A and Sarwoko S O 2015 Risk analysis of poultry feed production using fuzzy fmea *Science Direct: Procedia Manufacturing* **4** pp 270-281
[16] Septiani W, Marimin, Herdiyeni Y and Haditjaroko L 2016 Risk depedency chain model of dairy agro-industry supply chain using fuzzy logic approach *Supply Chain Forum: An Int. J.* **17** pp 218-230
[17] Ebrahemzadih M, Halvani G H, Shahmoradi B and Giahi O 2014 Assessment and risk management of potential hazards by failure modes and effect analysis (fmea) method in yazd steel complex *Open J. of Safety and Technology* **4** pp 127-135
[18] Pujawan and Geraldin L H 2009 Supply chain house of risk: a model risk management for proactive supply chain *Business Process Mgt. J.* **15** pp 953-67
[19] Camargo M E, Zanandrea G, Pacheco M T M, Malafia G C and Motta M E V 2013 Supply chain management operations reference (scor): study bibliometric *Int. J. of Ops. and Log. Mgt.* **2** pp 1-13
[20] Asrol M, Marimin and Machfud 2017 Supply chain performance measurement and improvement for sugarcane agro-industry *Int. J. of Sup. Chain Mgt.* **6** pp 8-21
[21] Franen S C, Franen B R, Hawkins N and Susel I 2013 *Median Aggregation of Distribution Functions* (California: Published Articles & Papers)
[22] Geraldin L H, Pujawan I N and Dewi D S 2007 Manajemen risiko dan aksi mitigasi untuk menciptakan rantai pasok yang robust *J. Tek. dan Rek. Teknik Sipil "TORSI"* pp 53-64
[23] Septiani W 2016 Rancang bangun model manajemen risiko rantai pasok agroindustri susu berbasis pengetahuan (Bogor: Institut Pertanian Bogor)
[24] Marimin and Maghfiroh N 2010 *Aplikasi Teknik Pengambilan Keputusan dalam Manajemen Rantai Pasok* (Bogor: IPB Press)
[25] Septiani W, Marimin, Herdiyeni Y and Haditjaroko L 2016 Method and approach mapping for agri-food supply chain risk management: a literature review *Int. J. of Sup. Chain Mgt.* **5** pp 51-64

IOP Conf. Series: Earth and Environmental Science **335** (2019) 012013 doi:10.1088/1755-1315/335/1/012013

Appendices

Appendix A. The possibility of risk event based on business process.

	Sub Process	Risk Event	Code	Si
Plan	Production capacity design	Incompatibility between capacity planning and processing realization	E1	4
	Raw materials control	The gap between planning and on-hand material	E2	4
	Adjusments of supply chain and cost planning	Supply chain incompatibility with cost budget	E3	4
	Perfect condition	The resulting yield doesn't meet the standard	E4	4
		Production target is not achieved	E5	4
Source	Raw material delivery's acceptance	The delay of raw materials from suppliers and plantations	E6	4
		Raw materials don't meet the capacity planning	E7	4
		The low quality of raw material supply	E8	4
		Raw materials that do not match/don't pass sorting process	E9	4
		The changes in the raw materials quality	E10	4
	Supplier selection and evaluation	Not evaluating supplier performance	E11	4
		Non-conformity of material sent by supplier	E12	4
Make	Production control	The raw material is broken so it can't be processed	E13	4
		Delays in the production implementation	E14	7
		The production process is inefficient	E15	5
		Engine failure (machine breakdown or lack of maintenance)	E16	8
		The output isn't perfect	E17	4
		Inhibition of production due to mechanical damage	E18	6
	Product quality checks	No quality checks are conducted during the process	E19	4
		The defect in products yield	E20	4
	Production activity	Production activities aren't done/stopped	E21	5
		The demand can't be fulfilled	E22	5
		The waste is generated in large quantities	E23	2
Deliver	Delivery selection	The lack of product delivery capacity	E24	4
		The lack of products in the distribution center	E25	3
		Delayed delivery of products to consumers	E26	4
		Damage to the product during transportation	E27	4
Return	Raw material returns	The quantity of raw materials is not suitable	E28	4
	Product returns	The stability of production is disrupted	E29	5
		The cost of the products reprocessing	E30	4
		Rescheduling delivery	E31	4

Appendix B. Risk event's potential impact.

Code	Risk Event	Potential Impact
E1	Incompatibility between capacity planning and processing realization	Unachieved production target, production process is disrupted
E2	The gap between planning and on-hand material	Production stability is disrupted, unachieved production capacity
E3	Supply chain incompatibility with cost budget	Ineffecient system, increasing of the total costs
E4	The resulting yield doesn't meet the standard	Low selling price, company losses
E5	Production target is not achieved	Unfulfilled orders, company losses
E6	The delay of raw materials from suppliers and plantations	Idle capacity
E7	Raw materials don't meet the capacity planning	Unmaximum holding capacities
E8	The low quality of raw material supply	The yield doesn't meet the standards/low quality
E9	Raw materials that do not match/don't pass sorting process	The yield doesn't meet the standards/low quality, CPO's price is low
E10	The changes in the raw materials quality	Products quality get drop and CPO's price is low
E11	Not evaluating supplier performance	Raw materials quality is low
E12	Non-conformity of material sent by supplier	Products quality get drop and CPO's price is low
E13	The raw material is broken so it can't be processed	Unachieved production target
E14	Delays in the production implementation	Orders are not met, delivery schedules' disruptions
E15	The production process is inefficient	Increased costs
E16	Engine failure (machine breakdown or lack of maintenance)	Unachieved production target, maintenance additional costs
E17	The output isn't perfect	Unfulfilled orders, low CPO's quality
E18	Inhibition of production due to mechanical damage	Raw materials' quality decrease, unachieved production target
E19	No quality checks are conducted during the process	Products quality below standard and rejected by consumers
E20	The defect in products yield	Products are rejected by consumers, low selling price
E21	Production activities aren't done/stopped	Raw materials accumulation, unachieved production target
E22	The demand can't be fulfilled	Consumer disclaimer
E23	The waste is generated in large quantities	Environmental damage, waste treatment costs
E24	The lack of product delivery capacity	Late order fulfillment
E25	The lack of products in the distribution center	Opportunity lost
E26	Delayed delivery of products to consumers	Consumer disclaimer
E27	Damage to the product during transportation	Products are rejected by consumers
E28	The quantity of raw materials is not suitable	Idle capacity
E29	The stability of production is disrupted	Idle capacity, unachieved production target
E30	The cost of the products reprocessing	Increasing of the total costs
E31	Rescheduling delivery	Sales delay, accumulating inventories

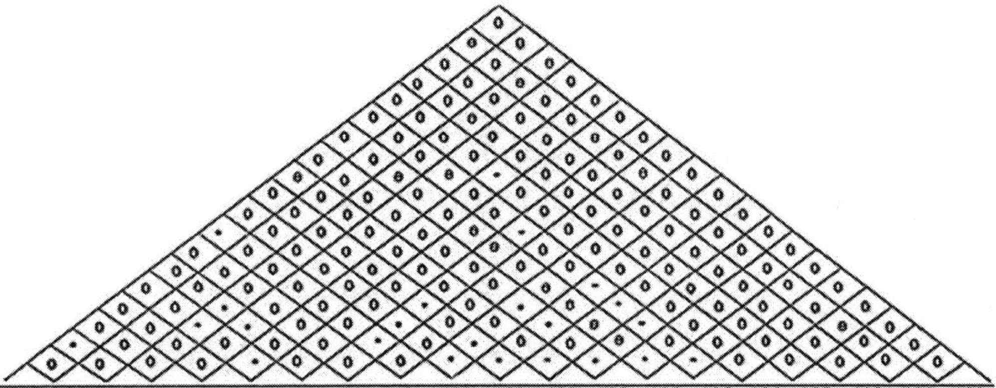

Risk Agent		A1	A2	A3	A4	A5	A6	A7	A8	A9	A10	A11	A12	A13	A14	A15	A16	A17	A18	A19	A20
	Oj	1	1	1	1.5	1	1	4.5	3	5	1.5	2	2.5	2	3	3	2.5	1.5	1	1	2
Risk Event	Si																				
E1	4	9		9			1		9	3											
E2	4	9		3		1															
E3	4	3	1	3	3		1		3					1							
E4	4							3		3	9		3	3	1	9				1	
E5	4	9		1			1	9	9	3	3	9									
E6	4				9	9													3		
E7	4	3		9			1		9												
E8	4					3							1	3						3	
E9	4				3	3	1					3	3	3							
E10	4	3						3					9	9						9	
E11	4											3									
E12	4	1				3													1		
E13	4										3		9	9							
E14	7			3				9		3	3										1
E15	5	9		1				3		3	3		3	1	9	3				1	
E16	8									9	9	3				3					
E17	4							3		3	1			3	3					1	
E18	6							3		9	9	1				1					
E19	4											3									
E20	4							1			1		3	3	1					1	
E21	5	3		1				3		9											
E22	5	1	1	1	1			3		9									1		
E23	2												1	1						1	
E24	4		9					3										1			1
E25	3	3	1	3				3		3								1			
E26	4											3						3	3		3
E27	4																3		3		
E28	4				1	1							3	9							
E29	5	1		1		1	3	9		3	3		3	1	1						
E30	4							1					1								
E31	4																	1	1		1
ARP		227	48	150	104	85	35	1224	360	1680	350	156	400	368	210	243	30	34.5	44	67	54
Rank		8	16	11	12	13	18	2	5	1	6	10	3	4	9	7	20	19	17	14	15

Appendix C. House of Risk 1 model result.

IOP Conf. Series: Earth and Environmental Science **335** (2019) 012014 doi:10.1088/1755-1315/335/1/012014

A design of competitive cocoa agro-industry supply chain system in Putat village, Gunung Kidul district

D M S Turnip[1], Y Arkeman[2] and Muslich[2]

[1]Graduate Program of Agro Industrial Technology, School of Graduate & Post Graduate Studies, Bogor Agricultural University, Indonesia e-mail: doris_monica@yahoo.com
[2]Department of Agro Industrial Technology, Bogor Agricultural University, Indonesia

Abstract. Inconsistency quality and availability of cocoa beans become a serious problem for cocoa agro-industry. In general, we can not trace whose cacao beans are processed in the industry. This is due to the lack of integrated supply chain consisting of farmer, collector, trader, exporter and industry. This study aimed to produce a design of competitive cocoa agro-industry supply chain system with dynamic system approach. Integrated partnership between Sidodadi farmer's group in Putat village, Patuk subdistrict, Gunung Kidul district with TTP Nglanggeran, Cokelat nDalem and local collector give a competitive value for cocoa agro-industry supply chain. Through this, fermented beans directly distributed to industry and processed. Cocoa supply in scenario increase 2.61% from existing, decreasing the balance of cocoa beans supply demand up to 0.7%, raising the first level of cocoa beans quality up by 2022, and lifting the accumulation of Farmers group's profit up to 29.31%. The design of this system is expected to inform cacao stakeholders about the importance of partnerships to improve cocoa agro-industry competitiveness.

Keywords: cocoa, dynamic system, supply chain system

1. Introduction

Cacao as one of the country's foreign exchange contributors has decreased production which disrupts the supply for cocoa industry. This condition caused by the old age of cacao plant, unsuitable cultivation pattern, pest attacks and climate change [1]. Post harvest handling especially fermentation has not been a priority, direct selling of wet beans, unwell established partnership between farmers and industry, prices fluctuate, and long of supply chain causes low quality of cacao beans [2]. Small land of each farmer (less than 1 acre) and economic needs are being a reason to sell wet and unfermented beans to local collector.

In global trading, our cacao beans has been exported to AS, Europe, Malaysia and Singapore produced as intermediate and end product [3]. It causes lack of beans, less utility of cocoa processing industry and low quality of cocoa products. This is one of reason we have to import cocoa products [4]. It shows a dilemma and contradiction in developing cocoa products with high value added than beans [5].

There are two kind of cocoa agro-industry, one is conventional which use blended fermented and unfermented beans to produce liquor, butter, powder and the other is bean-to-bar which use well fermented beans to produce chocolate, choco drink and nibs. These industries give an option for farmers to differ their quality of beans.

An integrated partnership between farmers' group and industry give a competitive value for cocoa agro-industry supply chain. An integrated partnership make a distribution line from producer to consumer directly [6]. Fermented beans produced by farmers' group are distributed to industry and processed. Product and consumer segmentation being an effective input for distribution management [7]. Cocoa supply and quality are one of cocoa competitive factor which must be focused and implemented on the model of small and medium scale of Agrokakao development system in production centre [8]. The goal of this research is continuing supply for cocoa industry, improving quality of beans, and fair of profit share to enhance the competitiveness of cocoa agro-industry. Integrity of supply management, inventory, demand, and price from upstream to downstream is known as a concept of supply chain management. It is a process for improving responsiveness and efficiency [9].

2. Literature

The accuracy of supply information at farmers, availability of inventory at collector/exporter, pengumpul/eksportir, prices fluctuate and batching order are four factor influences Bullwhip effect [10]. Coordination between demand and supply could be done through cooperation contracts such as biofuel supply chain in China [11]. [12] said that resources sharing has a positive influence and will increase the effectiveness of supply chain partnerships. The balance of supply and demand efficiently and effectively, followed by accurate information will improve agro-industry competitiveness [13].

Competitiveness factor including the availability of supply, quality consistency (water content, bean count, waste), fat content, flavour, price, infrastructure and logistic, policy and regulation [2]. [14] and [3] said the weakness of competitiveness of Indonesian cocoa caused by the low use of superior seeds, less of post harvest handling, lower quality of human resources and lack of infrastructure.

The balance of supply demand should be supported by distribution channel that can be differentiated in several level. Every actors who distributes product to end consumer form a trading channel [15]. [9] also said that distribution is one of important element of supply chain performance in order to fulfill consumer needs. [16] state that logistic infrastructure is very important for design of industrial distribution networks that are vertically integrated and depend on the source of raw material.

This research focus on design of cocoa agro-industri supply chain system in Putat village Patuk subdistrict Gunung Kidul district, Yogyakarta with dynamic system approach. [17] state that model is an abstract of real object or event with direct and indirect interrelation that show cause and effect, less complex than real world. Dynamic system is systems structure theories and approach which explain system complexity then analyse the changing in its behaviour [18]. It is dynamic, changes along within time and repeated iterations until optimal result are achieved [19]. Dynamic system has been used to evaluate oil supply chain in China for the overcapacity of crude oil refinary industry and energy security [20], the sustainable management of renewable energy involve the interaction of environmental, economic and social aspects [21], dynamic model simulation of CPO transport at PT. XYZ [22], Government program to increase corn production in order to have corn self-sufficiency which based on the consumption need in food and non-food [19].

The object of this research is Sidodadi farmer's group that distributed their fermented cacao beans to TTP Nglanggeran, Cokelat nDalem and local collector. The design of supply chain is built by cacao supply subsystem, cacao beans distribution subsystem, and quality improvement sub subsystem. The result of this research may give an input for policy makers and cocoa stakeholder in developing the competitiveness cocoa agro-industry through partnerships.

3. Methodology

Complexity problem of cocoa supply chain and interaction of all stakeholder with their own interest/need make a causal relationship among stakeholder and related factors. In order to accomodate the complexity and relationship along the period of time, a design of cocoa agro-industry supply chain system is built by dynamic system approach. According to [23], we can learn causal interaction between variables with feedback in the system in order to achieve the goal.

3.1. Need analysis, problem formulation and system identification

This step results actors who is getting involve, needs of each actor, contradictory needs between actors, and boundary of the systems. Variables are identified in this system boundary will interact one another to achieve the goals. Through this we can learn and analyse the behavior of the system [24].

3.2. Modelling

Interaction and relationship between variables are formulated in causal loop and flow diagram. Formulation of a dynamic hypothesis based on behavioral theory on the problem, assumption or even general description which influences modelers' understanding of real conditions [25]. The formulation of model is making structure specifications, decision rules, parameter estimation and consistency test of the goals and boundary that have been set previously [26]. Modeler can set a value of parameter and make experiments on model development, formulated ini mathematic equations [24]. The formulation of this system consist of cocoa supply subsystem, cacao beans distribution subsystem, and quality improvement sub subsystem.

3.3. Verification and validation

Verification to determine the suitability model structure of agro-industry supply chain system with real structure through expert information and scientific theory as well as simulation. Validation model is comparing the simulation results to real system (quantitative behavior pattern comparison).

Validation test use Mean Absolute Percentage Error (MAPE). MAPE test is one of relative measurements of percentage errors. This test is used to determine the suitability of result simulation to empirical data with 5 percent error is allowed [24], [26], [27].

$$MAPE = \frac{1}{n}\sum_{i=1}^{n}\frac{|X_m - X_d|}{X_d}\times 100\% \qquad (1)$$

X_m = result simulation, X_d = actual data, n = period/ number of data, i = 1,2,3,....n

Accuracy of model criteria with MAPE test are:

MAPE < 5% : precisely
5% < MAPE < 10% : precise
MAPE > 10 : not precise

4. Result

4.1. Need analysis, problem formulation and system identification

Literature study and field visit has been identified need analysis, problem formulation and system identification of cacao supply chain, shown in table 1.

Tabel 1. Need analysis.

No	Actor	Needs
1	Farmer	Cocoa seeds and plant rehabilitation
		Input for plantation
		Productivity improvement and cacao beans quality
		Cacao beans price
		Pests control
2	Local collector/trader	Continuity of cocoa beans supply
		Good quality of cocoa beans
		Minimum price of cocoa beans
3	Industry	Continuity of cocoa seed supply
		Good quality of cacao beans
		Minimum price of cacao beans

4	Government	Balance of raw material supply and demand Quality improvement of raw material Development of cocoa agro-industry Development of cocoa end product

Contradictory needs of actors form problem formulation :

1) The cocoa supply is low
2) The distribution flow of cocoa beans still long
3) The quality of cocoa beans varies

Interconnection needs analysis and problem formulation will form causal loop diagram, as shown below.

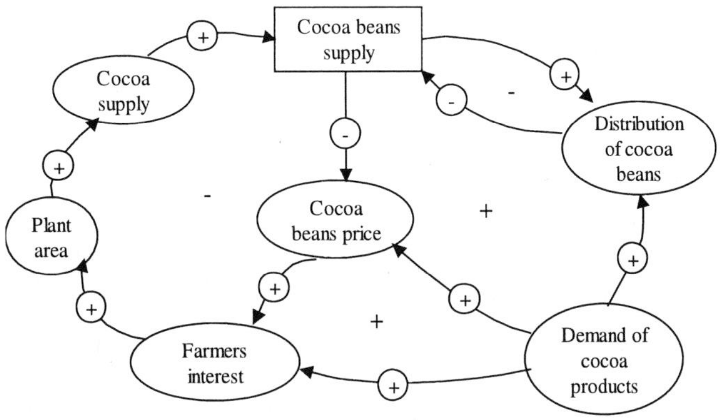

Figure 1. Causal loop diagram.

4.2. Cacao supply subsystem

The cacao supply of Sidodadi Farmers' group comes from existing plantation, pests control, maintenance and fertilization of plant, new plantation. This farmers' group also build partnerships to buy wet beans with other farmers group, then processes them into fermented beans. From wet cacao beans to fermented beans, yields 31.66 percent. The interactions flow of variables, parameters and stock for cacao supply in 13 years of simulation period result an increasing supply is 40.57 % which reduce the balance of cocoa beans supply demand about 18.62%, shown in figure 2.

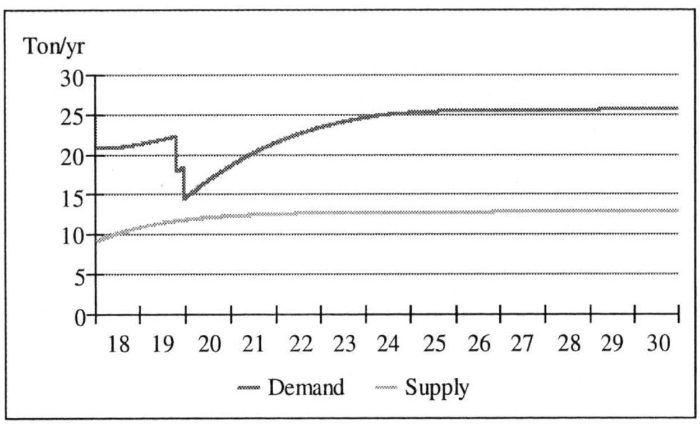

Figure 2. Supply demand of cocoa beans.

Result simulation shows an increasing flow of supply demand until 2024 and reaching the optimum level in 2025 to the end of simulation. The main factor of increasing cacao supply are fertilization, maintenance of plant and pests control. These are being a leverage factor in scenario.

4.3. Cacao beans distribution subsystem
The fermented beans distributed 50 percent to TTP Nglanggeran, 30 percent to local collector and 20 percent to Cokelat nDalem. This beans are processed into 3 in 1 choco drink, choco candy, and choco bar. The variables used are yield constant, cacao bean fractions, product demand fractions. For 3 in 1 choco drink and choco candy added with the trend of visitors parameter, average purchase of 3 in 1 choco drink, average purchase of choco candy, fraction of visitors who buy those products.

In second line, fermented beans are distributed to local collectors, the flow of beans processing does not continue because the industry is outside of Yogyakarta. The last line, fermented beans are processed into choco bars by Cokelat nDalem. The flow of distribution is shown in figure 3:

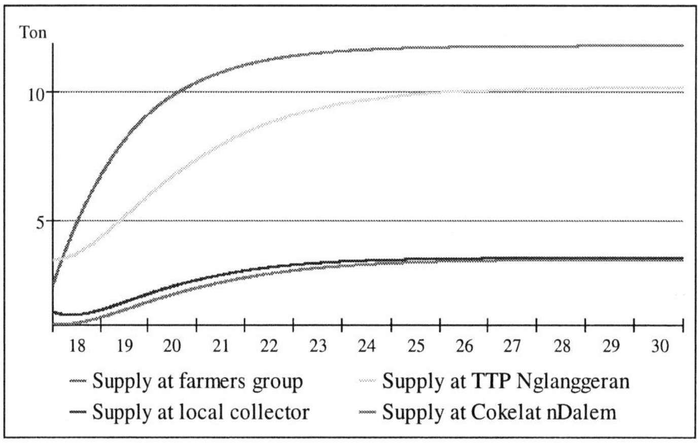

Figure 3. Cacao beans distribution.

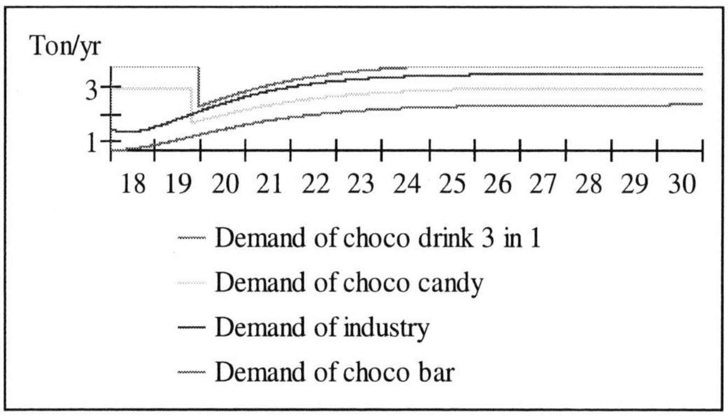

Figure 4. Demand of choco products.

The flow shown in figure 3 and figure 4 explain the same behaviour of supply and demand. The flow is reaching optimum level from 2025 to the end of simulation period, same as flow of cacao supply. Rate of increasing of cacao has a higher value until 2024, after that the pattern is moving on the optimum level. This pattern shows the increasing of cocoa beans distribution influenced by cacao supply and cocoa products.

Each product will benefit supply chain actors. Profit is influenced by cacao beans price, product price, production cost, distribution cost, post harvest cost, the amount of supply processed, number of products. The accumulation profit in the end of simulation period shows the highest is TTP Nglanggeran and the lowest is local collector. Local collector only sell the cocoa beans to industry without processing, it does not give added value. The accumulation profit of farmers' group is about 3.4 billion rupiahs.

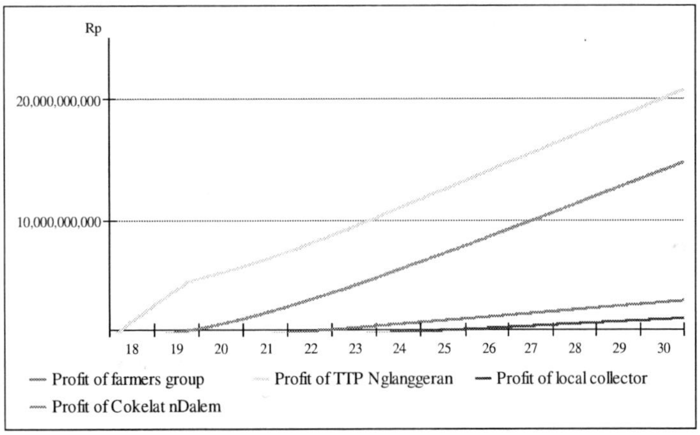

Figure 5. Profit of actor in cocoa supply chain

4.4. Quality improvement

Improvement quality of cacao will be achieved by implementing SOP of cultivation, harvest and post harvest. For this sub subsystem, parameter of management, input for plantations, technology, period of SOP implementation program, farmer interest in implementing SOP, and quality gap are connect and interact each other to achieve quality values. If farmers implement SOP consistently, it will improve the quality of cacao beans with time. However, this effort is often constrained by industry's impartiality to good quality beans and planting cacao is not the main livelihood. These factors will decrease the quality. Quality affects the price of cacao beans directly.

In this study, the quality of cacao beans is divided into 3 levels with 3 different prices. The price is become a leverage variable of farmer interest in implementing SOP. Cacao beans price and the quality affect price index. Best quality and good price index will increase cocoa competitiveness. This process is the goal of developing cocoa agro-industry to compete with foreign cocoa products.

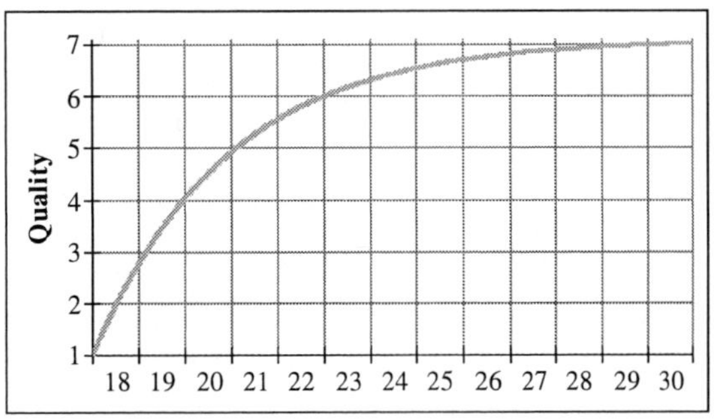

Figure 6. Improvement quality of cocoa beans.

Quality improvement during the 13-year simulation period, as shown in figure 6, the first level of quality is reached in 2031. Farmers' awareness in postharvest management with understanding fermented technology and appropriate uses of production input are the key factor to reach the highest level of quality. The effect of best quality on cacao beans price and cocoa product sales has increased farmers' interest in implementing SOP consistently.

4.5. Validation

Model of design of cocoa agro-industry supply chain that has been made, must be assessed the validity to determine the suitability of simulation results with real systems. The data used are the production and sales of Choco drink 3 in 1, Choco candy, and Choco bar from Januari to December 2017. The result of validity using MAPE method is shown below.

Tabel 2. MAPE result.

Product	Activity	MAPE result
Choco drink 3 in 1	Production	5,03 %
	Sale	8,06 %
Choco candy @ Rp 2000,-	Production	0,09 %
	Sale	4,11 %
Choco candy @ Rp 2500,-	Production	0,13 %
	Sale	2,74%
Choco bar	Production	9,06 %
	Sale	1,65 %

Result shows that all the process are valid. The model is reflecting the real system of cocoa agro-industry supply chain.

4.6. Scenario

According to the existing result, it need to make an effort in minimizing the balance of supply demand of cacao and improving the quality of cocoa beans. A leverage factor in cacao supply subsystem such as fertilization, maintenance of plant and pests control, losses and new plant area are changed. Also in sub subsystem of improvement quality, a leverage factor such as management, technology and production input are increased, shown in table 3.

Tabel 3. Scenario of leverage factors

Leverage factor	Existing	Scenario
Fertilization	90%/yr	100%/yr
Maintenance of plant	90%/yr	100%/yr
Pests control	0.65%/yr	2%/yr
Losses	10%	8%
New plant area	0	0.25 ha
Management	30%	50%
Technology	40%	50%
Production input	30%	50%

The result of cacao supply increase about 43.18%, decrease the balance of supply demand of cocoa beans about18.62%, and the highest quality level of cocoa beans is reached in 2020. Compare to existing condition, increasing of cacao supply up to 2.61% and decreasing of the balance of cocoa beans supply demand is 0.7%. It is giving the improvement profit of farmers' is about 4.4 billion rupiahs or equal to 29.31% than existing.

5. Conclusion

Integrated partnership between farmers' group and industry will form a traceability system of cocoa supply chain. The involvement of all actors and supply flows from upstream to downstream makes clear the flow of information on supply, demand and profit. Efforts that have been made by farmers are a key factor to increase the price and competitiveness of cacao beans.

The results of scenario should be carried out with operational regulation by government together with farmers' group and all related stakeholders. The operational regulation will be a guidelines to achieve the added value and competitiveness of the cocoa agro-industry.

References

[1] Utami A N 2016 *Struktur Pasar dan Daya Saing Produk Kakao Olahan Indonesia di Pasar Jerman* [Skripsi] (Bogor: Insitut Pertanian Bogor)

[2] Panlibuton H and Lusby F 2006 Indonesia cocoa bean value chain case study *USAID Microreport* **65**

[3] Rahmanu R 2009 *Analisis Daya Saing Industri Pengolahan dan Hasil Olahan Kakao Indonesia* [Skripsi] (Bogor: Institut Pertanian Bogor)

[4] Hamdani N S 2009 *Studi Kelayakan Pendirian Industri Pengolahan Kakao (Theobroma Cacao L) Skala Industri Kecil-Menengah (IKM) di Kabupaten Tanggamus, Lampung* [Skripsi] (Bogor: Institut Pertanian Bogor)

[5] Wicaksono R P K 2016 *Pengaruh Penerapan Bea Keluar Terhadap Daya Saing Produk Kakao Indonesia* [Skripsi] (Bogor: Insitut Pertanian Bogor)

[6] Kotler P 1996 *Manajemen pemasaran, analisis perencanaan, pengendalian* (Jakarta : Salemba 4)

[7] Rexhausen D, Pibernik R and Kaiser G 2012 Customer-facing supply chain practices-the impact of demand and distribution management on supply chain success *Journal of Operations Management* **30**(4): 269-281

[8] Syam H 2006 *Rancang Bangun Model Sistem Pengembangan Agroindustri Berbasis Kakao Melalui Pola Jejaring Usaha* [Disertasi] (Bogor: Insitut Pertanian Bogor)

[9] Chopra S and Meindl P 2013 *Supply chain management: strategy, planning,and operation 5th ed* (England: Pearson Education Limited)

[10] Lee H L, Padmanabhan V and Whang S 1997 The bullwhip effect in supply chains *Sloan Management Review* **93**-101

[11] Ye F, Li Y and Yang Q 2016 Designing coordination contract for biofuel supply chain in China *Resources, Conservation and Recycling.* **128**:306–314

[12] Rajagopal P, Zailani S and Sulaiman M 2009 Assessing the effectiveness of supply chain partnering with scalable partnering as a moderator *International Journal of Physical Distribution & Logistics Management.* **39(8)**: 649-668

[13] Setiawan A 2006 *Studi Peningkatan Kinerja Rantai Pasok Sayuran Dataran Tinggi Terpilih di Jawa Barat* [Tesis] (Bogor: Insitut Pertanian Bogor)

[14] Is I 2008 *Daya Saing Kakao Indonesia di Pasar Internasional* [Skripsi] (Bogor: Institut Pertanian Bogor)

[15] Mayangsari P 2009 *Kajian Jaringan Transportasi Multi Moda Manajemen Rantai Pasokan Produk Tomat dan Paprika di Jawa Barat* [Skripsi] (Bogor: Institut Pertanian Bogor)

[16] Blyde J and Molina D 2015 Logistic infrastructure and the international location of fragmented production *Journal of International Economics* **95(2)**:319-332

[17] Eriyatno 1998 *Ilmu Sistem* Jilid 1 *Meningkatkan Mutu dan Efektivitas Manajemen* (Bogor: IPB Press)

[18] Forester J W 1969 *Industrial Dynamics* MIT Pr

[19] Rachmawati L F 2016 *Rekayasa Model Sistem Dinamik Komoditas Jagung Untuk Mendukung Program Upaya Khusus di Jawa Barat* [Skripsi] (Bogor: Institut Pertanian Bogor)

[20] Pan L, Liu P and Li Z 2017 A system dynamic analysis of China's oil supply chain: over-capacity and energy security issues. *Applied Energy*. **188**:508-520.

[21] Saavedra R M M, Fontes C H and Freires F G M 2018 Sustainable and renewable energy supply chain: a system dynamics overview *Renewable and Sustainable Energy Reviews* **82(1)**:247-259

[22] Atikah N and Sutopo W 2014 Simulasi model dinamik pengangkutan crude palm oil (cpo) di pt. xyz untuk meminimalkan biaya transportasi pengadaan bahan *J@TI Undip* **9(2)**

[23] Hartrisari 2007 *Sistem Dinamik: Konsep Sistem dan Pemodelan Untuk Industri dan Lingkungan* (Bogor: SEAMEO Biotrop)

[24] Utami R 2006 *Simulasi Dinamika Sistem Ketersediaan Ubi Kayu (Studi Kasus di Kabupaten Bogor)* [Skripsi] (Bogor: Insitut Pertanian Bogor)

IOP Conf. Series: Earth and Environmental Science **335** (2019) 012014 doi:10.1088/1755-1315/335/1/012014

[25] Probowati B D 2011 *Perancangan Model Rantai Pasokan Untuk Agroindustri Kelapa Terpadu Dalam Skala Usaha Kecil* [Tesis] (Bogor: Institut Pertanian Bogor)
[26] Akbar A P 2014 *Aplikasi Sistem Dinamik Untuk Simulasi Model Produksi Biodiesel Terhadap Ketahanan Pangan Berbasis Crude Palm Oil* [Skripsi] (Bogor: Institut Pertanian Bogor)
[27] Budiman A D S 2016 *Analisis Kinerja Rantai Pasok Berkelanjutan Untuk Biodiesel Berbasis Minyak Jelantah di Kota Bogor Dengan Pendekatan Sistem Dinamis* [Skripsi] (Bogor: Institut Pertanian Bogor)

Development of smart TTI label based on kinetics diffusion of vegetable oils blends for cold supply chain monitoring

A Khairunnisa[1], N E Suyatma[1,2], D R Adawiyah[1,2]

[1]Food Science, Post Graduate School, Department of Food Science and Technology, Bogor Agricultural University (IPB), Bogor, Indonesia 16680 e-mail: nugrahaedhi@yahoo.com
[2]SEAFAST CENTER, Bogor Agricultural University (IPB), Bogor, Indonesia 16680

Abstract. Time-temperature control is a critical issue in cold supply chain, so it is very important to be tracked. Time-temperature indicator (TTI) label is devices that visually show the cold-chain accumulated history, from both time and temperature fluctuation effects. Diffusion model of TTI's indicator based on vegetable oils blend (VOB) has been investigated. The VOB indicators were made from canola oil (CA), soybean oil (SB), and olive oil (OV) that have blended with palm oil (PO). Nine formulas of indicator were designed in combination with the oils blendings i.e (A) PO:CA:SB, (B) PO:SB:OV, (C) PO:CA:OV and the combination of various ratio were (1) 50:40:10 %v/v, (2) 50:25:25 %v/v, and (3) 50:10:40 %v/v, then the blendings were added 0.25 %m/v red lake (C.I. 16255). The diffusion kinetics of this TTI indicators have tested on a waterproof photopaper material and storaged at 4, 18, 29, 37, and 40°C. The result present that the activation energy (Ea) values of indicators were about 28.925 – 40.646 $kJmol^{-1}$. The B3 and C indicators have an Ea value which have matched with Ea value of a commercial diffusion-based TTI. Based on Arrhenius equation, the diffusion model (x, m) for B3, C1, C2, and C3, respectively were $xB3^{0.5}=2t.e^{-4017(1/T)-4.4053}$, $xC1^{0.5}=2t.e^{-4124.3(1/T)-4.0196}$, $xC2^{0.5}=2t.e^{-4245.1(1/T)-3.7130}$, and $xC3^{0.5}= 2t.e^{-4888.9(1/T)-1.6641}$. Overall, the diffusion length of the indicators have significantly correlation with time-temperature changing and this diffusion model established can represented the diffusion behaviour of TTI's indicators.

Keywords: kinetics diffusion, smart TTI label, supply chain

1. Introduction

Diffusion-based of Time-Temperature indicator (TTI) is one of intelligent packaging innovation in a label formed. TTIs are devices that visually show the cold-chain accumulated history, from both time and temperature fluctuation effects. Temperature fluctuation makes the colored indicator diffuses irreversibly in a medium. This diffusion indirectly indicates the storage time and product quality decreasing (shelf life) based on temperature fluctuation [1].

The TTI and food packages is a system that used to monitors the condition of a product in real time, so it used to give an information about the transport and storage condition of the product and establishes the actual quality before consumption. Overall, the development of TTI head to decreases product waste phenomena and increases a more efficient management of products [2, 3, 4, 5].

The oils and some fatty acids (i.e isopropyl palmitic) have been used as indicator for diffusion based TTI [1, 5, 6]. Octaviasari [6] reports that using palmitic oil as TTI's indicator was results a typical diffusion length. The diffusible materials for TTI's indicator must have a specifically activation energy (Ea) value (33-50 $kJmol^{-1}$) [7]. The Ea value of palmitic oil as TTI's indicators was effected by the reaction (diffusion) rate. In order to decreases the palmitic oil's Ea value, rate of reaction must be increased. In the case of TTI, reaction rate of indicators was effected by the indicator's viscosity. The indicator's viscosity must be decreased to gets an indicator with low Ea. A blends methods between two or more oils which have different viscosities have reported as potential solution for decrease the viscosity and Ea values of final oil blends as TTI's indicator [8, 9]. Decrease the viscosities have effected by increased the unsaturation fatty acids chain length [9]. For that reasons, PO as potentially material must be blended with some vegetable oils which high in mono or poly-unsaturated fatty acids and low saturated fatty acids, i.e canola oil (CA), soybean oil (SB), and/or olive oil (OV).

The kinetics models of TTI needs to calculated. A systematic kinetic models of a TTI and a targetted product is a prerequisite for applicates the TTI [4]. By using reliable models of the product shelf-life and the kinetics both product and TTI, the effect of temperature can be monitored. For kinetics models, the TTI and product must be storaged under isothermal and/or dynamic temperature condition, for simulates the real storage [4, 10, 11]. The aims of this work were to developed a model for TTI's indicator diffusion prediction by investigated the indicator's viscosity and calculated the Ea values by the systematic kinetic models.

2. Experimental Section

2.1. Materials and Instrumentation

Four edible vegetable oils i.e palm olein (PO, PT SMART Tbk, Jakarta, Indonesia), canola oil (CA, Mazola), soybean oil (SB, Happy Soya Oil), and olive oil (OV, Bertolli) and C.I. 16255 red lake (PT Centra Lautan Pewarna) for coloring the indicators were used and purchased from a commercial source. Waterproof photopaper, 10 x 2.0 x 0.01 cm^3, respectively (Glossy Photopaper, Printech®), used as diffusion medium.

2.2. Procedures

2.2.1 Indicator preparation. Nine formulas of indicators prepared by blended the (A) PO:CA:SB, (B) PO:SB:OV, and (C) PO:SB:OV in (1) 50:40:10, (2) 50:25:25, and (3) 50:10:40 (%v/v solution) with heating at 40°C using magnetic stirrer, then 0.25% w/v C.I. 16255 red lake added to solution (modification from [12]). Nine (9) formulas of the indicators show in table 1.

Table 1. Formulation of diffusion-based TTI's indicators.

Indicators	Ratio Proportions (%v/v)				C.I 16255 (%m/v)
	PO	CA	SB	OV	
A1	50	40	10	-	0.25
A2	50	25	25	-	0.25
A3	50	10	40	-	0.25
B1	50	-	40	10	0.25
B2	50	-	25	25	0.25
B3	50	-	10	40	0.25
C1	50	40	-	10	0.25
C2	50	25	-	25	0.25
C3	50	10	-	40	0.25

2.2.2 Viscosity. The viscosity of indicators determined by viscometer Rotational Dial Reading (Brookfield Engineering Labs Inc, USA) at three different temperatures (18, 29, and 40°C) using 60 RPM (rotation per minutes), spindle number 1 for 29 and 40°C, while spindle number 2 and 3 used for samples at 4°C.

$$\text{Viscosity (mPas)} = \text{dial reading x correction factor}^{*} \quad (1)$$

*correction factor value based on spindles and speeds used

2.2.3. Indicator's diffusion length (x, cm). Photopaper medium diped in a 2 mL indicator and storaged at isothermal temperatures of 4, 18, 29, 37, and 40°C, respectively, using Leec compact incubator and B6200 Heraeus instrument Kelvitron® incubators for 30 hours. When the colored indicator diffuses through the medium, it changes from white to red medium (figure 1) (Modified from [6]). Data logger has used as the temperatures's recorder. Each temperature storage needs nine indicators and the experiments performed in triplicate.

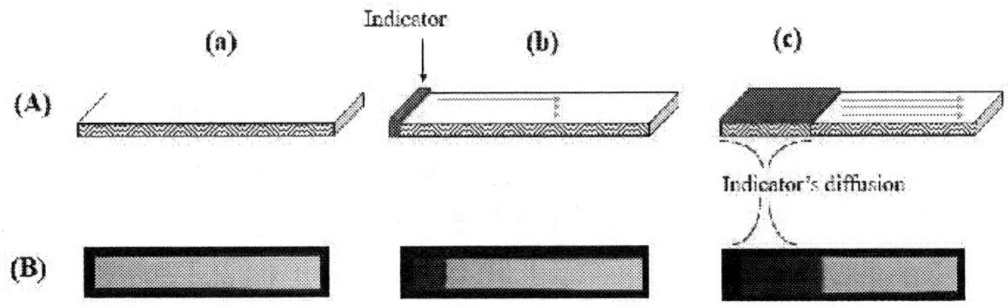

Figure 1. Schematic diagram of indicator's diffusion on photopaper medium before activation (a), on activation (b), and intermediate (c) from top view (A) and side view (B).

2.2.4. Indicators's diffusion kinetics reaction. Diffusion length value (x, cm) and storage time (t, h) which results from last analyses was used to calculate the diffusion rates (V, cmh^{-1}) (2) and diffusion coefficient value (D, cm^2h^{-1}) (4). The diffusion theory of this indicators have established by Fick's second law (3). An Arrhenius equation used to establishes the activation energy value (Ea, $kJmol^{-1}$) and a model of indicator's diffusion length prediction (5) by plotting the ln D and 1/T.

Table 2. Equations used for diffusion-based indicator kinetics.

	Models	Equations	References
(1)	Diffusion rate	$V = \frac{x}{t}$	[13]
(2)	Fick's second law	$\frac{\partial T}{\partial t} = D \frac{\partial^2 T}{\partial x^2}$	
(3)	D Value	$D = \frac{x^2}{2t}$	[1, 13, 14, 15]
(4)	Modified Arrhenius	$\ln D = -\left(\frac{Ea}{R}\right)\frac{1}{T} + \ln D_0$	

Where : J : *flux* per unit area ($gcm^{-2}s^{-1}$) ; T : absolute temperature (K) ; t : time (s) ; x : distance (cm or m) ; k : rate reaction constanta ; k_0, D_0 : frequency factor ; D : diffusion coefficient (m^2s^{-1}) ; Ea : activation energy ($kJmol^{-1}$) ; R : the universal gas constant (8.314 $Jmol^{-1}K^{-1}$)

2.2.5. Statistic analysis. The Microsoft Office Excel software used to conducted the data analysis. The curve fitting, regression analysis and analysis of variance (ANOVA) were performed according to SPSS statistic program (Version 16.0) software for Windows. Significant differences were determined using Duncan's multiple range test with P value less than 0.05 were considered statistically significant. All test were conducted in triplicate (n = 3).

3. Results and Discussion

3.1. Viscosities of Indicators

Viscosity (mPas) shows the resistance of fluid to flow, when internal friction was given on started [6]. Figure 2 shows that blended PO with two kinds of other edible vegetable oils was decreases the total viscosities of the oil blends. PO have a highest viscosity than other samples oils. The PO viscosity at 29°C was about 58.333±0.289 mPas. Viscosities of blended oils in A indicators at were about 52 – 54 mPas, viscosities of B indicators were 53.833 – 55.167 mPas, and viscosities of C indicators were about 56.833 – 58.167 mPas.

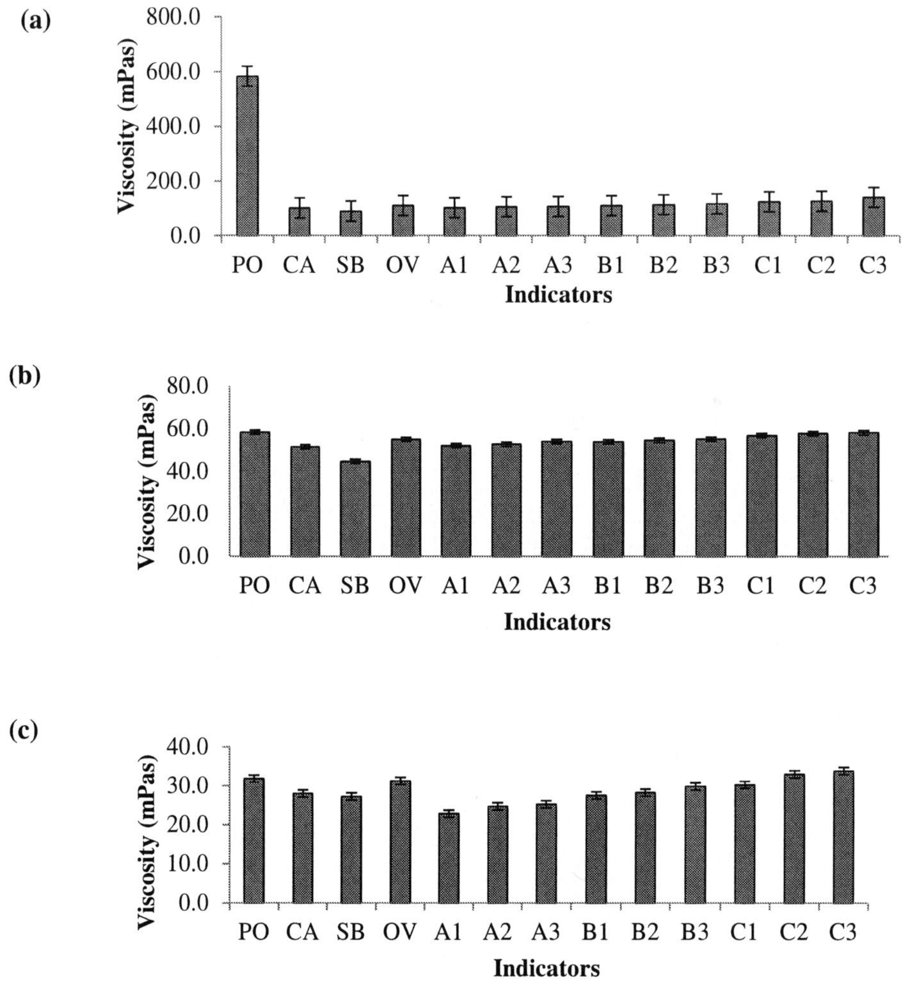

Figure 2. Viscosities of edible vegetable oils blends as TTI's indicators at (**a**) 18°C, (**b**) 29°C, and (**c**) 40°C.

Edible vegetable oils that have lower viscosity with higher ratios on oils blended method, have effected to decreases the viscosity of oils blends more. Other research reports that blended PO which have more saturated fatty acid (SFA) and high viscosity, with other edible vegetable oils which have unsaturated fatty acid more, can exactly modified the oil blends so it makes the final oil blends more stable when storaged and cooked process [6, 9, 16].

Fluid flow resistance is related to unsaturated and length of the fatty acid at triacylglycerol structures. Breaking of fatty acid in glycerol backbone from tryglyceride molecules was a factor that changes the oil viscosities. Large amount of polyunsaturated fatty acids (PUFAs) leads to low viscosity, conversely, large amount of SFA leads to high viscosity [8, 17]. Kim *et al.* [9] reported that oil viscosity tended to increase with increasing amount of saturated fatty acids and polymerization, so therefore relates to the melting point of the oil blends as well. Zhang *et al.* [18] also report that viscosities of oils influenced by oxidation reaction of the oils. When product of primer and secondary oxidation were formed, the viscosity increased. One of many methods that can decreases the oil viscosity without causing side effect is the blends method.

Other factor that affect the indicators viscosity was temperature storage. Figure 2 shows that temperature storage increased leads to decreases the indicator's viscosity [6]. Fasina and Colley [19] reported that oils viscosity at 35°C was about 10 to 15 fold of the viscosity at 180°C. Temperature increased leads to molecules modification that reduces the intermolecular pressure so the oil layers passed the other layers which have contributes to viscosty decreases [8, 9]. The diffusion of TTI's indicator have effected by the oils viscosities. Low viscosities of indicators have diffusion faster than high viscosities indicators, cause the rate diffusion retarded [20, 21].

3.2. Kinetics Diffusion of Indicators

3.2.1. Diffusion Rates of Indicators. Rates of diffusion (V, cmh^{-1}) determined from compared the indicator's diffusion length (x, cm) at the photopaper medium and the diffusion time (t, h). The results are shows in figure 3. The diffusion rates of indicators increases with increased temperature (T) and decreases as the ratio concentration of high viscosity oils increased. The diffusion rates of indicators at 40°C were fastest than other temperatures, otherwise the diffusion rates was lowest at 4°C. This indicators's diffusion rates were faster in the beginning but velocity decline with time. Increases the temperature storage changes energy of molecules, from potential to the kinetics energy. The kinetics energy breaks the molecules chain down, leads to molecules distance increases (viscosity decreases) [13].

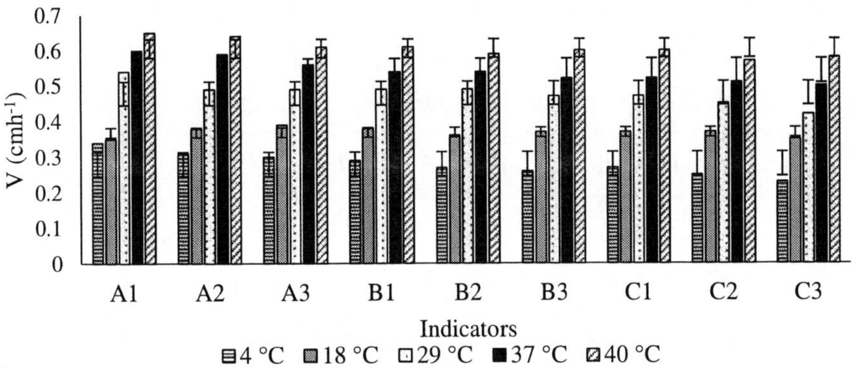

Figure 3. Diffusion rate (V) of indicators at isothermal temperatures.

3.2.2. Indicators's Diffusion Coefficient Value (D). Indicator's diffusion coefficient (D, cm^2h^{-1}) was determined from compared the squared of indicator's diffusion length (x^2, cm^2) at the photopaper medium and the diffusion time (t, h) [1, 13]. The length value squared causes D value higher than V value. However, D value have similar behaviour with V value. This values were depends on temperatures storage, the diffusion length, and indicators viscosities. As the decreases the indicators's viscosity, D value and V value increases with increased the temperature storage (as seen in figure 4) [15, 22]. Overall, the A indicators have highest D and V values, while C indicators have lowest D and V values.

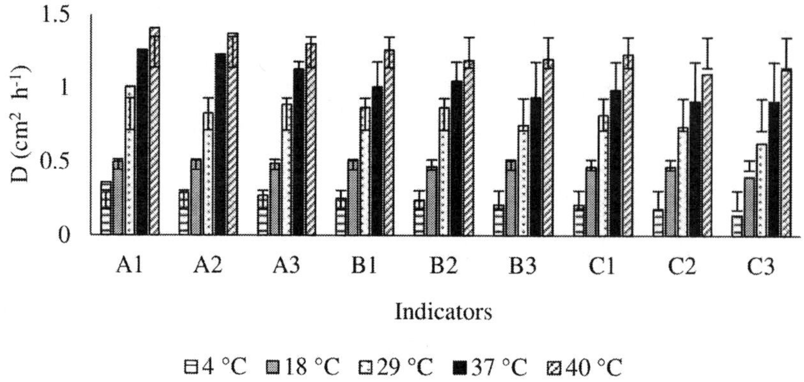

Figure 4. Indicators's diffusion coefficient (D) at isothermal temperatures.

3.2.3. Acvtivation Energy (Ea) of Indicator. This D values of the indicator at isothermal temperatures were used for calculates the activation energy value (Ea, kJmol^{-1}). A modified Arrhenius equation was used to calculated Ea value, based on linier gradient resulted by plots of ln D versus 1/T. The slope was used for ln D and 1/T converted to gets Ea value (R = 8.314 Jmol^{-1}K^{-1}). Increased the temperatures storage lead to ln D and Ea values decreases. Moreover, the ANOVA results of all the indicators showed significantly value (0.000) is less than probability value (0.05). Thus, the indicators were correlated with time and temperatures changes.

Table 3. Diffusion kinetics models and Ea values of indicators.

Indicators (%v/v)	Regression Equations	Ea ± SE (kJmol^{-1})	R^2	Diffusion prediction models (m)
A1	$\ln D = -3479.1\frac{1}{T} - 5.9526$	28.925±0.112	0.9627	$x = \sqrt{2t \times e^{-3479.1\frac{1}{T} - 5.9526}}$
A2	$\ln D = -3801\frac{1}{T} - 4.9609$	31.602±0.102	0.9943	$x = \sqrt{2t \times e^{-3801\frac{1}{T} - 4.9609}}$
A3	$\ln D = -3844.3\frac{1}{T} - 4.8515$	31.962±0.079	0.9918	$x = \sqrt{2t \times e^{-3844.3\frac{1}{T} - 4.8515}}$
B1	$\ln D = -3779\frac{1}{T} - 5.1065$	31.419±0.089	0.984	$x = \sqrt{2t \times e^{-3779\frac{1}{T} - 5.1065}}$
B2	$\ln D = -3932.9\frac{1}{T} - 4.6227$	32.698±0.070	0.9898	$x = \sqrt{2t \times e^{-3932.9\frac{1}{T} - 4.6227}}$
B3	$\ln D = -4017\frac{1}{T} - 4.4053$	33.397±0.105*	0.9793	$x = \sqrt{2t \times e^{-4017\frac{1}{T} - 4.4053}}$
C1	$\ln D = -4124.3\frac{1}{T} - 4.0196$	34.289±0.076*	0.9899	$x = \sqrt{2t \times e^{-4124.3\frac{1}{T} - 4.0196}}$
C2	$\ln D = -4245.1\frac{1}{T} - 3.713$	35.294±0.081*	0.9765	$x = \sqrt{2t \times e^{-4245.1\frac{1}{T} - 3.713}}$
C3	$\ln D = -4888.9\frac{1}{T} - 1.6641$	40.646±0.076*	0.9865	$x = \sqrt{2t \times e^{-4888.9\frac{1}{T} - 1.6641}}$

Where : D : diffusion coefficient (m^2/s) ; Ea : activation energy (KJ/mol) ; R^2 : linear regression ; x : distance (m) ; t : time (second) ; e : exponential ; T : absolute temperature (K) ; *selected indicator

The achieved Ea by the diffusion reaction of the indicators values ranged over 28.925-40.646 kJmol^{-1}. The Ea values from B3, C1, C2, and C3 were have similarity with an Ea values from commercial diffusion-based TTI that have Ea value range over 33 – 50 kJmol^{-1} [7, 23]. Overall, the indicators can be applied as TTI with diffusion-based systems. Based on several researchs, the diffusion-based TTI were applied for monitoring microbial quality for perishable food [1, 7].

Ea value of the indicators was the energy for diffusion reaction started. Estimation of Ea values should be preceded since Ea determines the accuracy of food safety and ensures the product quality [1]. The more higher of Ea value, the more sensitive the indicator (caused by temperatures changing). This Ea value was related to D and V value. The Ea value decreases with D and V value increased.

Table 3 shows the Ea values of indicators effected by the indicator's viscosity. Overall, A indicators which have lowest viscosities, results lowest Ea values than others. While, the C indicators which have highest viscosities results highest Ea values. Wang *et al.* [24] explaines that Ea values reflects the diffusion rates which effected by temperatures. Increases the Ea value relates to increased the effect of temperatures to the diffusion rates (temperatures sensible of the indicators increased). It indicates that C indicators (which have highest Ea) have more sensible than B and A indicators.

Diffusion prediction models used for established the indicators's diffusion length when the temperature and time of storage were known. Length of the indicator's diffusion indicates indirectly the product quality decreasing (can be used to established the actual shelf life of a specific product) [1]. The indicators have their diffusion reaction characteristics so this prediction model will be spesific just for one indicator respectively. For example, if the shelf life of a specific product was 25 hours (90.000 s) at 29°C (302 K), the diffusion length prediction were 6.06 cm, 6.16 cm, 5.88 cm, and 5.64 cm by using B3, C1, C2, and C3 indicator, respectively.

4. Conclusion

Viscosity of the indicators was affected by the fatty acids components and the temperatures storage. This viscosity relates to the V and D values, and even the Ea value of indicators. The more lowest viscosity, the more higher D value and the more faster diffusion rates, but the Ea will be more decreased. The Ea values of the indicators were about 28.925 – 40.646 $kJmol^{-1}$. The results shows that the Ea value of B3, C1, C2, and C3 have similarity with Ea value from a diffusion-based TTI commercial. Based on Arrhenius equation, the diffusion model prediction of this selected indicators was established and can be used for predicts the diffusion length of an indicator based on adjusment of spesific time and temperature values.

More researches needed to the develop the TTI prototype and studied more about application of this indicators to a spesific product. Fatty acids as indicators can be used for next researches to increases their stabilization at lower temperature storage.

References

[1] Kim J U, Ghafoor K, Ahn J, Shin S, Lee S H, Shahbaz H M, Shin H H, Kim S, and Park J 2015 *Food. Sci. Technol.* **67** 143-150
[2] Giannakourou M C, Kotsoumanis K, Nychas G J E, and Taoukis P S 2005 *Int. J. Food. Microbiol.* **102** 323-336
[3] Kuswandi B, Musa A, and Jayus J 2011 *Sens. Instrum. Food. Qual. Saf.* **1** 137-146
[4] Giannoglou M, Touli A, Platakou E, Tsironi T, and Taoukis PS 2014 *Innov. Food. Sci. Emerg. Technol.* **26** 294-301
[5] Pereira V A, de Arruda I N Q, and Stefani R 2015 *Food. Hydrocoll.* **43** 180-188
[6] Octaviasari R 2014 *Identifikasi Indikator dan Medium untuk Label Cerdas Pencatat Umur Simpan Produk* (Bogor - Indonesia: Institut Pertanian Bogor)
[7] Suppakul P, Kim D Y, Yang J H, Lee S B, and Lee S J 2017 *J. Food. Eng.*
[8] Siddique B M, Ahmad A, Ibrahim M H, Hena S, Rafatullah M, and Mohammad O A K 2010 *Grasas. Aceit.* **61** 423-429
[9] Kim J, Kim D N, Lee S H, Yoo S H, and Lee S 2010 *Food. Chem.* **118** 398-402
[10] Smolander M, Alakomi H L, Ritvanen T, Vainionpaa J, and Ahvenainen R 2004 *Food. Ctrl.* **15** 217-229
[11] Ellouze M and Augustin J C 2010 *Int. J. Food. Microbiol.* **138** 119-129

[12] Roiaini M, Ardiannie T, and Norhayati H 2015 *IFRJ.* **22** 1227-1233
[13] Setiawan J 2012 *Widyariset.* **15** 551-556
[14] Pratiwi Y K, Waluyo S, Warji, and Tamrin 2013 *JTPL.* **2** 59-66
[15] Zhou L, Nyberg K, and Rowat A C 2015 *Adv. Physiol. Educ.* **39** 192-197
[16] Myat M W, Abdulkarim S M, Ghazali H M, and Roselina K 2009 *J. Food. Agric. Environ.* **7** 175-181
[17] Hashempour-Baltork F, Torbati M, Azadmard-damirchi S, and Savage G P 2016 *Trends. Food. Sci. Technol.* **57** 52-58
[18] Zhang Q, Saleh A S, Chen J, and Shen Q 2012 *Chem. Phys. Lipids.* **165** 662-681
[19] Fasina O O and Colley Z 2008 *Int. J. Food. Prop.* **11** 738-746
[20] Rosyidah T 2005 *Aproksimasi Fungsi Suhu terhadap Koefisien Difusi pada Suatu Zat Cair* (Jember – Indonesia: Universitas Jember)
[21] Haryanto B 2008 *APLIKA.* **8** 1-7
[22] Agustina N, Waluyo S, Warji, and Tamrin 2013 *JTPL.* **2** 35-42
[23] Yan S, Huawei C, Limin Z, Fazheng R, Luda Z, and Hengtao Z 2008 *Food. Ctrl.* **19** 315-319
[24] Wang S, Liu X, Yang M, Zhang Y, Xiang K, and Tang R 2015 *Packag. Technol. Sci.* **28** 839-867

Acknowledgments
We thank Dr Nugraha Edhi Suyatma for the support to publish this research.

Mapping of palm oil technology based on patent information analysis

M H Wibowo[1,2], A M Fauzi[3], F M Taqi[4]

[1] Business School, Bogor Agricultural University, Jalan Pajajaran Bogor 16151, Indonesia e-mail: hendra@apps.ipb.ac.id
[2] Directorate of Innovation and Entrepreneurship, Bogor Agricultural University. Gedung AH Nasoetion Lantai 5, Kampus IPB Dramaga Bogor 16680, Indonesia
[3] Department of Agroindustrial Technology, Bogor Agricultural University, Gedung Fakultas Teknologi Pertanian Lantai 2, Kampus IPB Dramaga Bogor 16680, Indonesia
[4] Department of Food Science and Technology, Bogor Agricultural University, Gedung Fakultas Teknologi Pertanian Lantai 2, Kampus IPB Dramaga Bogor 16680, Indonesia

Abstract. Patent system has been implemented globally to protect intellectual property in technology. Patent information analysis has been used for various purposes, including technology mapping, technology forecasts, business intelligence, patent portfolios, and others. Indonesia is the largest producer and exporter of palm oil in the world thus palm oil is one of Indonesia's largest foreign exchange earning commodities, but Indonesia's palm oil research activities are still lagging behind Malaysia. This study aims to map the conditions and development of palm oil technology in the world and Indonesia based on patent information analysis that expected to be an input for researchers & policy maker (government, university, R&D institution) in structuring of research and development strategic planning to produce applicative researches which suitable to industry needs. Patent information analysis of palm oil in this study includes technology life cycle, technology classification, patent applicants, market size of technology, and experts of palm oil technology. Mapping of palm oil technology based on patent information analysis in this study shows the development of palm oil technology both in the world and Indonesia are in the stage of growth. In the period 1997-2016, technologies that were widely developed were production of oil and fat from palm oil raw materials and enzyme utilization for synthesis of organic compounds, processing of wastewater and fertilizers from palm oil waste, cultivation or nursery of oil palm, herbal preparations, and utilization of oil palm biomass for particle board. Data on patent applicants for palm oil in the world shows that 83% of patent applicants came from companies, where the Malaysian Palm Oil Board (MPOB) is the largest producer of palm oil technology in the world.

Keywords: patent analysis, patent information, technology mapping, palm oil

1. Introduction

Palm oil industry is an important part in the Indonesian economy which currently places Indonesia as the largest palm oil's producer and exporter in the world. Palm oil industry had become one of the 22 main economic activities included in the Masterplan of Acceleration and Expansion of Indonesian Economic Development (Masterplan Percepatan dan Perluasan Pembangunan Ekonomi Indonesia/ MP3EI) 2011-2025 [1]. Palm Oil Analytics [2] showed that Indonesia's palm oil production was 34.520.000 tons, almost double the production of Malaysian palm oil (Table 1). Indonesia contributed

48% and Malaysia 37% of the total volume of world palm oil production, meaning that 85% of the world market for palm oil controlled by Indonesia and Malaysia [3]. Palm oil is also one of Indonesia's export commodities and Indonesia's largest foreign exchange earner in addition to tourism, crude oil and gas. The export value of Indonesian palm oil in 2016 was USD 16,277,282,000 [4].

Research and development which directly influenced the development of national palm oil industry is still very much required. The Palm Oil Plantation Fund Management Board (Badan Pengelola Dana Perkebunan Kelapa Sawit / BPDPKS) [3] stated that research activity is the foundation of downstream industry required as the spearhead of industrial development, which Indonesia still possess a big task to pusue its backwardness over Malaysia – who have been the benchmark in terms of palm oil industry development and Crude Palm Oil (CPO) downstream technology.

Tabel 1. The Top Ten Palm Oil Producing Countries in 2016 [2].

No	Country	Production (000 ton)
1.	Indonesia	34,520
2.	Malaysia	17,320
3.	Thailand	2,300
4.	Colombia	1,280
5.	Nigeria	970
6.	Ecuador	560
7.	Honduras	545
8.	Papua New Guinea	522
9.	Ghana	520
10.	Guatemala	515

The condition of palm oil technology is an essential matter to be comprehended as an effort of policy making, research and development implementation as well as building a cooperation between research and business. The information included in the patent document can be utilized to comprehend the condition of palm oil technology development also as a tool for business intelligence as well as development strategic planning and investment in the research. Statistically, patent had been used as an indicator of technology activity, together with other data used in the innovation field decision making analysis, such as the role of intelectual property in economic performance, entrepreneurship, as well as science and technology system [5]. Patent analysis information report designed to efficiently solves problems in the decission making of various field of technology based on the data and information [6].

The patent system has been implemented globally to protect intellectual creations in the field of technology; and the technology generally used by the industry was the patent based technology. Patent documents contained technical information of a certain field of technology which is applicative. An invention under current condition is not a truly new technology in the world, the fact shows that most of the new idea developed from existing ideas. The patents found across the United States of America 80% to 90% of them were estimated as a development of existing patents [7].

Many benefits can be obtained from patent documents information, among others: to avoid research and development duplication; to reveal the experts and institution/company in the world who produce the most number of technology in a certain field; to find out technology development trend of a certain field; to discover innovative activities and business development direction conducted by the competitor (business intelligence); and assist business planning and decision making such as technology partnership, merger and acquisition [8,9]. This study aims to map the condition and development of palm oil technology in the world and Indonesia based on patent information analysis. The result of palm oil technology mapping in this study is expected to be an input for researchers and policy maker (university, R&D institution, government) in structuring of research and development strategic planning to produce applicative researches which suitable to industry needs.

2. Patent Information Analysis

Patent is one type of legal protection for intellectual property (IP) in the field of technology, either in the form of process or method, formula or composition, product, instruments or machines. Patent is an exclusive right granted by the State to the inventors as the result of their invention in technology field for a certain period of time to perform the invention mentioned themselves or giving approval to other party to implement it [10].

Patent document utilization can be categorized into three, namely: 1) patent document as scientific source of information; 2) information utilization of patent document in research and development activities; and 3) patent document for business benefit [9]. Up to date and complete information regarding one technology field contained in the patent document can be used as one of the refferences. Eusebi and Silberglitt [11] revealed that many researchers have analyzed bibliography data related with technical and patent papers to obtained information regarding the latest technology. Patent data is a valuable source of information which enable decision maker to: (1) assess their own technology portfolios compared to the competition; (2) assess the attractiveness of technology, especially new technology that has the potential to be a threat or vice versa to be an opportunity for business; (3) recognize company's strategic alteration in a competitive environment; (4) identify and evaluate external resources for cooperation purposes (i.e Management and Accounting, research and development cooperation, etc); (5) assessing patent status in business fields that can be explored; (6) evaluate important market partner, especially customer and vendor, to determine whether companys' research and development strategy in line with customers' needs; and (7) to upgrade human resources management related inventor or leading experts in a certain field of technology (specific) [8].

WIPO [12] devided patent information main users group as follow: industry; research and development (R&D) institutions; government authorities; small and medium enterprises (SMEs); inventor/individual; professionals in patent field, for example patent agency, researchers, data bank producer; educational institutions and university students. There are many designation used to analysed patent information, namely patent map analysis, patent portfolio analysis, etc. Fabry et al [13] used patent portfolio analysis as a tool utilized to identify research and development and business opportunity. According to Trippe [6], based on his report, patent information analysis can be mentioned with several terminology, namely patent landscape, patent map, watch or alerts, freedom-to-operate / clearance, patentability / prior art, validity, general statistics.

Several studies related to patent information analysis have been conducted for various field of technology and the methods used can be view in table 2. The aspects studied in patent information analysis described in table 3.

Table 2. Patent information analysis for various field of technology and the method used.

No	Author	Method	Study Result
1.	Suprijadi [14]	Zipf Curve and Bibliometric Analysis	Determining renewable energy technology sectors based on patent information. According to patent data analysis collected from international data, identified as many as 202 technology sectors and three technology sector groups namely group categorized as trivial, interesting and noise group.
2.	Jun and Lee [15]	Text mining technique, statistical inference and neural networks method	The model that can be implemented in forecasting the development of various technology field (emerging technology forecasting/ETF). An example of the technology being studied is the nanotechnology.
3.	Poranki [16]	*Delphi, Analogy Methods, Extrapolation, Statistics,* dan *Causal Relations*	Examine the potential of new technology growth based on patent analysis for flexible electronic technology case. Several innovation based new technology considered not mature enough to show a high growth, despite potential. This study will not

No	Author	Method	Study Result
			only help in technology evaluation, but also provide a better understanding of emerging technologies, especially flexible electronic applications.
4.	Kim [17]	Quantitative data analysis, ANOVA or t-test technique	The road map generated providing valuable information for non woven tissue industry regarding potential market, condition when tissue watered, including challenge, strengthness, opportunity and solution. Study result providing a very valuable information for research invesment, development and manufacture facility.
5.	Sandal and Kumar [18]	Text mining and quantitative analysis, namely patent maps and patent network analysis	The tracer study and patent document analysis to solve R&D problems in the field of medicine. The patent analysis combination narrow the patent search result to 14 of the most relevant patent documents (drug delivery, inhalation and fentanyl).
6.	Kim et al [19]	Text-mining, term frequency - inverse document frequency (TF-IDF), topic model, cross-impact analysis, association rule mining, key word anayisis, social network analysis, and applicant analysis	Developing a hybrid model patent analysis for sustainable technology management requirement in humanoid robot industry. Cross-impact combination analysis result described in Weighted network graph for main cross-impact matrix and bipartite network graph for technology company matrix. The model developed is useful to extend sustainable technology management strategy, because it's possible to comprehend the technology specifically.
7.	Garland [20]	Text mining and TRIZ TF	The application of both method combination, text mining and TRIZ TF will increase technology forecasting process period (longer time). Text mining and TRIZ TF method provides different type of information. Text mining method is an exact method to identify the trend and technology tracking from time to time, and to identify technology development stage but limited in predicting the developing technology for the future. TRIZ TF method is an exact method to determine technology system maturity, understanding the technology component which contributes to and identifies future possibility but limited in forecasting a more specific technology.
8.	Fujii et al [21]	Logarithmic Mean Divisia Index (LMDI)	Studying research and development strategy for biology technology (biochemistry and biotechnology; medicine and disease; food material and agriculture) based on patent data analysis from Japanese manufacture company.

Tabel 3. The aspects studied in patent information analysis.

No	Author, year and research title	Method	Examined Aspects
1.	Egan [22]	Descriptive Statistic	Patent quotation (sitation), commercialization strategy, triumph possibility and the value of succes.
2.	Fabry et al [13]	Patent portfolio analysis	Patent activity, patent strength and patent quality presented in the form of graph and spider diagram.

No	Author, year and research title	Method	Examined Aspects
3.	Shuang and Yan [23]	Patent analysis	Technology development trend, area distribution, patent applicant, technology classification and technology distribution.
4.	She et al [24]	Patent information analyisis	Applicant rank (top ten patent applicant), patent classification (top 10) in accordance to International Patent Classification (IPC), patent type or variety (invention patent and new practical patent), patent legal status (valid or invalid).
5.	Altuntas et al [25]	S Curve *technology life cycle* (TLC)	Technology life cycle, technology diffusion velocity, patent strength and patent expansion.

3. Methodology and Limitations

3.1. Research design and methodology

Research design used is secondary data content analysis from written document, which is bibliography patent document data. The method used in this research presented in Figure 1. Patent data obtained from data base that accessed by online and categorized into two groups, namely 1) world patent data and 2) Indonesia patent data. World patent data harvested with the tool of Patent Inspiration software basic lisence version which contained EPO database consisting two data base, namely: 1) EPO (European Patent Office) and 2) patent inspiration data base. EPO data base contained patent bibliography data (title, abtsract, applicant, inventor, quotes/sitation, classification code) of more than 102 countries which is updated every week. Patent inspiration data base contains complete text (claim and description) from the main search authorities (WO, EP, US, CA, etc).

World patent data harvesting in Patent Inspiration software conducted based on the keywords "palm oil" or "oil palm" or "palm kernel" with search filter as follow:

- *field: title*
- *show only one per family: order by text content*
- *patents without empty title or abstract*

Searching with the above key words in the field title conducted to obtained more focused search result and the certainty that patent documents obtained are palm oil technology. Total palm oil patent application in the world generated from Patent Inspiration database since 1904 to September 28, 2017 were 695 patent documents.

Indonesia patent data obtained from the data base found in Directorate General of Intelectual Property (DGIP) Ministry of Law and Human Rights of Republic Indonesia (Kemenkumham RI), namely patent data which can be accessed from http://e-statushki.dgip.go.id/. Based on the search in May 16, 2017 with the keyword "sawit", the total palm oil patent application listed in Indonesia since 1989 to 2016 were 331 patent documents.

Bibliography data which contained in patent document resulted from patent data harvesting, next would be analized quantitatively. The quantitative analysis of world patent information using the available tool in form of patent information analysis software namely Patent Inspiration Basic License version from AULIVE (http://www.patentinspiration.com/) and Microsoft Excell program. Indonesia patent information quantitative data analized using Microsoft Excell program. The analized aspects based on the information within patent documents in this research are as follow:

1. Technology Life Cycle (TLC) of palm oil, conducted by building S curve based on cumulative amount of palm oil patent registration per year from 1997 to 2016. TLC analysis used to explore the changes and development of palm oil technology.
2. Classification of palm oil technology, conducted by classifying palm oil technology categories based on number of International Patent Classification (IPC) code distribution. IPC analysis used to reveal most produced fields of technology. In this research, the big top ten classification of palm oil technology would be collected based on IPC main group.

3. Patent producer (technology) of palm oil, have analized major palm oil producer countries, the big 10 palm oil patent applicants and research collaboration between company and higher education institution. The primary palm oil patent producer countries determined based on the distribution of patents number per applicant country. Primary patent applicant conducted by counting the number of patents produced by every applicant and identifying the top 10 patent applicants as well as the percentage of share from total palm oil patents. Besides that, the sharing of patent applicants can be viewed based on company group, academic and company-academic joint venture by calculating the number of patent registered by applicant company, academic (higher education institution) and company-academic joint venture. Collaboration or cooperation intertwined between company and higher education institution viewed based on joint patent application.
4. Market size technology, conducted to reveal the destinated area of application or implementation of a technology and identifying the market size of a patent. Generally the countries which become the destination of patent registration analized based on prefix country, while the market size of a certain patent can be identified based on the number of patent family. In this study, the top 10 patent have been identified to have the biggest patent family in the field of palm oil technology and its market area countries.

At the final stage, a matrix of palm oil technology mapping was build which obtained from the quantative analysis result of bibliography data in accordance with the aspects studied. Matrix structuring will make the palm oil technology development easier to be review based on patent information analysis.

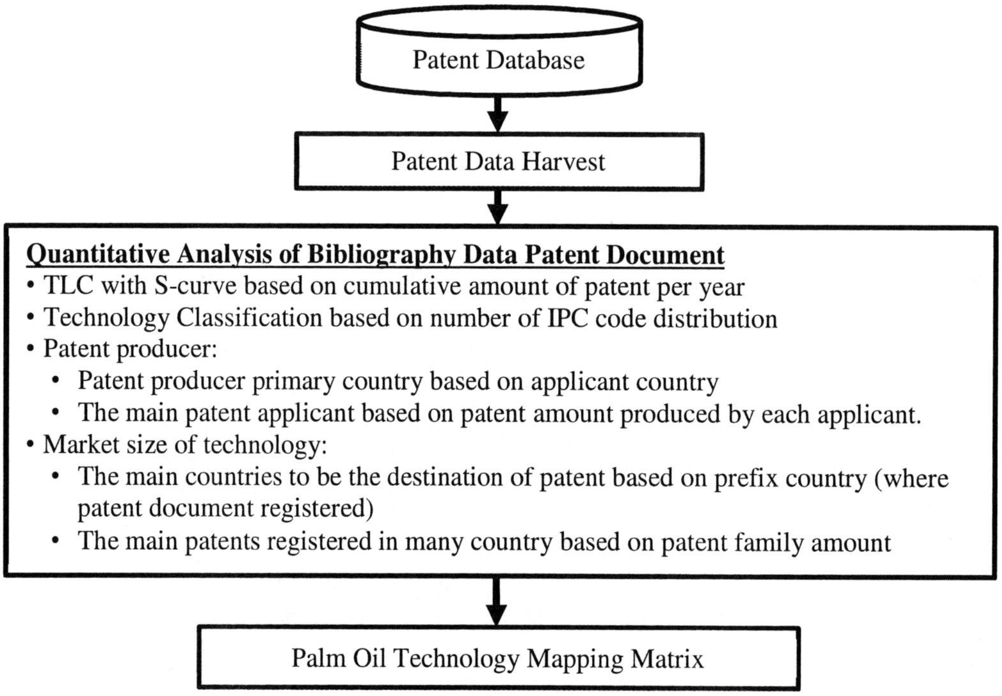

Figure 1. Research methodology.

3.2. Limitiations

Indonesian patent documents contained in Patent Inspiration database was in Bahasa Indonesia, so it was not included in worlds' palm oil patent data processing. Analysis of Indonesian patent data not carried out by Patent Inspiration Software are applicant country analysis, prefix country analysis, joint patent application analysis, and patent family analysis.

IOP Conf. Series: Earth and Environmental Science **335** (2019) 012016 doi:10.1088/1755-1315/335/1/012016

The searching using keywords "palm oil" or "oil palm" or "palm kernel" on the field filter "title" were affect a narrower scope of search results, which is specifically to palm oil technology starting from its cultivation to palm oil raw material processing (CPO and PKO), not including the technology to produce palm oil derivative products or palm oil products utilization in various other field of technology.

4. Result and Discussion

4.1. Palm oil technology life cycle

Patent data have become the source of information that can be used to explore technology changes and development [26-28]. Patent data in terms of technology trend can be used to predict technology life cycle (TLC) with the S-curve model approach build by utilizing cumulative number of patent registration each year [25-28]. Ernst [26] explained the four TLC stages as presented in figure 2, namely new technology (emerging), growth, maturity and saturation.

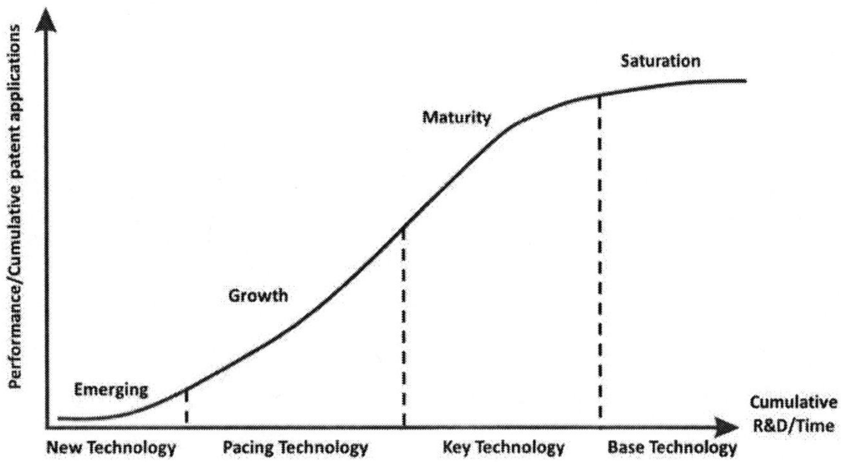

Figure 2. The TLC S-curve concept [26].

The first stage of technology development exhibit the effect of competition and integration of a product or a lower process [29] and patent request application growth still indicated a small number [30], on the other hand the number of patent application continue to increase [29]. During the growth stage, competition effect high [29] and the growth characteristic is exponential or rapidly [30]. On maturity stage, some technology will be appear as primary technology and having a high competitive impact in the market, the number of patent application relatively stabilized and tends to decrease [29]. The saturation stage indicates no further growth existed [30] and the number of patent application shows a significant decrease trend [29].

Based on TLC concept, palm oil technology development generally still being on the growth stage. The S-curve graph of worlds' palm oil technology development (Patent Inspiration database) and Indonesia (DGIP database) exhibit an increase in the cumulative number of patent registration since 1997 to 2016 (Figure 3 and Figure 4). Generally, the trend of worlds' palm oil technology development as well as Indonesia were being at the growth stage, meaning still provides very good opportunity for invesment and conducting palm oil research and development. Lee [31] stated that on the growth stage there will be found many technology purchase and sale transaction, technology licensing process will be more aggresive and intellectual property (IP) protection will be required to prevent IP patent violations.

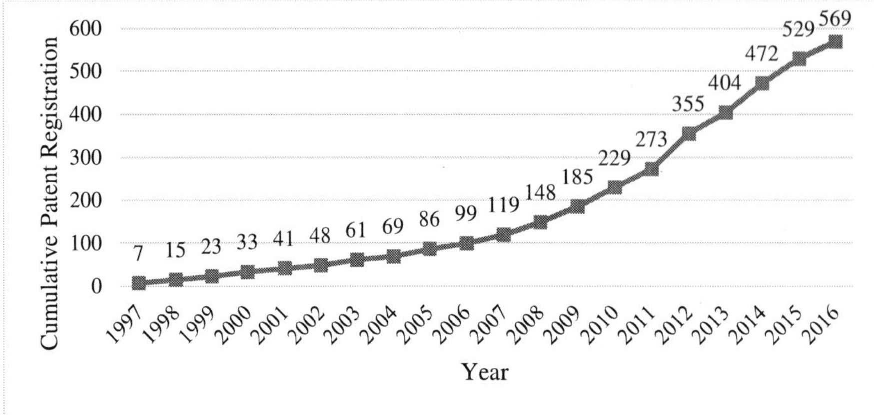

Figure 3. The number of cumulative palm oil patent registration worldwide based Patent Inspiration database.

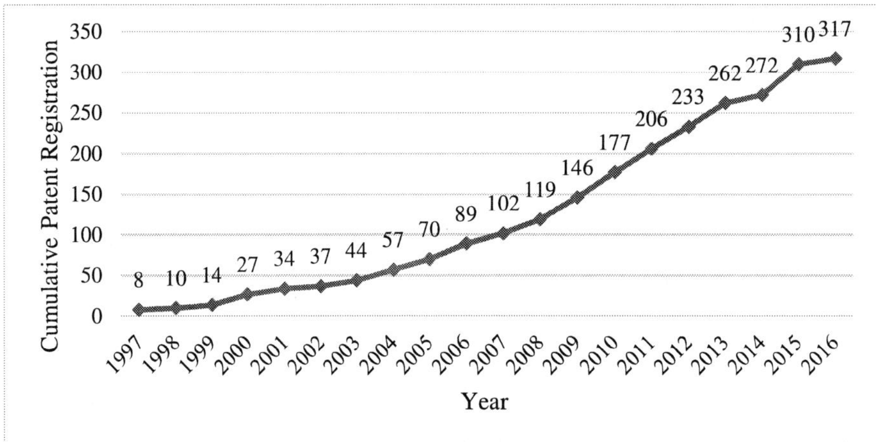

Figure 4. Cumulative number of Indonesia palm oil patent registration based on DGIP database.

4.2. The most developed palm oil technology group

The classification of palm oil technology conducted based on *International Patent Classification* (IPC) distribution. IPC is grouping/classifying field of patent internationally which was established by WIPO as a tool to obtain similar patent documents classification internationally [9]. One of IPC purpose as the basic of structuring technology statistical data which enable technology development assessment in various area [32]. IPC classification consists of four stages namely *section (1st level, 1 digit), class (2nd level, 3 digits), sub class (3rd level, 4 digits),* and *main group* atau *sub group* (*4th level, > 4 digit*), meaning the higher the level, the more specific the classification, however one patent can consists of several IPC.

Technology classification based on IPC section and class will generate an unduly wide scope of technology. Some researches related to technology classification based on IPC, generally classify technology based on their sub class and/or IPC main group. Altuntas et al [25] and She et al [24] were using IPC sub class to classify technology, meanwhile Shuang and Yan [23] using IPC main group; while Leydesdorff et al [33] using IPC class and sub class. Ini this research palm oil technology classification was classified based on IPC main group.

Based on IPC main group, it was discovered that as many as 358 palm oil technology groups in the world which the top 10 technology classification presented in Table 4. Palm oil classification in

IOP Conf. Series: Earth and Environmental Science **335** (2019) 012016 doi:10.1088/1755-1315/335/1/012016

Indonesia based on IPC main group quite diverse, there were 393 IPC main group code of palm oil in Indonesia which the top ten technology classification presented in Table 5.

Table 4. Top ten palm oil classification in the world based on IPC main group.

IPC Code	Number of Patent	IPC Code Description
C11B1/00	82	Production of fats or fatty oils from raw materials
A23D9/00	54	Other edible oils or fats, e.g. shortenings, cooking oils
C11B3/00	49	Refining fats or fatty oils
C11C3/00	38	Fats, oils or fatty acids obtained by chemical modification of fats, oils or fatty acids, e.g. by ozonolysis
C02F1/00	25	Treatment of water, waste water, or sewage
C11B7/00	24	Separation of mixtures of fats or fatty oils into their constituents, e.g. saturated oils from unsaturated oils
C12P7/00	24	Preparation of oxygen-containing organic compounds
A61K36/00	19	Medicinal preparations of undetermined constitution containing material from algae, lichens, fungi or plants, or derivatives thereof, e.g. traditional herbal medicines
A23D7/00	18	Edible oil or fat compositions containing an aqueous phase, e.g. margarines
A01H1/00	18	Processes for modifying genotypes

Table 5. Top ten palm oil technology classification in Indonesia based on IPC main group.

IPC Code	Number of Patent	IPC Code Description
A23D9/00	7	Other edible oils or fats, e.g. shortenings, cooking oils
C11B1/00	7	Production of fats or fatty oils from raw materials
C11B1/04	6	Production of fats or fatty oils from vegetable raw material
C11B7/00	6	Separation of mixtures of fats or fatty oils into their constituents, e.g. saturated oils from unsaturated oils
A23D9/02	5	Other edible oils or fats, e.g. shortenings, cooking oils, characterised by the production or working-up
B27D1/04	5	Joining wood veneer with any material; Forming articles thereby; Preparatory processing of surfaces to be joined, e.g. scoring, to produce plywood or articles made therefrom; Plywood sheets
C05F5/00	5	Fertilisers from distillery wastes, molasses, vinasses, sugar plant, or similar wastes or residues
C12N15/82	5	Mutation or genetic engineering; DNA or RNA concerning genetic engineering, vectors, e.g. plasmids, or their isolation, preparation or purification; Use of hosts therefor, for plant cells
A61K36/889	4	Medicinal preparations of undetermined constitution containing material from algae, lichens, fungi or plants, or derivatives thereof, e.g. traditional herbal medicines, (Arecaceae, Palmae or Palmaceae (Palm family), e.g. date or coconut palm or palmetto)
B27N3/00	4	Manufacture of substantially flat articles, e.g. boards, from particles or fibres

The top ten palm oil technology classification based on IPC main group indicated specifically the technology that widely developed in 1997 to 2016 are the technology related to production of fats or fatty oils from palm oil raw materials and enzyme utilization for the synthesis of organic compounds (C11B1/00, A23D9/00, C11B3/00, C11C3/00, C11B7/00, A23D7/00, C11B1/04, A23D9/02, C12P7/00); treatment of liquid waste (C02F1/00) and fertilizer from palm oil waste (C05F5/00); palm oil cultivation and nursery (A01H1/00, C12N15/82); preparation of medicine from herbal ingredients (A61K36/00, A61K36/889); and palm oil biomass utilization for particle board (B27D1/04, B27N3/00).

The study conducted by WIPO indicated the most active technology which relevant to palm oil production in 2010 to 2015 is cultivation and nursery as well as pressing process and palm oil refining, while the trend of waste processing and explotation heading to mattress/cushion development, composite material, adsorbent, filter and biomass fermentation [34].

Trend of the top 10 palm oil technology areas based on main group of IPC in 1997-2016 shows that processing technology to produce fats or fatty oils from palm oil is increasing (Figure 5). This means that the processing technology to produce fats or fatty oils from palm oil is still potential to develop by universities or R&D institutions.

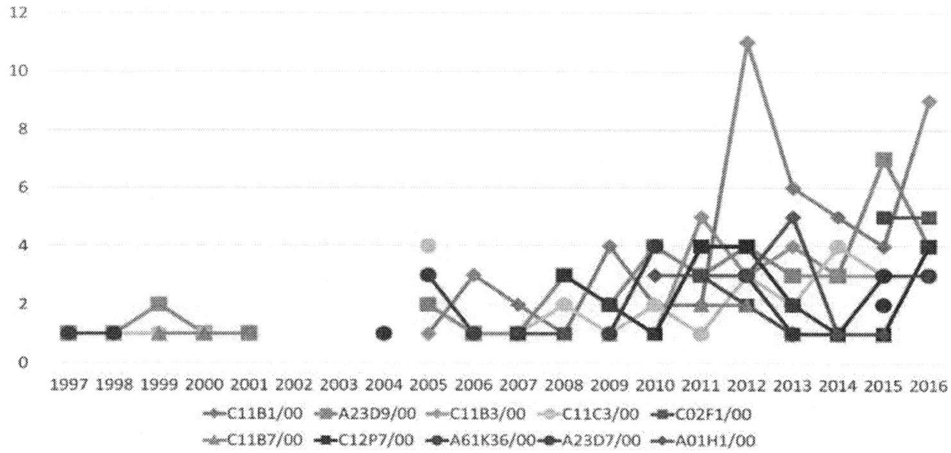

Figure 5. Trend of the top 10 palm oil technology areas based on main group of IPC in 1997-2016.

4.3. The biggest palm oil patent producer and applicant in the world and Indonesia

4.3.1. The countries dominating palm oil technology in the world and Indonesia. The distribution of palm oil patent application based on applicant country intend to reveal the producer of palm oil technology countries. This analysis will give a general description of the countries which become research and development center of a certain technology. Figure 6 describe that Malaysia is the biggest palm oil technology producer in the world, followed by Japan, China and so on. Malaysia being the biggest palm oil technology developer in the world is reasonable, considering they are the second best producer of palm oil in the world, with 37% from total world volume of palm oil production [3].

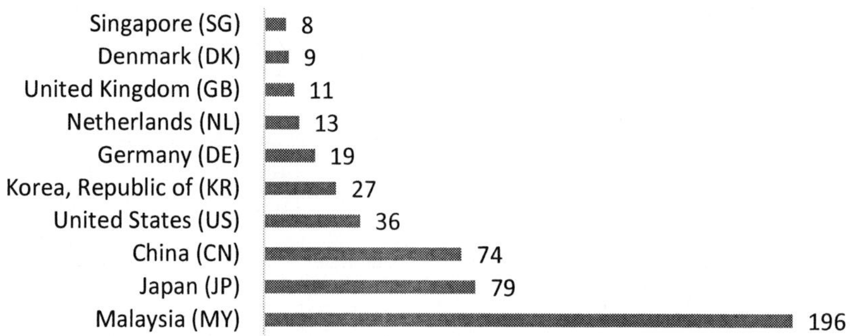

Figure 6. Top ten countries of palm oil patent producer in the world (applicant country).

Palm oil patent application data in Indonesia indicates most of palm oil patent applicant registered in Indonesia (51.70%) came from abroad, palm oil patent applicant who trully comes from Indonesia only as much as 48.30% (Figure 7). The biggest countries who registered palm oil patent to Indonesia were

Malaysia (32,39%) and Japan (6.82%). The data shows that most palm oil technology implemented in Indonesia came from abroad, especially Malaysia and Japan. BPDPKS [3] stated that Indonesia still have a big task to pursue its backwardness from Malaysia, which have been the benchmark all this time in terms of palm oil industry development and CPO downstream technology.

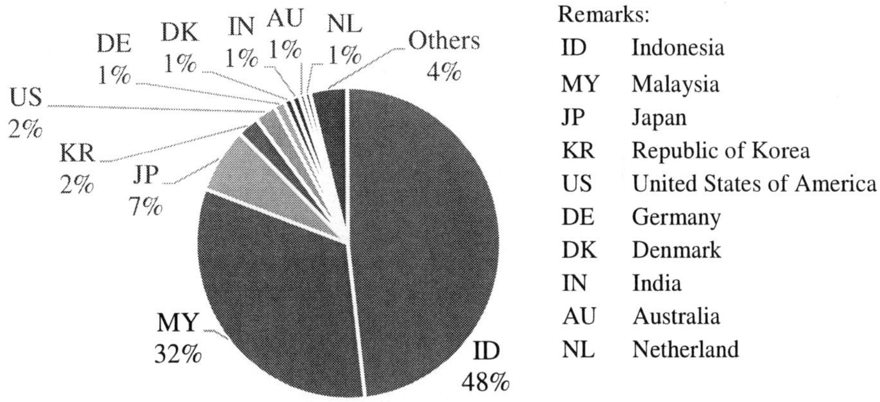

Figure 7. Top ten (10) palm oil patent applicant countries in Indonesia.

4.3.2. Applicants from companies dominated palm oil patent in the world and Indonesia. Patent applicant analysis will show who the main patent applicant in the field of technology being studied. The biggest patent applicant in a technology field expressed that those applicants are the main technology provider and also potential user of the technology at once. Patent applicant analysis of palm oil in the world studied in this research categorized into three categories, namely category of company, academic and individual.

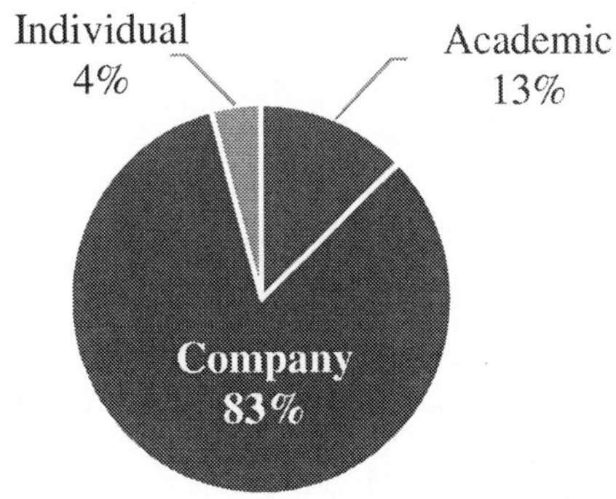

Figure 8. Percentage of palm oil patent applicant based on category of company, academic and individual.

Palm oil patent applicant data show that 83% of palm oil patent applicant came from company and 13% came from academic (Figure 8). The biggest palm oil patent applicant is Malaysian Palm Oil Board (MPOB) as many as 64 patent or owned a share of 5.83% from total applicants (Figure 9). The company has an interest in protecting its technology with a patent system because patents as intellectual assets play an important role in business performance and economic growth [35] and allow business cooperation between companies [36].

IOP Conf. Series: Earth and Environmental Science **335** (2019) 012016 doi:10.1088/1755-1315/335/1/012016

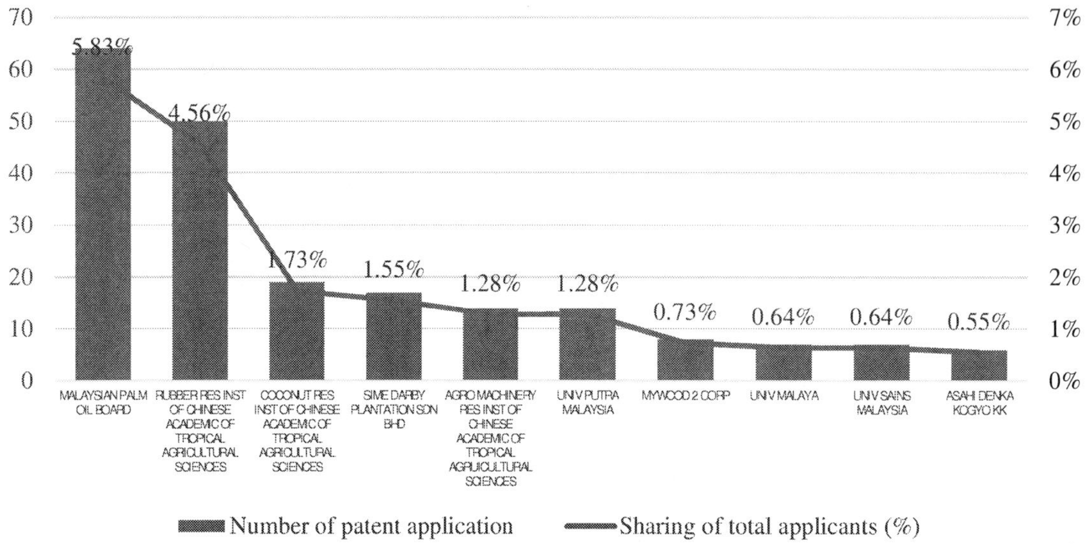

Figure 9. Top ten palm oil patent applicant in the world.

Patent data registered in Indonesia presented that the top ten palm oil patent applicant placed two patent applicant from Malaysia with patent amount reaching 53.85%, meanwhile patent application from eight patent applicant who came from Indonesia only reached 46.15% (Table 6). The number of patent application from MPOB reached 47.86%, showing a significant difference compared to patent application from other patent applicant. Those data trully define the statement of BPDPKS (2016) that Indonesia still have a big task to pursue its backwardness over Malaysia, who have been the benchmark in terms of palm oil industry development and CPO downstream technology.

Tabel 6. Top ten biggest palm oil patent applicant in Indonesia.

No	Name of Patent Applicant	Number of Patent
1.	Malaysia Palm Oil Board (MY)	56
2.	Bogor Agricultural University (ID)	13
3.	Indonesian Institute of Sciences - LIPI (ID)	10
4.	Yusuf Abdul Gani (ID)	9
5.	Sime Darby Malaysia Berhad (MY)	7
6.	Research Institution of Riau University (ID)	5
7.	University of Indonesia (ID)	5
8.	Agency for the Assessment and Application of Technology - BPPT (ID)	4
9.	Bayu Himawan, ST SE MM MM-SI MComm (ID)	4
10.	LPPM Sebelas Maret University (ID)	4

Based on the data analysis of patent applicant, it can be assumed that the institution who has most of palm oil patent application has strong applicative research in the palm oil sector, for example MPOB in Malaysia and Bogor Agricultural University (Institut Pertanian Bogor/IPB) in Indonesia. Other data that support the strength of palm oil research in an institution is non-patent publication (Scopus indexed publications), which shows that MPOB is among the top 10 palm oil publication producers in the world (596 publications). While in Indonesia, IPB is the largest Scopus indexed publication producer in the field of palm oil (152 publications) [37]. These institutions can be used as a reference for sources of palm oil technology or for research collaboration.

4.3.3. Joint patent application of higher education institution and company. Collaboration or cooperation intertwined between higher education institution with a company can be seen on joint patent application data. Joint patent application between higher education institution with a company indicates an ongoing research cooperation process between higher education institution with industry, furthermore expected to become technology shifting from higher education institution to industry or vice versa. Figure 10 presented that only 21 patent application submitted jointly between higher education institution and company, or only as much as 3.02% of 695 palm oil patent application. The data indicates that research cooperation between higher education institution with industry still needs more improvement.

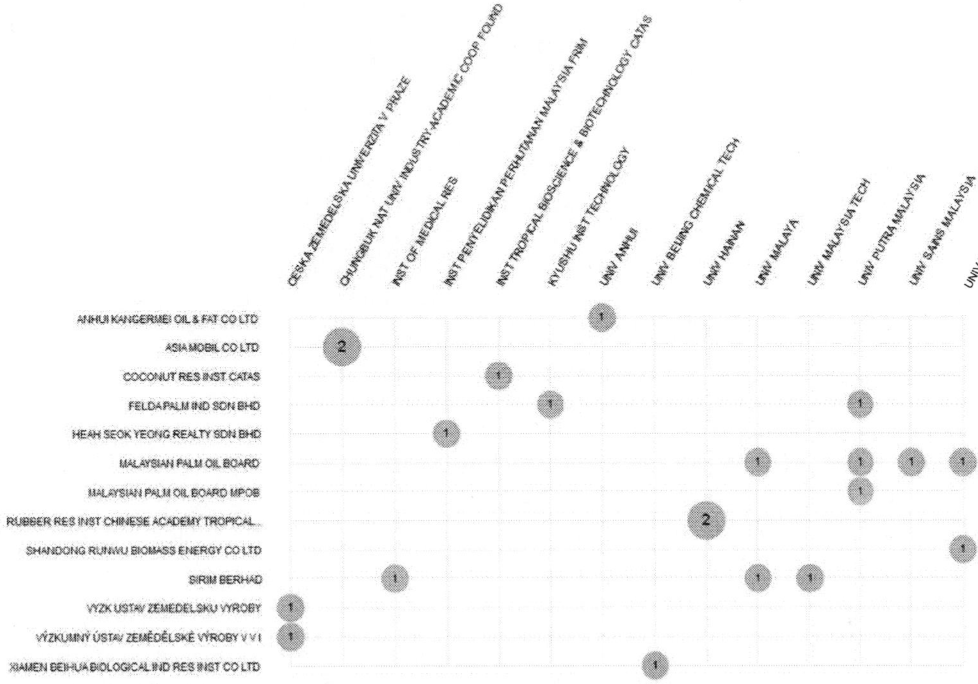

Figure 10. Joint patent application between higher education institution with company.

4.4. Market size of palm oil technology

Patent rights are territorial, which only applies in the territory of the country where the patent is registered. Thus, to obtain patent protection in several countries or regions, patent owners must submit patent applications in each of countries or regions [38]. Filing patents in several countries is a business decision that requires careful consideration that is closely related to the target of implementation of the invention. There are several factors that need to be considered for filing patent applications in several countries, including demand and markets for inventions and the possibility of invention can be implemented in the destination country [39].

4.4.1. Palm oil patent application destination country. Analysis of patent registration country origin (Prefix Country) conducted by discovering the application destinated area or targeted area of technology implementation. Based on prefix country analysis, discovering that China, Malaysia, United States of America, Japan and United of Kingdom are the main countries to be the destination of palm oil technology implementation in the world (Figure 11). Those countries are the targeted or market share of palm oil technology implementation.

Patent registration through WIPO of 18% is patent registration through Patent Cooperation Treaty (PCT) system, which enable to apply patent registration to many countries at once or the same time by a single registration internationally. Therefore, it would be difficult to identify one by one the countries which become patent registrations' destination through WIPO or PCT system, so that the patent registration through WIPO is grouped separately.

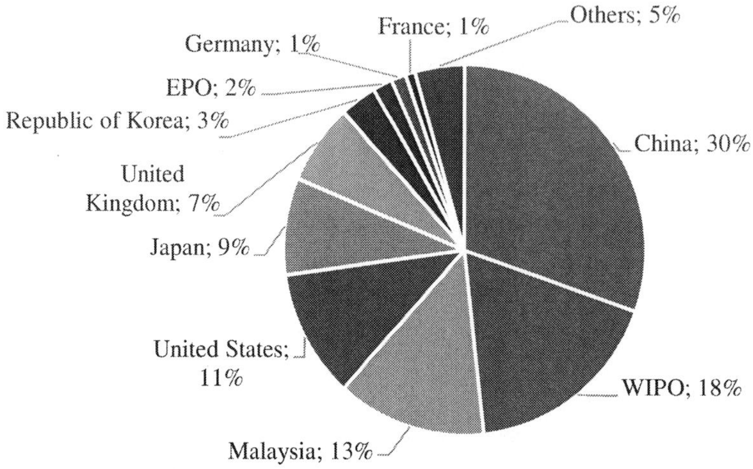

Figure 11. Palm Oil patent application destination country.

4.4.2. Patents with the biggest market size. Patent family is the sum of all patents registered in various countries either directly or indirectly connected with the same application which can be use to identify patents' value [40]. Patent family sum analysis can be use to measure market size of each invention or patent [41]. Formally, patent family covers patent application submitted in various country which interconected to each other or several priority application [5]. Putnam [42], Harhoff et al [41] as well as Van Pottelsberghe and Van Zeebroeck [43] discover a positive correlation between patent value with the number of countries in which the patent protected for the same invention. Patent family could become a simpler indicator to measure market size of a technology. Table 7 presented 10 patents that possess the biggest patent family in the field of palm oil and the countries of its marketing area.

Table 7. Top ten palm oil family patents with the highest number of family and market area.

Patent Publication Number	Family Size	Patents' Title	Market Area
US5932275A	24	Palm kernel oil blends	United States of America, Eropa, Australia, Czech Republic, Jepang, Kanada, Cina, Brazil, New Zealand, South Africa, Singapura, Rusia, Jerman, Spanyol, India, Malaysia, Argentina, Austria, Kolombia, Hungaria, Norwegia, Turki, Ukraina
US2005069620A1	15	Trans free hard palm oil fraction, trans free non-hydrogenated hard structual fat and fat blends and methods	Eropa, United States of America, Kanada, Australia, Jepang, Cina, Jerman, Singapura, Spanyol, Austria, Denmark, Malaysia
US4004041A	14	Production of liquid edible oil from palm oil or similar oils	Belgia, Jerman, Belanda, Swedia, Jepang, Perancis, Inggris, Kanada, Italia, Swiss, Australia, Denmark, Israel

Patent Publication Number	Family Size	Patents' Title	Market Area
US2009155434A1	13	Process for production of highly enriched fractions of natural compounds from palm oil with supercritical and near critical fluids	Eropa, WIPO, Cina, Jepang, Meksiko, Jerman, Brazil, Malaysia, Austria, Spanyol, Hongkong,
US2010138951A1	12	Methods of producing haploid and doubled haploid oil palms	WIPO, Australia, Eropa, Cina, African Regional Intellectual Property Organization (ARIPO), Colombia, Costa Rica, Ecuador, Brazil, India
WO9213453A1	12	Herbicides for citrus, sugarcane, oil palm and the like	Australia, Cina, Eropa, Afrika Selatan, Jepang, Selandia Baru, Jerman, Spanyol, Kanada, Meksiko
WO2012108756A1	11	Optimising the utilisation of renewable energy from biomass resources in the palm oil industry	Inggris, Costa Rica, Dominican Republic, Meksiko, Peru, Ekuador, African Regional Intellectual Property Organization (ARIPO), Brazil, Guatemala
US5567670A	10	Herbicides for citrus oil palm, rubber and other plantation crops	WIPO, Australia, Cina, Afrika Selatan, Eropa, Brazil, United States of America, Kanada, Israel
US2014303389A1	10	Palm oil enriched in unsaturated fatty acids	WIPO, Australia, Kanada, Eropa, Cina, Jepang, Meksiko, Rusia, Brazil
WO2013012311A1	10	Method and apparatus for energy-efficient operation in the sterilization of oil palm fruit	ARIPO, Inggris, Kolombia, Dominican Republic, Meksiko, Costa Rica, Ekuador, Peru, Brazil

4.5. Palm oil technology mapping matrix

Palm oil technology condition and development map in the world and Indonesia can be organized briefly with a matrix in accordance to the analyzed aspects (Table 8). The matrix organization will simplify the observation of palm oil technology development based on patent information analysis.

Tabel 8. Palm oil technology matrix map based on patent information analysis.

No.	Analyzed Aspects	Map of Palm Oil Technology Condition and Development	
		World	Indonesia
1.	Technology Life Cycle	The trend of palm oil technology development is increasing and being at the growth level	The trend of palm oil technology development is increasing and being at the growth level
2.	Technology classification	• There are 358 technology group based on IPC main group code • The most developed technologies are the technology of production process of fats or fatty oils from palm oil raw materials and enzyme utilization for the synthesis of organic compounds; processing or treatment of palm oil wastewater; processes for modifying genotypes; and preparation of medicine from herbal ingredients	• There are 393 technology group based on IPC main group code • The most developed technologies are the technology of production process of fats or fatty oils from palm oil raw materials and enzyme utilization for the synthesis of organic compounds, fertilizer from palm oil waste, mutation or genetic engineering, preparation of medicine from herbal ingredients, and palm oil biomass utilization for particle board.

No.	Analyzed Aspects	Map of Palm Oil Technology Condition and Development	
		World	Indonesia
3.	Patent Producer	• The biggest palm oil technology producer in the world is Malaysia • 83% palm oil patent applicant came from company and 13% of them came from academic • The biggest palm oil patent applicant is Malaysian Palm Oil Board (MPOB) (28.20%) • Collaboration between higher education institution with company is only 3.02% of 695 palm oil patent application.	• 51.70% palm oil patent applicant registered in Indonesia came from abroad • The biggest countries who submit the most registration of palm oil patent to Indonesia are Malaysia (32.39%) and Japan (6.82%). • 47.86% palm oil patent application in Indonesia submitted by Malaysian Palm Oil Board (MPOB)
4.	Market size of technology	• China, Malaysia, United States of America, Japan and United Kingdom are the primary countries to be the destination of palm oil technology implementation in the world • Palm kernel oil blends (US5932275A) is the patent with the biggest market area registered in 24 countries.	Data not available

5. Conclusion

Palm oil technology mapping result based on patent information analysis in this research show palm oil technology both in the world and Indonesia being on the increasing trend or growth level, meaning still offering very good atmosphere for invesment and conducting palm oil research and development. The most developed technologies in 1997 to 2016 are the technology of production process of fats or fatty oils from palm oil raw materials and enzyme utilization for the synthesis of organic compounds; treatment of liquid waste and fertilizer from palm oil waste; palm oil cultivation and nursery; preparation of medicine from herbal ingredients; and palm oil biomass utilization for particle board. This study recommends the processing technology to produce fats or fatty oils from palm oil to be developed by universities or R&D institutions.

Malaysia through *Malaysian Palm Oil Board* (MPOB) become the biggest palm oil technology producer in the world (28.20%), meanwhile in Indonesia most (51.70%) of palm oil patent applicants came from abroad. Palm oil patent applicant data in the world shows that 83% of palm oil patent applicant came from company (industry) group and the low number of joint patent application of Higher Education Institution-industry (3.02%) indicates that research cooperation of Higher Education Institution-industry still need more improvement. Prefix country analysis indicates that China, Malaysia, United States of America, Japan and England are the primary countries to be the destination of palm oil technology implementation in the world.

Mapping result of palm oil technology condition and development in the world and Indonesia which is conducted in this research can be positive input for Higher Education Institution, R&D institution as well as policy maker in organizing policy, conducting research and development activity, and developing research and business cooperation. Technology technical information contained in patent document can be utilized as references to generate applicative researches in accordance to industry requirements.

In this research, searching limitation on the field filter "title" affect narrower search result scope, which is specifically for technology cultivation, preparation of palm oil raw material (CPO and PKO) and waste treatment, not including the technology which produces palm oil derivative products or utilization of palm oil product in various other fields of technology. Therefore, further research still need to conduct with a wider scope of search, which based on market research report by Technavio which describing the existance of shifting phenomena from palm oil technology requirement heading to derivative research and palm oil utilization, for example oleochemistry industry and cosmetic products. Future research needs to compare the patent information analysis with scientific publications analysis and case studies of recent developments from R&D institutions or universities.

Reference

[1] Regulation of The President of Republic Indonesia Number 32 year 2011 regarding Masterplan of Acceleration and Expansion of Indonesian Economic Development (Masterplan Percepatan dan Perluasan Pembangunan Ekonomi Indonesia / MP3EI) 2011-2025

[2] Essential Palm Oil Statistics 2017 (Internet) (Referenced February 7th, 2018) available on http://www.palmoilanalytics.com/files/epos-final-59.pdf

[3] Palm Oil Plantation Fund Management Board (Badan Pengelola Dana Perkebunan Kelapa Sawit / BPDPKS) 2016 Technical Guidance of Proposal Application for Palm Oil Research Grant

[4] BPS-Statistics Indonesia 2016 Indonesia Palm Oil Statistic

[5] Organisation for Economic Co-operation and Development (OECD) 2009 OECD Patent Statistics Manual (Paris: OECD publishing)

[6] Trippe A 2015 Guidelines for Preparing Patent Landscape Reports (Geneva: World Intellectual Property Organization)

[7] Yubas M 2004 Product Idea to Product Success: A Complete Step-by-Step Guide to Making Money from Your Idea (Cardiff: Broadword Publishing)

[8] Ernst H 2003 Patent information for strategic technology management *World Patent Information* **25** 233-43

[9] Wibowo M H 2010 Practical Guidance of Patent Searching (Bogor: Bogor Agricultural University)

[10] Republic of Indonesia Law Number 13 of 2016 concerning Patents

[11] Eusebi C A, Silberglitt R 2014 Identification and Analysis of Technology Emergence Using Patent Classification, The RAND Corporation

[12] World Intellectual Property Organization 2000 WIPO Patent Information Services for Developing Countries (Geneva: World Intellectual Property Organization)

[13] Fabry F B, Ernst H, Langholz J, Koster M 2006 Patent portfolio analysis as a useful tool for identifying R&D and business opportunities—an empirical application in the nutrition and health industry *World Patent Information* **28** 215-225

[14] Suprijadi J 2016 Proceedings of the National Conference on Mathematical Research and Learning (Surakarta) p 819-826

[15] Jun S, Lee S J 2012 Emerging Technology Forecasting Using New Patent Information Analysis *International Journal of Software Engineering and Its Applications* **6(3)** 107-116

[16] Poranki S S 2010 A Framework to Evaluate Growth Potential of Emerging Technologies – A Case of Flexible Electronics (Dissertation) (New York: Binghamton University)

[17] Kim M J 2009 Industry Technology Roadmap for the Flushable Pre-moistened Nonwoven Wipes Industry (Dissertation) (Raleigh-North Carolina: North Carolina State University)

[18] Sandal N, Kumar A 2016 Searching and Analysing Patent Document to Solve R&D Problems *DESIDOC Journal of Library & Information Technology* **36(2)** 65-96 DOI 10.14429/djlit.36.2.9504

[19] Kim J, Lee J, Kim G, Park S, Jang D 2016 A Hybrid Method of Analyzing Patents for Sustainable Technology Management in Humanoid Robot Industry *Sustainability* **8** 474

[20] Garland G M 2013 Technology Forecasting (TF) using Hybrid Text mining, TRIZ TF for Research and Development Planning: Forecast for Nonwovens Air Filtration Media (Dissertation) (Raleigh-North Carolina: North Carolina State University)

[21] Fujii H, Yoshida K, Sugimura K 2016 Research and Development Strategy in Biological Technologies: A Patent Data Analysis of Japanese Manufacturing Firms *Sustainability* **8(4)** 351

[22] Egan E J 2012 The Economics of Patent Citations: Startup Commercialization Strategy, Value, and Success (Dissertation) (Berkeley: University of California)

[23] Shuang F, Yan G 2015 The Analysis and Prospect of Development of Fresh Cut Flower Industry Based on the Patent Analysis *Agricultural Science and Technology* **16(9)** 2064-2069
[24] She L, Zhang L, Wang L 2016 Patent Information Analysis of Front-end Industry Chain in China's Tobacco Industry *Agricultural Science and Technology* **17(2)** 438-442
[25] Altuntas S, Dereli T, Kusiak A 2015 Forecasting technology success based on patent data *Technological Forecasting and Social Change* **96** 202–214
[26] Ernst H 1997 The Use of Patent Data for Technological Forecasting: The Diffusion of CNC-Technology in the Machine Tool Industry *Small Business Economics* **9(4)** 361-381
[27] Andersen B 1999 The hunt for S-shaped growth paths in technological innovation: a patent study *Journal of Evolutionary Economics* **9** 487–526
[28] Trappey C V, Wu H Y, Dutta F T, Trappey A J C 2011 Using patent data for technology forecasting: China RFID patent analysis *Advanced Engineering Informatics* **25(2011)** 53–64
[29] Wang Y H, Trappey A J C, Trappey C V 2015 Life Cycle Analysis of the Optical Disc Industry Market Innovation and Development *Innovation: Management, Policy & Practice* **17**(2) 196-216
[30] Dutta F T, Trappey A J C, Trappey C V, Wu H Y 2009 An exploratory RFID patent analysis *Management Research News* **32(12)** 1163-1176
[31] Lee S H 2010 An analysis of technology market from the perspective of technology life cycle *African Journal of Business Management* **4(17)** 3641-3654
[32] World Intellectual Property Organization 2017 Guide to the International Patent Classification version 2017 (Geneva: World Intellectual Property Organization)
[33] Leydesdorff L, Kogler D F, Yan B 2017 Mapping Patent Classifications: Portfolio and Statistical Analysis, and the Comparison of Strengths and Weaknesses *Scientometrics* **112(3)** 1573-1591
[34] Wang H, Coleman A, Dhall K, Singh M 2016 Patent Landscape Report on Palm Oil Production and Waste Treatment Technologies (Geneva: World Intellectual Property Organization / WIPO)
[35] Kamiyama S, Sheehan J, Martinez C 2006 Valuation and Exploitation of Intellectual Property (STI Working Paper 2006/5 Statistical Analysis of Science, Technology and Industry - OECD)
[36] Sugimitsu K 2017 Intellectual Property as a Marketing Tool *Journal of Intellectual Property Association of Japan* **13(3)** 4-14
[37] Scopus (Internet) (Referenced September 18th, 2017) available on https://www.scopus.com
[38] Hidayah K Protection of Patent Rights in the Study of Islamic Law and the Role of Muslims in the Field of Science and Technology *de Jure: Jurnal Syariah dan Hukum* **4** 85-95
[39] Nanjappan P 2011 Creating a Business Plan to Generate Knowledge-based Income *Palm Oil Engineering Bulletin* **99** 15-18
[40] Gkika D A, Kontogoulidou C, Nolan J W, Mitropoulos A C, Vansant E F, Cool P, Braet J 2016 Nano-patents and Literature Frequency as Statistical Innovation Indicator for the use of Nano-porous Material in Three Major Sectors: Medicine, Energy and Environment *Journal of Engineering Science and Technology Review* **9(5)** 24-35
[41] Harhoff D, Scherer F M, Vopel K 2003 Citations, family size, opposition and the value of patent rights *Research Policy* **32(8)**1343-1363
[42] Putnam J 1997 *The Value of International Patent Rights* (Dissertation) (New Haven-Connecticut: Yale University)
[43] Van Pottelsberghe B, van Zeebroeck N 2007 A Brief History of Space and Time: The Scope-Year Index as A Patent Value Indicator Based on Families and Renewals (Makalah Diskusi) Centre for Economic Policy Research No. 6321

Quality characteristic determination of red dragon fruit noodle with Kano Model and QFD Method

M E Isharyani[1,2], **Y Wijaya**[1] **and F D Sitania**[1]

[1] Department Industrial Engineering, Mulawarman University, Samarinda, East Kalimantan, Indonesia, e-mail: murianiemelda@gmail.com

Abstract. Red dragon fruit noodles in Samarinda, East Kalimantan have quality attributes that differ from other food commercial products. Therefore, the determination of quality attribute of red dragon fruit noodles need to be investigated. This paper discusses the use of Kano model and Quality Function Deployment (QFD) method for determining the quality parameters of red dragon fruit noodles. Kano questionnaire and Voice of Customer (VOC) questionnaire were used to categorize consumer-defined attributes and the level of importance of consumer defined attributes. The result obtained that 14 attributes are in One Dimensional category, 1 attributes in Must Be category and 13 attributes in Indifferent category. It is finally concluded that the consumer prefer red dragon fruit noodles with soft texture, thoroughly cooked, easily carried, fast and practical served and served in hot temperature. Technical parameters that need to be exist are technical presentation and the amount of wheat flour.

Keywords: red dragon fruit noodle, Kano model, quality function deployment

1. Introductions

Every human need food to get energy needed for their daily activities. 70% of human health depends on the dietary habit and food. Balanced meal will help human health. Health dietary habit related with carbohydrate, fat and protein. Carbohydrate not only obtained from rice, but can be found in fruits and vegetables [1].

One of the potential of East Kalimantan's natural resource is red dragon fruit. The production of red dragon fruit in East Kalimantan is very abundant. The benefit of East Kalimantan's geographical condition made the red dragon fruit production not depends on the season. The other advantages of this fruit are sweeter taste and larger size compare to other red dragon fruit that produced outside East Kalimantan [2]. Red dragon fruit is rich in various vitamin and mineral substances that help increase endurance and beneficial to human body metabolism. Red dragon fruit is very good for the circulatory system, reduce emotional pressure and neutralize toxic in the blood [3]. Red dragon fruit processed products that are suitable to be marketed for the area of East Kalimantan have been studied by [4], namely red dragon fruit noodles and ice cream that have the potential to be developed with franchise cart marketing systems. Therefore, red dragon fruit combined with noodle as one source of carbohydrate would create healthy food for human consumption.

The combination of noodle and dragon fruit has been made by Jember University student at East Java in 2016. The research utilized dragon fruit in Banyuwangi, East Java, in the noodle to provided added value while utilized unsold dragon fruit [5]. Nowadays, there is no commercial product of red dragon fruit noodle in East Kalimantan. Therefore, research needed to create red dragon fruit noodle with preferred quality characteristic by East Kalimantan consumer, which is certainly different from the noodle produced previously.

Kano model and Quality Function Deployment (QFD) method are considered to determine quality attributes of red dragon fruit noodles based on detailed categorized level that could result from these methods. The integration approach of Kano's model and QFD involves a series of activities, from classifying costumer attributes to evaluate the priority analysis of technical characteristics [6][7].

Various researchers have recommended some techniques on how to QFD can be enhanced, whereby surveys, focus group, questionnaire and interviews can be used as the traditional ways of collecting customer requirements and to identify the degree of importance for each customer requirements [8][9]. The Kano model is used to categorize product attributes based on the ability of the product to satisfy the consumer. QFD then connects consumer needs with the technical parameters to create preferred quality characteristic of red dragon fruit noodle. These results compile House of Quality (HOQ) that show information about the needs of consumers with technical parameters to crate preferred consumer product. Technical parameters used in this paper are considered from [10] which did noodle research based on 10 parameters.

2. Research methodology

The design and development of red dragon fruit noodle is based on data from the Kano questionnaire and voice of costumer (VOC) questionnaire. Kano questionnaire are used to classify consumer attributes, while the VOC questionnaire is to obtain the level of importance of consumer attributes. These questionnaires were distributed by using purposive sampling techniques because the selected sample is believed to represent a particular population where in this case are people who have consumed products similar to red dragon fruit noodle. The product is segmented to all ages, therefore the number of samples is determined using the following [7] formula:

$$n = \frac{(Z_{\alpha/2})^2 (pq)}{(SE)^2} \quad (1)$$

With n=sample size, $Z_{\alpha/2}$ =criticism value, α=significance level, p=variance value, q=1 – p, and SE=sampling error.

Given that the population cannot be known with certainty, it is assumed that the population is unlimited. Therefore, the value of p and q using the highest value, with each of the value is about 0.5, with a confidence level of 95% and an error rate of 10%. From the calculation using the formula (1), it was found that the number of respondents needed was 97 people. The number of questionnaire distributed is 107 questionnaires to anticipate if there is an invalid questionnaire.

Valid questionnaires ready to process after going through the validity and reliability test. Validity test is intended to find out whether the statement items in the questionnaire can measure the interest of consumers, meanwhile reliability test to determine the consistency of measurement results or the extent to which these measurements can be trusted despite repeated measurements [8]. After the results of the questionnaire distribution are declared valid and reliable, then the data is processed through two main stages:

2.1. Grouping consumer attributes into the Kano Model

Kano model is a model used to categorize the attributes of a product based on how well the product is able to give effect to customer satisfaction. Kano model provide a linear view of the results provided by the performance of a product or service to customer satisfaction that can be used to identify attributes that potentially lead to customer satisfaction or dissatisfaction [9]. The Kano model can group product attribute needs into six categories:

2.1.1. Must-be. Attributes belonging to this category are considered the basic criteria of a product. The customer will be dissatisfied if the attribute performance is not met. Customer assumes that the attributes of this product are the attributes that must be present in the product, but if the attribute performance is met then it will not affect the increase of customer satisfaction.

2.1.2. One dimensional requirements. The level of customer satisfaction is directly proportional or linear with the level of fulfillment of the availability of attributes of a product. Customer satisfaction will decrease if the attributes in this product are not met, but customers will be satisfied if the attributes in the product are given. Attributes in the one dimensional requirements category will be explicitly demanded by the customer.

2.1.3. Attractive. Customer satisfaction will increase accordance with the improvement of attribute performance given to the product, but the decrease in attribute performance will not caused decreased of customer satisfaction. Customers will feel satisfied if the attributes on this product continue to be improved, but customers will not feel disappointed if the attributes in this category are not in the product. Customer are not demanding and not expect fulfillment of attributes in this category. Giving attribute in this category will give more positive effect to customer and improve product image in customer mind compared to competitor product. Some customers will even pay more to be able to get products that have this attribute.

2.1.4. Indifferent. Attributes in this category are seen as a neutral attribute of attributes for customers, so they will not have any effect on changes in customer upgrade or reduction.

*2.1.5. Reverse.*The customer will be dissatisfied if the attribute in this category is given, otherwise customer satisfaction will appear if the attributes does not exist.

2.1.6. Questionable. The attribute needs in this category can not be translated clearly because there is a contradiction between the customer's answer and the paired questions. This category indicates there is confusion and misunderstanding between the questions raised and the answers given.

2.2. Preparation of House of Quality (HoQ)

HOQ is the most recognized and widely used form of Quality Function Deployment (QFD) method, which would generate quality characteristics that need to be in the red dragon fruit noodle that will be developed. QFD is a way to improve the quality of a product or service by understanding the needs of consumers and then connecting it with technical provisions to produce a good or service at every stage of making goods and services produced. The dissemination of quality function in QFD is a plan tool used to help businesses focus on the needs of their customers when compile specification design and fabrication [10]. HOQ translate customer requirements, based on marketing research and benchmarking data, into an appropriate number of engineering targets to be met by a new product design. According to [11], it is "a kind of conceptual map that provides the means for inter functional planning and communication." There are six basic steps performed to build HOQ:

2.2.1. Idenfity WHAT costumer wants. It considered as consumer attributed which is obtained based on questionnaires. Consumer attributes used are the attributes of consumer who have been declared valid and reliable. Furthermore, the level of importance of relative consumer attributes, raw weight and normalized raw weight are calculated with formula as shown below.

$$\bar{X} = \frac{\sum x}{n} \tag{2}$$

With $\bar{X}$=importance to customer, $\sum X$ = total of score, and n=number of respondents.

$$Raw\ Weight\ =\ impotance\ to\ customer\ \mathrm{x}\ improvement\ ratio\ \mathrm{x}\ sales\ point \tag{3}$$

$$Normalized\ Raw\ Weight\ =\ \frac{Raw\ Weight}{\sum Raw\ Weight}\ x\ 100 \tag{4}$$

$$Wweighted\ Iimportance\ =\ \sum(Normalized\ Raw\ Weight\ \times\ relationship\ level) \tag{5}$$

However, competitor data excluded due to red dragon fruit noodle is still classified as a new product. This makes the value of raw weight determined by normalized level of importance to consumer attributes.

2.2.2. Idenfity HOW the product will satisfy the customer. It refers to identifying specific product characteristics, features or attributes and showing how they will satisfy customer wants.

2.2.3. Identify relationships between how's. It would construct the interrelationship matrix which its main function is to establish a connection between the customer's product requirements and the performance measures designed to improve the product. After setting up the basic matrix, it is necessary to assign relationships between the customer requirements and the performance measures. These relationships are portrayed by symbols indicating a strong relationship, a medium relationship or a weak relationship. The symbols in turn are assigned respective indexes such as 9-3-1. When no relationship is evident between a pair, a zero value is always assigned.

2.2.4. Develop importance ratings. It refers to using the customer's importance ratings and weights from the relationships in the matrix to compute our importance rating. The calculation of the relative importance of the consumers with the absolute importance level using the Likert scale is done by calculating the average of the questionnaire answers given by the consumers.

2.2.5. Evaluate competing products or services.

2.2.6. Determine the desirable technical attributes. The absolute importance for each technical requirement is calculated from the cell value and the customer importance rating. Numbers are then added up in their respective columns to determine the importance of each technical requirement.

3. Discussions

3.1. Qualification of consumer preference with Kano model

Respondent data from Kano questionnaire were calculated into each category in the Kano model. The result show 14 attributes that are categorized as One Dimensional, which are 1) great and interesting appearance, 2) the sweet and salty taste in red dragon fruit noodle blend together, 3) has chewy texture, 4) served in hot temperature, 5) the colour is interesting and appetizing, 6) has natural colour, 7) has soft texture, 8) thoroughly cooked, 9) tasted delicious, 10) the packaging represent red dragon fruit, 11) boiled cooked, 12) served with gravy, 13) can be brought easily, and 14) fast and practical serving process. Attributes that falls into this category means that these attributes result in satisfaction when fulfilled and dissatisfaction when not fulfilled. Therefore, it is important to have these attributes in red dragon fruit noodle.

Moreover, attribute the similarity in thickness of red dragon fruit noodle categorized in Must Be, meanwhile 13 of the rest attributes fall into the Indifferent category. Must be category means this attribute must be included in red dragon fruit noodle; otherwise consumers would be very dissatisfied. The indifferent category considered insignificant because these attributes refer to aspects that are neither good nor bad, and would not result in either customer satisfaction or customer dissatisfaction.

3.2. Consumer attributes

Consumer attributes in the questionnaire were compiled based on previous similar studies, namely research on corn noodle [5], strawberry chili chips [12] and guavagua dodol [13]. Only 15 of 34 consumer attributes are considered in building HOQ, since these 15 attributes belonging to the One Dimensional and Must Be categories. These attributes are then calculated normalized relative importance score based on the average score of the questionnaire from the respondents. Top 5 priorities considered from consumer attributes that have the highest normalized relative importance score. Based on the result of calculations using equation (3) and (4) as shown in Table 1, the 5 priorities respectively were soft texture of red dragon fruit noodle, thoroughly cooked, can be brought easily and has packaging design that represent red dragon fruit.

Table 1. Normalized importance level relative.

No	Consumer Attribute	Normalization
1	The appearance of red dragon fruit noodle good and interesting	6,529

2	The thickness of red dragon fruit noodles is relatively same	6,265
3	Sweet and salty flavor in red dragon fruit noodle balance	6,312
4	Red dragon fruit noodle has a chewy texture	6,762
5	Red dragon fruit noodle served with hot temperature	6,823
6	The color of red dragon fruit noodle is interesting and appetizing	6,591
7	The color of red dragon fruit noodle looks natural	6,545
8	The texture of the red dragon fruit noodle is soft	7,103
9	red dragon fruit noodle thoroughly cooked	6,963
10	red dragon fruit noodle tasted good	6,762
11	The packaging design describe red dragon fruit	6,110
12	Red dragon fruit noodle cooked by boil	6,808
13	Red dragon fruit noodle served with gravy	6,606
14	Red dragon fruit noodle can be brought easily	6,948
15	The process of serving red dragon fruit noodlesquickly and practically	6,870

3.3. Compiling technical response and relationship matrix

Technical parameters or technical responses obtained based on the translation of consumer attributes. These technical parameters would be a reference in making red dragon fruit noodle. The determination of the technical parameters refers to the research of [5]. There are 9 technical parameters related to consumer attributes, which are the amount of wheat flour, number of red dragon fruit, number of eggs, the amount of salt, volume of water, number of STTP (baking soda), mixing process, length of drying time and presentation techniques.

The strength of relationship between consumer attributes and technical responses as seen in figure 1, includes 25 relationship that are relatively strong, 9 relationship that are classified as intermediate and 2 relationships that are classified as low relations. Each of relationship type is indicated by a particular symbol as follows:

◉ = *Strong relationship with weight is equal to 9*
O = *intermediate relationship with weight is equal to 3*
Δ = *low relationship with weight is equal to 1*
Blank = *no relationship with weight is equal to 0*

3.4. Technical correlation determination

This stage is carried out to determine the relationship between technical parameters with one another. The influence between these technical parameters is classified into strong positive, medium positive, medium negative and strong negative influences. There are 4 strong positive relations and 9 medium positive relations between each other technical parameters. This correlation can be seen as a roof of HOQ in figure 1.

3.5. Technical matrix determination

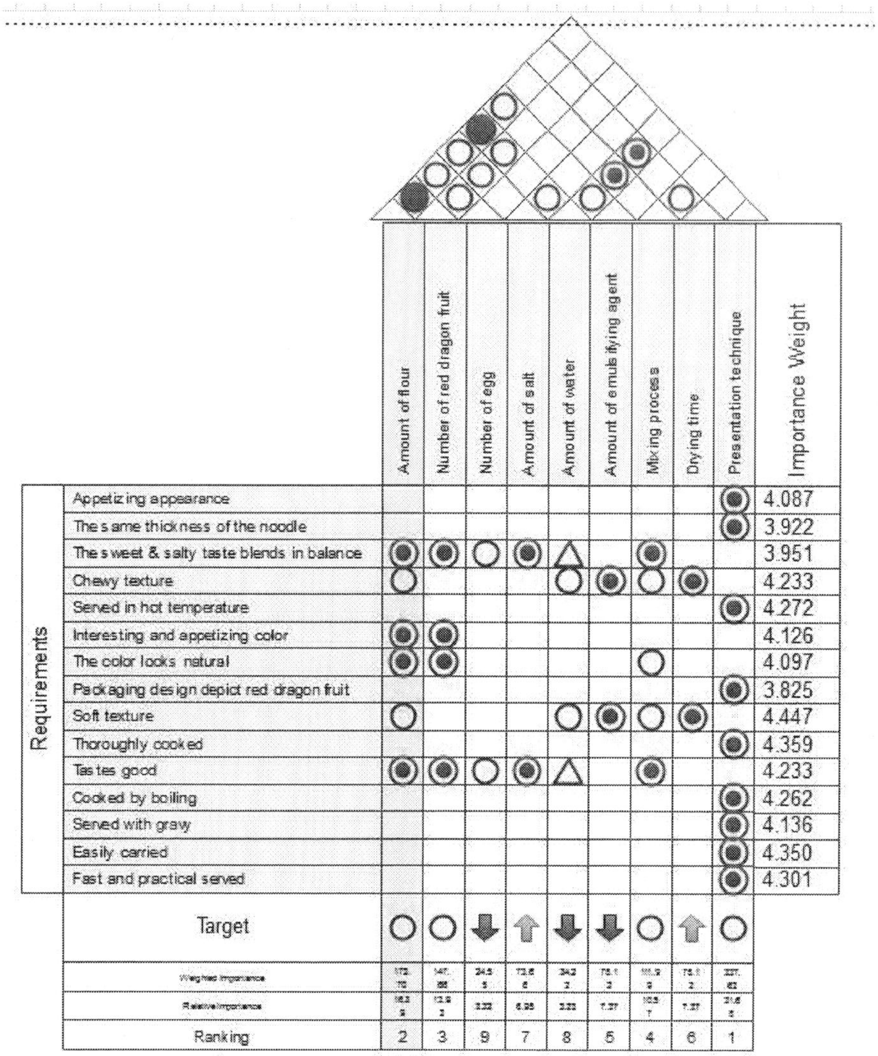

Figure 1. HOQ of red dragon fruit noodle.

3.5.1. Weight importance. The weight of the technical parameters is obtained by using equation (5), which represent importance to consumer value and the linkage weight (relationship) between the consumer attribute and the technical parameters. The calculation of weight importance for all technical responses placed in the bottom of HOQ, as seen in figure 1. The weight value of each technical parameter is then used as a guide in ranking the priority to determine the characteristics of red dragon fruit noodles. The priority ranking from highest to lowest weight can be sorted as technical presentation as first priority, amount of wheat flour as second priority, followed by number of eggs. Therefore, red dragon fruit noodle favored by the consumer is soft texture noodle which presented in a cooked, hot, quick and easy to carry which is related to the amount of wheat flour and technical presentation.

3.5.2. Direction of development. This direction of development was obtained from interviews with a researcher from Balai Riset dan Standardisasi Industri Samarinda (Baristand Samarinda) who had made research on red dragon fruit noodle in 2016. The results show that 2 of technical parameters need to improved, 3 of them need to be reduced and 4 of them need to meet the appropriate target values to satisfy consumer desires. Those 2 technical parameters that need to be improved are the amount of salt and length of drying time. Meanwhile, 3 of technical parameters that need to be reduced are the number of eggs, the volume of water and amount of STTP (baking soda). Furthermore, 4 technical parameters that need to meet certain target values are the amount of wheat flour, the number of red dragon fruit, the mixing process and presentation techniques.

All the steps above are done to get the HOQ components; therefore after all the data in HOQ has been determined, HOQ could be built as seen in figure 1.

4. Conclusions

Red dragon fruit noodle that must be made are red dragon fruit noodles that are preferred by consumers, which are has soft texture of noodle, served in a mature, hot, fast serving process and easy to carry (practical). These quality characteristic of red dragon fruit noodle related to the amount of wheat flour and the presentation techniques. The amount of wheat flour associated with soft texture of the red dragon noodle, meanwhile the technical presentation associated with noodles that served in cooked, hot, fast serving process and easy to carry.

HOQ is only the first stage of the QFD method, therefore further research is needed covering all stages in QFD. The level 2 QFD matrixes is a used during the Design Development Phase, which could discover the assemblies, systems, sub-systems and components have the most impact on meeting the product design requirements and identify key design characteristics. The level 3 QFD is used during the Process Development Phase where examined which of the processes or process steps have any correlation to meeting the component or part specifications. Last but not least, the level 4 QFD would have list all the critical processes or process characteristics in the "What" column on the left and then determine the "How" for assuring quality parts is produced and list them across the top of the matrix.

References

[1] Nainggolan S Y 2017 Lupakan super food ini pola makan sehat dan seimbang di 2016 http://rona.metrotvnews.com/read/2016/01/05/467069/lupakan-super-food-ini-pola-makan-sehat-dan-seimbang-di 05-11-2017
[2] Dinas Pertanian 2017 Buah naga kaltim miliki sejumlah keunggulan http://dispertan.kaltimprov.go.id/berita-buah-naga-kaltim-miliki-sejumlah-keunggulan.html 02-09-2017
[3] Leong A I 2017 Manfaat buah naga dan kandungan gizi buah naga http://doktersheat.com/manfaat-buah-naga-dan-kandungan-gizi-buah-naga/ 05-11-2017
[4] Abdi F N, Isharyani M E, Kuncoro D K R, Sitania F D, and Saptaningtyas W W E 2018 Pengembangan Produk Olahan Buah Naga Merah Kaltim yang Cocok Dipasarkan Dengan Sistem Usaha Gerobak Waralaba *Journal Industrial Services* **3 No.2** 27-32
[5] Iim 2017 Manfaatkan buah naga yang berlimpah jadi pahlawan pangan loka https://www.unej.ac.id/manfaatkan-buah-naga-yang-berlimpah-jadi-pahlawanpangan-lokal/?lang=id., 04-10-2017.
[6] S F Liu and M H Lee 2013 Research on Prospective Innovation Design of Smart Electric Vehicle *Advance Industrial Engineering Management* 2 No. 2 63-67
[7] W Xiong, Y Yu, and J Wang 2015 An Improved Algorithm for Product Conceptual Design based on Quality Function Deployment *Appl Math. Inf. Science* 297 No. 1 289-297
[8] H Wu, Y Tang and J Shyu 2010 An Integrated approach of Kano's model and Importance-Performance Analysis in identifying key success factors *African Journal Business Management* 4 No. 15 3238-3250
[9] L Ho, T Peng, S Feng, and T Yen 2013 Integration of Kano's model and SERVQUAL for enhancing standard hotel customer satisfaction *Africa Journal Business Management* 7 No. 23 2257-2265
[10] Suyaningrat I B, Rutiani E, and Kurniawati I 2010 Application of Quality Function Deployment (QFD) method for quality improvement of corn noodle product *Agrotek* **4 No. 1** 8-17
[11] McClave J T and Sincich T 2010 *Statistik untuk bisnis dan ekonomi* (Jakarta: Erlangga)
[12] Suliyanto 2005 *Analisis Data dalam Aplikasi Pemasaran* (Bogor: Ghalia Indonesia)
[13] Kano N 1984 Attractive quality and must-be quality *The Journal of the Japanese Society for Quality Control* **14 No.2** 39-48
[14] Ginting R 2010 *Perancangan Produk* (Yogyakarta: Graha Ilmu)

IOP Conf. Series: Earth and Environmental Science **335** (2019) 012017 doi:10.1088/1755-1315/335/1/012017

[15] Hauser J R, Clausing D 1988 The House of Quality *Harvard Businees Review* 66: 63-73
[16] Wulandari S 2015 Peningkatan kualitas produk keripik sambal stroberi pada usaha kecil menengah (UKM) rizqia dengan menggunakan metode Quality Function Deployment *e-Proceeding Engineering* **2 No.3** 7516
[17] Dwirini E, Rohayati Y, and Fashanah A 2015 Analisis kebutuhan produk dodol guavagua menggunakan integrasi model Kano dan product quality di UKM Barokah Alam Ciwidey *e-Proceeding Engineering* **2 No.2** 4070

IOP Conf. Series: Earth and Environmental Science **335** (2019) 012018 doi:10.1088/1755-1315/335/1/012018

Isolation of collagen from chicken feet with hydro-extraction method and its physico-chemical characterisation

O Suparno[1] and N B Prasetyo[1]

[1]Department of Agroindustrial Technology, Faculty of Agricultural Engineering and Technology, Bogor Agricultural University, Darmaga Campus, Bogor 16680, Indonesia e-mail: ono.suparno@ipb.ac.id

Abstract. Chicken slaughterhouses generate a large volume of chicken feet as a waste or by product. Nowadays, the chicken feet have not been optimally utilised. One of the ways to increase added value of the chicken feet is to utilize collagen contained in it. The objectives of this study were to obtain the best treatments in pre-treatment and hydrolysis stages of collagen extraction from chicken feet using hydro-extraction method, and to characterise the physico-chemical properties of the isolated collagen. Collagen extraction consisted of three stages of pre-treatment, hydrolysis, and hydro-extraction. Pre-treatment stage used NaOH concentrations of 0.5, 1, and 2 M, and soaking times of 2, 4, 6, and 8 hours. The hydrolysis used CH_3COOH concentrations of 0.1, 0.3, and 0.5 M, and soaking times of 1, 2, and 3 hours. The best condition of pre-treatment was 0.5 M NaOH concentration with 6 hours of soaking time, while the best condition of hydrolysis was 0.5 M CH_3COOH concentration with 1 hour soaking time. FTIR spectroscopy showed that the isolated collagen amino acids consisted of amide A, amide B, amide I, amide II, and amide III groups. The dominant amino acid contents were glycine, L-serine, L-proline, L-glutamic acid, and L-alanine.

Key words: chicken feet, hydro-extraction, hydrolysis, collagen, pre-treatment.

1. Introduction

Collagen is an extracellular matrix of connective tissue which is abundant, about 30% of the total protein. Collagen consists of 26 types; one of the collagen types is type I collagen which is the main component in tendon, bone and skin tissue [1]. Type II collagen is present in the cartilage and the conjunctiva of the eye. Type III collagen is found in blood vessels. About 80-90% of collagen types that are in the human body are type I, II, and III. Differences in collagen types are characterized by bond complexity, structural uniformity, connection variants (slice), presence of non-helix domains, assembly and function. The collagen structure unit is a rod-shaped tropocollagen and consists of three polypeptide units that play a role in the formation of triple helix structure. Collagen consists of 3 large and repetitive polypeptide chains. Amino acid composition and physicochemical characteristics of collagen are varies and dependent on tissues [2]. The amino acid composition of collagen found in chicken feet is dominated by glycine, glutamic acid, proline, and hydroxyproline [3]. The use of collagen is quite extensive both in the fields of biomedicine, cosmetics, and food. Collagen has been widely applied for biomedical purposes, pharmaceutical, food, drug, and cosmetics industries. Collagen of animal skin or hide can be converted to leather by a tanning process to produce a variety of products [4, 5, 6, 7, 8]. Applications of collagen in biomedicine are as a homeostatic agent, regeneration in bone tissue replacement, membrane oxygenator, contraception (barrier method), implant, and drug delivery system. Utilisations in the field of cosmetics are as active ingredients to prevent the occurrence of premature aging (ant aging) either in the form of lotions or creams. The food industry utilises collagen as an emulsifier and foaming agent. Collagen has several advantages including easily absorbed in the body, low antigenicity, high water affinity, non-toxic, biocompatible, biodegradable, relatively stable, and can be prepared in various forms as needed [9].

Collagen found on the market today generally comes from cow hide and pig skin. The use of collagen from these materials has constraints from religious aspects [3]. An alternative source of collagen is needed to meet the collagen demands. This causes high potential of chicken feet as an alternative in the selection of raw materials for making collagen. This study uses chicken feet as a raw material for making collagen. The use of chicken feet is due to the abundant in Indonesia and their protein content.

The chicken feet are a waste from chicken slaughterhouse with a large volume of waste. Based on animal husbandry and animal health statistics data in 2017, there were 1.698 billion broilers in Indonesia [10]. During this time the potency of chicken feet has not been utilised optimally; chicken feet only became a waste of the chicken slaughterhouse. One way to increase the added value of chicken feet is to utilise their collagen.

The collagen isolation process can be carried out using the existed methods of acid-soluble collagen and pepsin-soluble collagen [11]. Collagen can be extracted chemically or with combination between chemical and enzymatic processes [12]. However, both extraction techniques have disadvantages, so other collagen isolation techniques are needed. Hydro-extraction is a method of isolating collagen using water as a solvent to extract collagen. Determination of NaOH and CH_3COOH concentrations, and also the duration in isolation of collagen is necessary to shorten the isolation time and to minimise the chemicals used in the collagen extraction. Physico-chemical characteristics of collagen isolated by hydro-extraction from chicken feet are also important to understand in product development from slaughterhouse by products.

The objectives of this study were to obtain the best treatments in pre-treatment and hydrolysis stages of collagen extraction from chicken feet using hydro-extraction method, and to characterise the physico-chemical properties of the isolated collagen.

2. Material and Method

2.1. Material and Equipment

Main material used in this study was chicken feet. The materials used to produce collagen were distilled water, NaOH, and acetic acid. The materials used in quantitative protein analysis were Bradford's solution (10 mg Coomassie Brilliant Blue (CBB G-250), 5 ml 95% ethanol, 10 ml of 85% orthophosphoric acid solution and distilled water) and bovine serum albumin (BSA) standards. Analysis for swelling and level of collagen solubility used filter paper and 5 M NaCl. Proximate analysis used distilled water, kjeltabs grains (catalyst tablets), H_2SO_4, NaOH, HCl, H_3BO_3 containing a mixture of 0.1% bromocresol green indicator and methyl red 0.1% with a ratio of 2:1, NaOH-$Na_2S_2O_3$ and hexane solvent 96%. Other ingredients included ingredients for characteristic analysis of collagen.

Equipment used in this study were cooler box, analytic scale, thermometer, UV-Vis spectrophotometer, water bath shaker, and freeze dryer, Erlenmeyer glass, measuring cup, and stirring rod. The equipment for the proximate, protein content, fat content, and ash content analyses. FTIR (Fourier transform infrared) type Bruker infrared spectrophotometer was used for the functional groups analysis and ultra-performance liquid chromatography (UPLC) was used for the amino acid test.

2.2. Method

2.2.1. Chemical preparation and characterisation of chicken feet. The chicken feet was transported to the laboratory under freezing conditions, then thawed, and cleaned all remaining dirt by washing with water. Chicken feet were cut to a size of approximately 0.5-1 cm using a knife or scissors and stored at freezing temperature until the chicken feet were used. The proximate analysis of the clean chicken were carried out including water, protein, ash and fat contents [13].

IOP Conf. Series: Earth and Environmental Science **335** (2019) 012018 doi:10.1088/1755-1315/335/1/012018

2.2.2. Collagen isolation (modification for Liu et al. [14], modification from Sukkwai et al. [15] and Nur'aenah [16]). Collagen isolation process with hydro-extraction method consists of three stages, namely pre-treatment with NaOH solution, hydrolysis with acetic acid solution, and extraction with distilled water. The first stage is the process of pre-treatment with NaOH solution aimed to eliminating non-collagen proteins and other impurities such as fats, minerals, pigments and odours. The NaOH concentrations used were 0.5, 1, and 2 M with a soaking times of 2, 4, 6 and 8 hours, then every 2 hours the NaOH solution was replaced with a new NaOH solution, but still used the same chicken feet. The ratio between chicken feet and NaOH solution was 1:10 (b/v). NaOH solution from the immersion process was tested quantitatively by the Bradford analysis to determine the best NaOH concentration and soaking time. The chosen chicken feet were neutralized by distilled water before were used for the next stage. The second stage was hydrolysis with a solution of acetic acid (CH_3COOH) to change the structure of collagen fibres to facilitate the extraction process. The concentrations of acetic acid used were 0.1, 0.3, and 0.5 M with soaking times of 1, 2, and 3 hours. The ratio between chicken feet and acetic acid solution was 1:10 (b/v). The samples were washed using distilled water until neutral. The third stage was extraction by hydro-extraction method. The extraction was carried out at 40°C for 2 hours using a water bath shaker with a speed of 150 rpm. The ratio between chicken feet and distilled water was 1:2 (b/v). Extraction result was in the form of collagen solution, then the solution was dried with a freeze dryer to obtain dry collagen and the yield was calculated.

2.2.3. Characterisation of collagen. Collagen characterization consisted of yield analysis [17], functional group analysis with FTIR [18], and amino acid analysis [19, 20].

2.2.4. Proximate analysis [13]. Chemical characterization of chicken feet raw material was carried out by proximate analysis consisting of water, protein, ash, and fat content analyses.

3. Results and Discussion

3.1. Chemical characteristics of chicken feet

The chemical composition of material needs to be analysed for the identification of chemical content contained in the material and as indicator of material quality. Chemical composition analysis used in this study was a proximate analysis. The proximate analysis results of chicken feet were presented in table 1. Potency of chicken feet can be seen from the collagen content of 5.64-36.38 % of total protein [21]. The presence of ash and fat in materials can interfere effectiveness of collagen extraction [22]. Extraction disturbances occur because fat and ash can block the penetration of water to extract collagen from chicken feet, so it needs to be removed at the pre-treatment stage.

Table 1. Proximate compositions of chicken feet.

Proximate composition	Composition (% w/w)	Composition (% w/w)
Water	64.22±0.66	65.08±0.90[1]
Ash	6.95±0.44	5.98±0.37[2]
Protein	18.09±0.07	17.42±0.73[2]
Fat	8.22±0.19	12.04±0.44[2]

[1]Hashim et al. [3], [2]Liu et al. [21]

3.2. Pre-treatment of the chicken feet with NaOH solution

Collagen extraction process begins with a pre-treatment process which aims to remove non-collagen protein. The alkaline solutions that can be used to remove non-collagen proteins are NaOH and $Ca(OH)_2$ [23]. The alkaline solution (NaOH) can break down most of the telopeptide regions of collagen molecules which cause slight swelling of materials [24]. The telopeptide regions are the ends of the open triple helix chain (short non-helix structures) [25]. The condition of swelling can cause easily released of non-collagen proteins and other impurity components such as fat and pigments in the collagen matrix [26]. This causes the NaOH solution used to soak the chicken feet to turn brownish. The concentration of dissolved protein found in NaOH solution because of the influence of concentration and soaking time could be seen in Figure 1. Figure 1 showed the concentration of protein (mg/l) non soluble collagen measured every 2 hours for 8 hours. The samples used at a concentration of 2 M were damaged which were indicated by the materials that became soft, so that the non-collagen protein value increased sharply from the second to forth hours. This results in macro particles being extracted in the measurement of dissolved proteins.

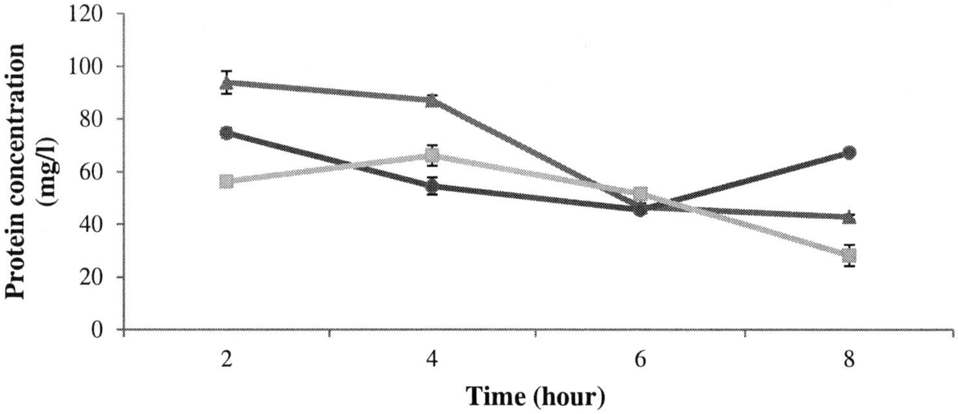

Figure 1. The concentrations of protein in the NaOH solution residues; NaOH 0.5 M, NaOH 1 M, NaOH 2 M.

The results of the split plot analysis showed that the differences in soaking time and their interaction with soluble protein levels were significant ($p < 0.05$). NaOH concentration had no significant effect on soluble protein levels ($p > 0.05$). The data were then tested by Duncan's Multiple Range Test (DMRT). The use of 0.5 M NaOH concentration for 6 hours was the best treatment from the pre-treatment stage which showed the lowest soluble protein concentration of 0.046 mg /ml. The concentration of dissolved protein showed that non collagen protein has been maximally extracted. This concentration was chosen because lower concentrations can reduce the risk of collagen component damage [27]. Hydrolysis is the process of breaking chemical bonds with the addition of water, where water is broken down into the form of H^+ and OH^-, generally were used in breaking down certain polymers [28].

The efficiency of the pre-treatment stage in NaOH solution is influenced by several factors including time, temperature, NaOH concentration, and raw materials used [8]. The use of alkaline solutions in the pre-treatment process is more effective in the process of extracting non-collagen proteins and only causes low level of collagen loss [23]. The extraction of non-collagen compounds occurs due to the breakage of some of the bonds between fibres in the structure of collagen in alkaline conditions [29]. The duration of soaking and alkali ratio will cause depolymerisation and deacetylation. NaOH has a role in separating strands from collagen fibres [24].

3.3. Chicken feet hydrolysis with CH_3COOH

Neutral pH causes decrease of collagen solubility [30]. The hydrolysis stage is carried out with acids which cause changes in the structure of collagen fibres to facilitate the extraction process of collagen in the next stage. Hydrolysis using acetic acid solution causes the material expanding and collagen solubility will occurs [31]. Based on the results of the study, the swelling increased along with the increase in acetic acid concentration and soaking time (figure 2). The biggest swelling (27.39% was found at 0.5 M acetic acid concentration with a 3 hour soaking time, whereas the lowest (2.61%) was found at 0.1 M acetic acid concentration with 1 hour soaking time. The variance analysis showed that the differences in acetic acid concentrations affected the swelling significantly ($p<0.05$), while the differences in soaking times and their interactions did not significantly influence the swelling (p> 0.05). Nur'aenah [16] stated that the swelling increases when the higher concentration of acetic acid used and the longer of soaking time. Acetic acid serves as a catalyst that helps the performance of water in the process of hydrolysis, so that it has a large influence on the results of extraction [28].

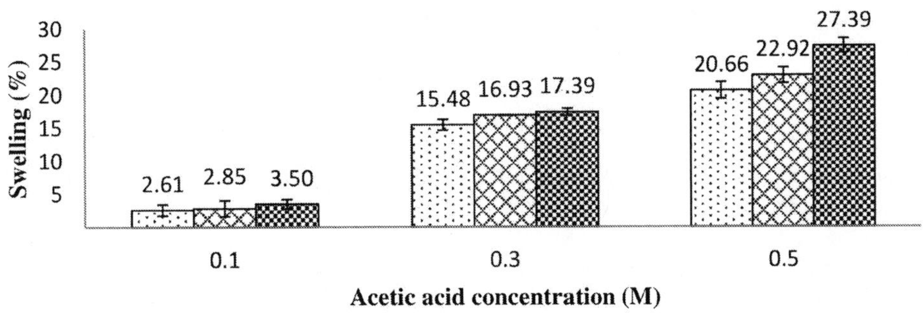

Figure 2. Swelling of chicken feet on differences of acetic acid concentrations and soaking times; soaking time of 1 hour, soaking time of 2 hours, soaking time 3 hours.

During the soaking process with acetic acid, collagen fibres will have a swelling process, resulting in a decrease in the internal cohesion of the collagen fibres [32]. During swelling, the cross-linking structure of the collagen molecule undergoes an opening and acetic acid acts as a catalyst between the bonds [28], so that water can enter the space of the collagen fibrils which causes swelling of the chicken feet (figure 3). Swelling of collagen in acidic solution is influenced by the osmotic pressure difference between protein and solution to reach the point of balance of osmotic pressure [33]. Increased collagen swelling at low pH is caused by the breakdown of the reticular tissue and elastin tissue by breaking intermolecular interactions and hydrogen bonds. Breaking intermolecular interactions can reduce the cohesion of protein molecules found in chicken feet, to facilitate the extraction process. Collagen protein changes in the form by swelling, because of the interaction between collagen fibrils and acidic solutions. This interaction is indicated by an increase in the percentage of weight on chicken feet after soaking in a certain period. The magnitude of swelling indicates more space in collagen fibrils can be entered into the solution.

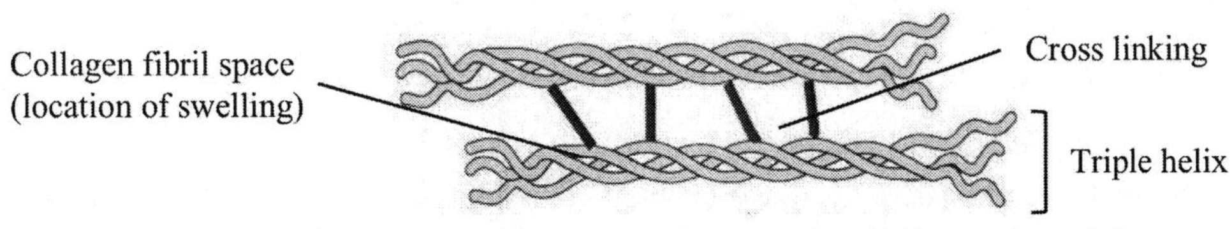

Figure 3. Swelling of collagen on the triple helix structure [34].

The results of variance analysis of collagen solubility showed that the differences in the soaking time of acetic acid affected significantly the level of collagen solubility ($p<0.05$). Acetic acid concentration and the interaction of acetic acid concentration and soaking time did not significantly influence the level of collagen solubility ($p>0.05$). The data was then tested further by DMRT. Figure 4 showed the level of solubility of chicken feet collagen. The highest level of collagen solubility (1.04 %) was found in 0.5 M acetic acid concentration with 3 hour soaking time, which is equivalent to losing 0.011 gram. The lowest level of collagen solubility (0.46 %) was at 0.5 M acetic acid concentration with 1 hour soaking time equivalent to 0.005 gram loss. Wulandari et al. [31] reported that the concentration of acetic acid determines the pH value of the solution, so that it regulates the level of charge of collagen which affects electrostatic interactions, the degree of solubility, and the extractability of collagen from the tissues.

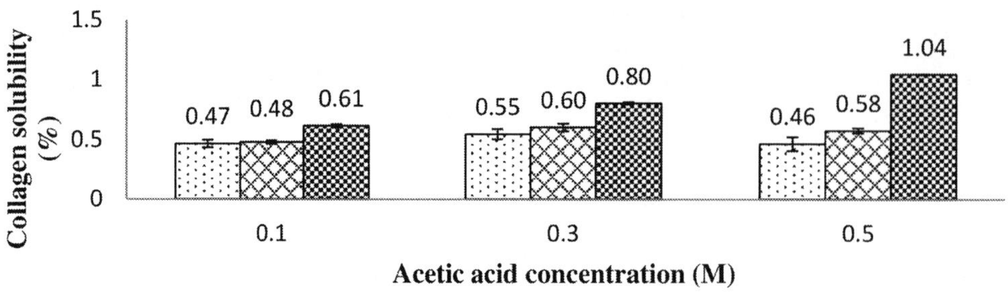

Figure 4. The degree of collagen solubility on the chicken feet during soaking by acetic acid; ▢ soaking time of 1 hour, ▩ soaking time of 2 hours, ▧ soaking time of 3 hours.

Collagen is a hydrogel so that increases in solubility causes an increase in the swelling [30]. Swelling of the skin structure affects the separation of collagen fibre structure and can interfere the non-covalent bond of collagen which ultimately can facilitate extraction [24]. Therefore, in the hydrolysis process, the swelling factor is more important than the level of collagen solubility. Concentration of 0.5 M is the concentration level which gave the highest swelling compared to the concentration of 0.1 M and 0.3 M, and had insignificant soaking times. The lowest collagen solubility was at a acetic acid concentration of 0.5 M with a 1 hour soaking time with a large swelling. Based on the variance analysis on swelling and the level of collagen solubility, the best treatment chosen at the hydrolysis stage was 0.5 M acetic acid concentration with 1 hour soaking time.

3.4. Hydro-extraction

Before extracting collagen, chicken feet washed first with distilled water. The aim was to decrease the pH until neutral. Chicken feet extraction was carried out with distilled water, with a ratio of 1:2 (b/v). Extraction was carried out at 40°C for 2 hours using a water bath shaker with a speed of 150 rpm.

3.4.1. Yield. The results (table 2) showed that the yield of collagen dry base from chicken feet extracted by hydro-extraction method was lower compared to collagen yield obtained using enzymatic methods, sodium chloride-soluble collagen (SSC), and acetic acid-soluble collagen (ASC). Collagen extraction is currently used by pepsin enzyme (PSC) and acid method (ASC). Both methods have several disadvantages such as requiring a long time, a lot of chemicals and high production cost. Hydro-extraction method is an extraction method that has the advantage of producing collagen with a relatively fast production, requires little equipment, can be produced continuously, small amount of waste, and lower production costs. The difference in yield caused by differences in the extraction method used, the concentration of the solution used to remove non-collagen protein, the type of raw material used, temperature differences, and the duration of extraction process [35]. The amount of collagen wasted during the pre-treatment and washing process can cause a decrease in yield [36]. Low yields occur due to collagen leaching during the washing process or denaturation by acid during the hydrolysis process [37]. The hydro-extraction process has a low reaction speed and low ability to penetrate into a tissue.

Table 2. Yield of collagen resulted by the hydro-extraction and other extraction methods.

Source of collagen and method of extraction	Yield(% w/w)
Chicken feet, hydro-extraction	0.14
Chicken feet skin, sodium chloride-soluble collagen (SSC)	1.13[1]
Chicken feet skin, acetic acid-soluble collagen (ASC)	14.49[1]
Chicken feet, papain enzyme	18.16[2]
Chicken feet, pepsin enzyme	22.94[2]

[1]Zhou et al. [38], [2]Hashim et al. [3]

IOP Conf. Series: Earth and Environmental Science **335** (2019) 012018 doi:10.1088/1755-1315/335/1/012018

The chicken feet has a dense or less porous physical form which causes low level of swelling, so that absorption of water is low and the ionization reaction with water during the extraction process is reduced. The chicken feet have a thick structure that requires extracting material that is able to penetrate into the material aggressively. The small number of cross links opened during the swelling process can reduce yield [39]. Soluble collagen in acetic acid cause reduce of collagen yield [31].

3.4.2. Functional group. The functional group of a compound can be identified by the FTIR (Fourier Transform Infrared) spectroscopic analysis by detecting infrared light as a source of electromagnetic radiation. If certain infrared frequencies are absorbed by a sample of organic compounds, vibration will occur in the molecule of the compound [40]. Characteristics of chicken feet collagen functional groups detected by FTIR are shown in table 3.

The functional group analysis with FTIR can be used to confirm the existence of a triple helix structure, which is known from the value of the amide III absorption ratio (1241.01 cm^{-1}) with an absorption peak at 1450 cm^{-1} which is close to 1 [41]. The intensity ratio between the amide III band and the 1450 cm^{-1} band is 1.16. FTIR analysis results showed that the value in accordance with the statement, so this indicates that the chicken feet collagen had not been degraded into gelatine because of the presence of the triple helix structure. The results of functional group analysis with FTIR is shown in figure 5.

Table 3. The functional groups of chicken feet collagen.

Amide type	The wave number of absorption peaks (cm^{-1})			Information
	This research data	Collagen literature	Absorption area standards	
Amide A	3426.65	3308[1]	3400-3440[2]	N-H stretch
Amide B	2924.37	2925[4]	2915-2935[3]	Asymmetrical stretch of CH_2
Amide I	1637.13	1630[1]	1600-1700[4]	C=O stretch
Amide II	1409.51	1548[1]	1550-1600[4]	N-H bend and C-N stretch
Amide III	1241.01	1242[1]	1229-1301[5]	N-H bend and C-N stretch

[1]Zhou et al. [38], [2]Liu et al. [42], [3]Coates et al. [43], [4]Hashim et al. [3], [5]Kong et al. [44]

3.4.3. Amino acid. The amino acid composition determines the type and physical characteristics of collagen. The amino acid composition of collagen from chicken feet was expressed in mg/kg (table 4). The largest amino acids in collagen found in chicken feet extracted with pepsin soluble collagen method were glycine (18.91%), glutamic acid (12.43%), proline (10.75%), and alanine (8.88%) [3]. The difference in amino acid content can occur due to differences in the species of material used and the extraction method used. Glycine is the most dominant amino acid found in collagen, and all types of collagen are characterized by tripeptide (Gly-X-Y) repetition, where X is proline and Y is hydroxyproline responsible for forming triple helix structures [18]. Collagen with the amino acid glycine, alanine, and proline in sufficiently high amounts, and contains little amino acids histidine and tyrosine, and does not contain cysteine is a characteristic of type I collagen [45]. High levels of proline and hydroxyproline are rarely found in proteins other than collagen [46]. Hydroxyproline is an amino acid that has a function as a heat stabilizer from collagen [18].

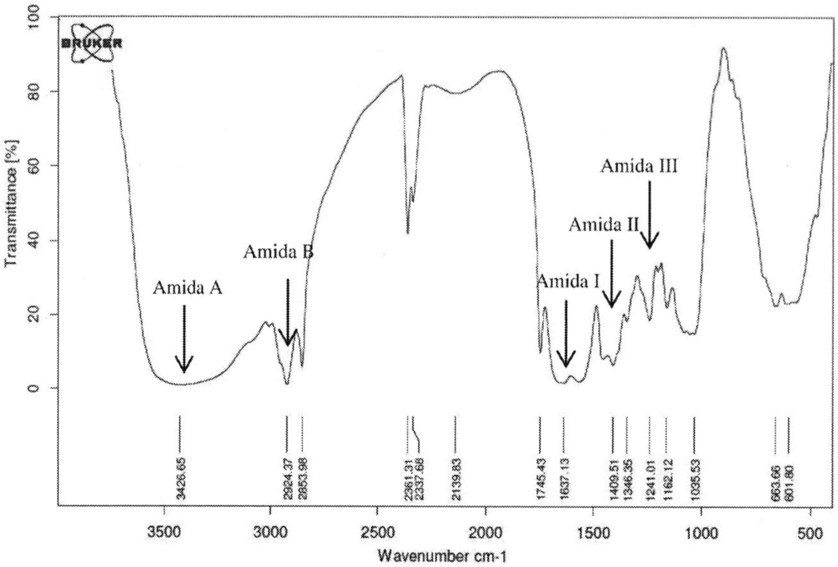

Figure 5. The FTIR spectrum of chicken feet collagen.

Table 4. The amino acid composition of chicken feet collagen resulted by the hydro-extraction and enzymatic extraction.

Amino acids	Hydro-extraction (mg/kg)	Enzymatic extraction (%)[1]	Enzymatic extraction (%)[2]
L-Glutamic acid	107.98	10.07	12.43
L-Phenylalanine	50.31	4.69	2.54
L-Isoleucine	-	0.00	2.50
L-Valine	24.63	2.30	3.07
L-Alanine	96.10	8.96	8.88
L-Arginine	93.97	8.76	7.34
Glycine	253.36	23.62	18.91
L-Lysine	42.31	3.95	5.37
L-Aspartic acid	72.03	6.72	7.63
L-Leucine	22.80	2.13	4.56
L-Tyrosine	21.14	1.97	1.01
L-Proline	113.65	10.60	10.75
L-Threonine	50.04	4.67	2.54
L-Histidine	-	0.00	-
L-Serine	124.14	11.58	3.02
L-Cysteine	-	0.00	-
L-Methionine	-	0.00	1.34

[1] Zhou et al. [38], [2] Hashim et al. [3]

4. Conclusion

Based on this study, it can be concluded that the use of 0.5 M NaOH for 6 hours was the best condition for pre-treatment in chicken feet collagen extraction. The best treatment of hydrolysis was the concentration of 0.5 M acetic acid with 1 hour soaking time. The collagen yield was 0.14%. FTIR spectroscopy showed amide groups of collagen consisted of amide A, amide B, amide I, amide II, and amide III. The highest amino acid contents were glycine, L-serine, L-proline, L-glutamic acid, and L-alanine which are the composition of amino acids which build collagen structure.

References

[1] Gelse K, Poschlb E and Aigner T 2003 Collagens-structure, function, and biosynthesis *Advanced Drug Delivery Reviews* **55 (1)** 1531-1546
[2] Hema G S, Shyni K, Mathew S, Anandan R and Ninan G A 2013 Simple method for isolation of fish skin collagen-biochemical characterization of skin collagen extracted from Albacore tuna (*Tunnus alalunga*), Dog Shark (*Scoliodon Sorrakowah*), dan Rohu (*Labeo Rohita*) *Annals of Biol Res.* **4(1)** 271-278
[3] Hashim P, Ridzwan M S M and Bakar J 2014 Isolation and characterization of collagen from chicken feet *International Journal of Bioengineering and Life Sciences* **8(3)** 250-254
[4] Covington A D, Evans C S, Lilley T H and Suparno O 2005 Collagen and polyphenols: new relationships and new outcomes Part 2 Phenolic reactions for simultaneous tanning and coloring *Journal of the American Leather Chemists Association* **100(9)** 336-343
[5] Suparno O 2005 Phenolic reactions for leather tanning and dyeing PhD Thesis (Leicester: University of Leicester)
[6] Suparno O, Covington A D and Evans C S 2007 Novel combination tanning using diphenols and oxazolidine for the production of high stability leather *Journal of the Society of Leather Technologists and Chemists* **91(5)** 188-192
[7] Suparno O, Kartika I A and Muslich 2009 Chamois leather tanning using rubber seed oil *Journal of the Society of Leather Technologists and Chemists* **93(4)** 158-161
[8] Suparno O 2010 Optimization of chamois leather tanning using rubber seed oil *Journal of the American Leather Chemists Association* **105(6)** 189-194
[9] Lee C H, Singla A and Lee Y 2001 Biomedical application of collagen *International Journal of Pharmaceutics* **22** 1-22
[10] Ministry of Agriculture 2017 *Statistics of Animal and Animal Health* 2017 (Jakarta: Directorate General of Livestock and Animal Health, Indonesian Ministry of Agriculture)
[11] Aberoumand A 2012 Comparative study between different methods of collagen extraction from fish and its properties *World Applied Sciences Journal* **16(3)** 316-319
[12] Karim A A and Bhat R 2009 Fish gelatin: properties, challenges, and prospects as an alternative to mammalian gelatins *Food Hydrocolloid* **23** 563-576
[13] AOAC [Association of Official Analytical Chemists] 2005 *Official Method of Analysis of the Official Analytical Chemists* (Virginia: The Association of Official Analytical Chemists, Inc.)
[14] Liu D, Wei G, Li T, Hu J, Lu N, Regenstein J M and Zhou P 2015 Effects of alkaline pre-treatments and acid extraction conditions on the acid-soluble collagen from grass carp (*Ctenopharyngodon idella*) skin *Food Chemistry* **172** 836-843
[15] Sukkwai S, Kijroongrojana K and Benjakul S 2011 Extraction of gelatin from bigeye snapper (*Priacanthus tayenus*) skin for gelatin hydrolysate production *International Food Research Journal* **18(3)** 1129-1134
[16] Nur'aenah N 2013 Extraction and characterisation of collagen and collagen nanoparticles from stingray (*Pastinachus solocirostris*) skin as cosmeceutical raw materials [MSc Thesis] (Bogor: Bogor Agricultural University)
[17] Djailani F, Trilaksani W and Nurhayati T 2016 Optimisation of extraction and characterisation of collagen from swimming bubbles of *Cunang* fish by acid-hydro-extraction method *Jurnal Pengolahan Hasil Perikanan Indonesia* **19(2)** 156-167
[18] Muyonga J H, Cole C G B and Duodu K G 2004 Characterisation of acid soluble collagen from skins of young and adult Nile perch (*Lates niloticus*) *Food Chemistry* **85** 81–89
[19] Nollet L M L 1996 *Handbook of Food Analysis* (New York: Marcel Dekker)
[20] Waters 2012 *Acquity UPLC H Class and H Class Bio amino Acid Analysis System Guide* (Milford: Waters Corporation)
[21] Liu D C, Lin Y K and Chen M T 2001 Optimum condition of extracting collagen from chicken feet and its characteristics *Asian-Australasian Journal of Animal Sciences* **14(11)** 1638-1644
[22] Shon J, Ji-Hyun E, Hwang S J and Jong-Bang E 2011 Effect of processing conditions on

functional properties of collagen powder from Skate (*Raja kenojei*) skins *The Journal of Food Science Biotechnology* **20(1)** 99-106

[23] Zhou P and Regenstein J M 2005 Effect of alkaline and acid pre-treatments on alaska pollock skin gelatin extraction *Journal of Food Science* **70(6)** 392-396

[24] Jaswir I, Monsur H A and Salleh H M 2011 Nano-structural analysis of fish collagen extracts for new process development *African Journal of Biotechnology* **10(81)** 18847-18854

[25] Fratzl P 2008 *Collagen Structure and Mechanics* (Potsdam: Springer)

[26] Cho S M, Gu Y S and Kim S B 2005 Extracting optimization and physical properties of yellowfin tuna (*Thunnus albacares*) skin gelatin compared to mammalian gelatins *Food Hydrocolloids* **19** 221–229

[27] Huang C Y, Kuo J M, Wu S J and sai H T 2016 Isolation and characterization of fish scale collagen from tilapia (*Oreochromis sp.*) by a novel extrusion–hydro-extraction process *Food Chemistry* **190** 997-1006

[28] Mardina P, Prathama H A and Hayati D M 2014 Effect of hydrolysis time and sulfuric acid catalyst concentration on furfural synthesis from rice straw *J. Konversi* **3(2)** 1-8

[29] Hinterwaldner R 1977 Raw material *In:* Ward AG dan Courts A (editor) *The Science and Technology of Gelatin* (New York: Academic Press)

[30] Kartika I W D 2017 Characterisation and biocompatibility of swimming bubble of *cunang* fish (*Muraenesox talabon*) collagen as a cell culture scaffold biomaterial [MSc Thesis] (Bogor: Bogor Agricultural University

[31] Wulandari, Suptijah P and Tarman K 2015 Effectivity of alkali pre-treatment and acetic acid hydrolysis on the characteristics of collagen from cork fish skin *JPHPI* **18(3)** 287-302

[32] Ockerman H W and Hansen CL 2000 *Animal by Product Processing and Utilization* (NewYork: CRC Press)

[33] Bowes J H and Kenten R H 1949 The swelling of collagen in alkaline solutions *J. Biochem.* **46(1)** 1-8

[34] Marsaid A L 2010 Characterisations of chemical, physical, and terminal properties of gelatin extract from tuna (*Thunnus sp.*) bone in a variety of acid solutions for soaking *J. Tek. Kimia* **1(1)** 1-15

[35] Potaros T, Raksakulthai N, Runglerdkreangkrai J and Worawattanamateekul W 2009 Characteristics of collagen from nile tilapia (*Oreochromis niloticus*) skin isolated by two different methods *Kasetsart Journal* **43** 584-593

[36] Ratnasari I, Yuwono S S, Nusyam H and Widjanarko S B 2013 Extraction and characterization of gelatin from different fresh water fishes as alternative sources of gelatin *International Food Research Journal* **20(6)** 3085-3091

[37] Jamilah B and Harvinder K G 2002 Properties of gelatins from skins of fish-black tilapia (*Oreochromis mossambicus*) and red tilapia (*Oreochromis nilotica*) *Food Chemistry* **77** 81-84

[38] Zhou C, Li Y, Yu X, Yang H, Ma H, Yagoub AEA, Cheng Y, Hu J and Otu PNY 2016 Extraction and characterization of chicken feet soluble collagen *Food Science and Technology* **74** 145-153

[39] Shyni K, Hema G S, Ninan G, Mathew S, Joshy C G and Lakshmanan PT 2014 Isolation and characterization of gelatin from the skins of skipjack tuna (*Katsuwonus pelamis*), dog shark (*Scoliodon sorrakowah*), and rohu (*Labeo rohita*) *Food Hydrocolloids* **39** 68-76

[40] Nagarajan M, Benjakul S, Prodpran T, Songtipya P and Kishimura H 2012 Characteristics and functional properties of gelatin from splendid squid (*Loligo formosana*) skin as affected by extraction temperatures *Food Hydrocolloids* **29** 389-397

[41] Silva T H, Silva J M, Marques A L P, Domingues A, Bayon Y and Reis R L 2014 Marine origin collagens and its potential applications *Marine Drugs* **12(12)** 5881–5901

[42] Liu D, Liang L, Regenstein J M and Zhou P 2012 Extraction and characterisation of pepsin-solubilised collagen from fins, scales, skins, bones and swim bladders of bighead carp (*Hypophthalmichthys nobilis*) *Food Chem.* **133** 1441–1448

[43] Coates J 2000 Interpretation of infrared spectra, a practical approach. *Encyclopedia of Analytical Chemistry* (Chicester: John Wiley & Sons, Ltd.)
[44] Kong J and Yu S 2007 Fourier transform infrared spectroscopic analysis of protein secondary structures *Acta Biochimica et Biophysica Sinica* **39(8)** 549-559
[45] Nalinanon S, Benjakul S, Kishimura H and Osako K 2011 Type I collagen from the skin of ornate threadfin bream (*Nemipterus hexodon*): Characteristics and effect of pepsin hydrolysis *Food Chemistry* **125** 500-507
[46] Katili A S 2009 Structure and function of collagen proteins *Jurnal Pelangi Ilmu* **2(5)** 19-29

Acknowledgments
The authors thank to Department of Agroindustrial Technology of Bogor Agricultural University for its financial support and facilities for conducting the research.

IOP Conf. Series: Earth and Environmental Science **335** (2019) 012019 doi:10.1088/1755-1315/335/1/012019

Antioxidant activity and phytocemical analysis from black rice bran

Amirotul Muniroh[1], Slamet Budijanto[1,3], Bambang Pontjo Priosoeryanto[2]

[1)]Food Science and Technology Department, Faculty of Agricultural Technology, Bogor Agricultural University, Bogor 16680, Indonesia e-mail: amirotul.muniroh@gmail.com
[2)]Division of Pathology; Department of Veterinary Clinics, Reproduction and Pathology (CRP), Faculty of Veterinary Medicine, Bogor Agricultural University, Bogor 16680
[3)]Center for South East Asia Food and Agricultural Science and Technology (Seafast Center), Bogor Agricultural University, Bogor 16680, Indonesia

Abstract. Black rice bran is a one of potential antioxidant component source. Black rice bran bioactivity is better than white rice bran. However, the high bioactivity of black rice bran is not accompanied by its utilization as a functional food ingredient. The ability of R. oryzae to degrade lignocellulose and polysaccharide matrix through its enzyme activity is proven to increase antioxidant activity on rice bran variety of white rice. The capability of these molds on the black rice bran matrix is expected to maximize the potential functional properties of antioxidants. This study aims to determine the content of phytochemicals and antioxidants from black rice bran and the comparison between before and after being given fermentation treatment with addition of *Rhyzopus oryzae.* In this study, black rice bran was given two different treatment that is non-fermentation and fermentation using *Rhizopus oryzae* for 96 hours. After the cultivation period, each sample was extracted with methanol solvent. Further phytochemicals including flavonoid, alkaloid, tannin, saponin, quinone, steroid and triterpenois were evaluated by qualitative phytochemical tests while the antioxidant capacity was evaluated by DPPH assay. The results showed that the fermented black rice bran samples contained flavonoids, saponins and triterpenoid while non-fermented black rice bran sample contained flavonoids, tannins and steroids. Antioxidant activity of fermented black rice bran (67,37%) was more effective to inhibit free radicals than non fermented black rice bran (43,93%). According to the result, studies are needed to produce a black rice bran functional food product that can be used as a source of food containing bioactive compouns and antioxidants.

Keywords: Black Rice Bran, *Rhizopus oryzae*, Phytocemical, DPPH

1. Introduction

Black rice bran is a one of potential antioxidant component source. Black rice bran bioactivity is better than white rice bran. However, the high bioactivity of black rice bran is not accompanied by its utilization as a functional food ingredient. Fermentation process have been studied for many decades. a Simple technique for the production of bioactive compounds was solid state fermentation. It is helps reduce the environmental impact of their disposal and economically viable due to the use of agro-industrial residues [8,11]. Degradation of structural components of the cell wall during fermentation is proven to be able to free phenolic bound and increase free phenolic components. The high antioxidant capacity in fermented black rice bran samples is thought to be a consequence of the increase in free phenolic components such as ferulic acid, anthocyanin, and oryzanol. According to [6,9], ferulic acid as the main phenolic compound in black rice bran have been reported to act as radical scavengers against free radical. The ability of R. oryzae to degrade lignocellulose and polysaccharide matrix through its enzyme activity is proven to increase antioxidant activity on rice bran variety of white rice. The capability of these molds on the black rice bran matrix is expected to maximize the potential functional properties of antioxidants. This study aims to determine the content of phytochemicals and antioxidants from black rice bran and the comparison between before and after being given fermentation treatment with addition of *Rhyzopus oryzae.*

2. Material and Methods

2.1. Fermentation of the rice bran

2.1.1. Preparation of the rice bran. Black rice bran (cempo ireng variety) as the main ingredient, was obtained in the milled dry grain from Cigudeg farmers, Bogor, Indonesia. The grain was peeled by milling it with huller machine and then put on a Satake polisher machine for 1.5 minutes every 200 grams. From this process, two outputs will be obtained rice and mixture of rice bran. The mixture of rice bran was sieved with a 20 mesh sieve. The obtained black rice bran is ready to proceed to the next stage.

2.1.2. Preparation of Inoculum. The fungus R. oryzae from IPB culture collection were cultured in PDA (potato dextrose agar) media. R. oryzae culture used in the fermentation process is a culture that is 7 days old and has entered the sporulation phase so that the mold isolate must be refreshed before use. The fungus culture refresher technique is one ose of each mold culture from PDA media as long as it is scratched on to oblique the new PDA media. The new PDA media was incubated for 7 days at 30 °C. After 7 days, the mold inoculum is ready to be used as a spore suspension at the stage of the fermentation process. The spore were spread by adding 10 mL of an sterile disttilled water then the spores counted to reach 106 spores/mL. In the fermentation process, the suspension solution is ready to be inoculated into the black rice bran substrate [10].

2.1.3. Fermentation process. The fermentation technique used in this study is solid media fermentation. The fermentation process begins with a petri dish which is entered by 50 grams of black rice bran and added with 25 mL of water (50% of the weight of the substrate) [10]. Bran substrate was sterilized at 121°C for 15 minutes using autoclave. Furthermore, the substrate is cooled at 30°C and the substrate is ready for the inoculation stage. Fifteen percent of spores (106 spores/mL) from each liquid R. oryzae culture were inoculated into petri dishes containing sterile rice bran then stirred evenly with a sterile spatula. The inoculated rice bran was incubated at 30°C for 4 days (96 hours) [12].

2.2. Phytochemical Analysis

2.2.1. Flavonoid. Black rice bran was added distilled water, then heated at 100°C for 3-5 minutes. After cold, the solution is filtered. Ethanol was added to the filtrate, Mg powder, amyl alcohol and concentrated HCl. the filtrate mixture is shaken homogeneously. A positive test for flavonoids is indicated by the orange color in the solution.

2.2.2. Alkaloid. Black rice bran added NH_3 and $NHCl_3$, then shaken homogeneously and filtered. The filtrate is added with H2SO4 2 M, then shaken homogeneously. The acid layer (the top) is pipetted in another test tube. Alkaloid positive test is indicated by the formation of precipitated orange to brown red after penetrated by Dragendorf reagent, white precipitate after Meyer reagent drops or deposition of light brown to yellowish after Wagner reagent drops. The standard used is Tapak Dara leaves.

2.2.3. Tannin. Black rice bran was added distilled water, then heated at 100°C for 3-5 minutes. After cold, the solution is filtered. The resulting filtrate was added with $FeCl_3$ 10%. Blackish green solution indicates tannins.

IOP Conf. Series: Earth and Environmental Science **335** (2019) 012019 doi:10.1088/1755-1315/335/1/012019

2.2.4. Saponin. Black rice bran was added distilled water, then heated at 100°C for 3-5 minutes. After cold, the solution is filtered. The filtrate is shaken vigorously. The positive saponin test is indicated by the formation of a stable foam for 30 seconds.

2.2.5. Quinon. Black rice bran was added metanol, then heated at 100°C for 3-5 minutes. After cold, the solution is filtered. The filtrate was added NaOH 10%. The positive quinon test is indicated by the red color in solution.
Steroid and triterpenoid. Black rice bran was added etanol, then heated at 100°C for 3-5 minutes. After cold, the solution is filtered. The filtrate is heated until the extract is produced. The extract was added diethyl eter and shake homogeneously. The solution is poured into pourcelain, was added H_2SO_4 and CH_3COOH anhydrous. A positive test for steroid is shown when it color was blue or green. While purple or orange color was positive result for triterpenoid.

2.3. Antioxidant activity of black rice bran
2.3.1. Extraction process. The extraction techniques used is maceration. Black rice bran were place in an erlenmeyer and added methanol solvents. During extraction process, erlenmeyer was covered by alumunium foil. This process was carried out for 3 hours with orbital platform shaker innova 2300 (New Brunswick Scientific, USA) at room temperature. After that, for 15 minutes, black rice bran were resuspend in an ultrasonic bath (Branson 8510) with distilled water. Furthermore, the extract obtained was filtered through filter paper (Whatman no. 1). After centrifugation (10 m, 1791 g, room temperature), filtrate was separated and put into a 100mL flask. The extract has been evaporated with Rotavapor (R-300 BUCHI, Switzerland) at temperature 50°C until all solvent is evaporated.

2.3.2. DPPH assay. The antioxidant activity of black rice bran was measured by the reduction of free radicals 1.1-diphenyl-2-pikrihidrazil (DPPH) according to the method described [1]. The DPPH radical scavenging capacity of samples was defined as Trolox Equivalent Antioxidant Capacity (mg TEAC per g dry weight samples). The DPPH radical scavenging capacity assay was performed by 96 well plate method. Each sample (20 μL) were transferred into 96 well plate and added working solution DPPH (180 μL). Furthermore, the sample incubated at the dark room temperature for 40 min . The absorbance was measure by Epoch microplate spectrophotometer (BioTek, USA) at 515 nm. The percentage of DPPH radical scavenging was measured by following equation:

$$\%\text{Scavenging}=\left[1-\left(\frac{A_{\text{sample}}-A_{\text{blank}}}{A_{\text{control}}-A_{\text{blank}}}\right)\right]\times 100 \quad (1)$$

This method begin with is based on the transfer of electrons from one antioxidant substance to a free radical, DPPH, which loses its purple color upon reduction, becoming yellow.

2.4. Statistical
One way Analysis of Variance (ANOVA) test was used to determine significant differences between variable. Differences with a probability value of <0.05 were considered significant and all data were reported as mean ± sd.

3. Result and discussion
3.1. Phytochemical Analysis
Phytochemical compounds are natural bioactive compounds produced by secondary metabolites found in plants that act as nutrients and natural fibers that can prevent disease. Phytochemical compounds commonly found in plants, namely alkaloids, flavoniods, quinones, tannins, polyphenols, saponins, steroids and triterpenoids [5]. In this study, analysis of black rice bran phytochemicals was carried out based on [5]. Identification carried out is the test of alkaloids, flavonoids, saponins, tannins, quinones, steroids and triterpenoids. Fermented black rice bran is shown positive result to flavonoid, saponin and

IOP Conf. Series: Earth and Environmental Science **335** (2019) 012019 doi:10.1088/1755-1315/335/1/012019

triterpenoid compounds. Meanwhile unfermented black rice bran is shown positive result to flavonoid, tannin, and steroid compounds. Flavonoids are commonly found in plants, bound to sugar as glycosides and aglycone flavonoids (sugar-bound flavonoids) are present various forms of structure [7]. It is known that antioxidant activity from plants is due to their presence phenol compounds. The antioxidant effect of this compound is caused by capture free radicals through donor hydrogen atoms from flavonoid hydroxyl groups. Antioxidants are compounds that protect cells against damage caused reactive oxygen. The imbalance between antioxidants and reactive oxygen resulting in oxidative stress, which causes cell damage [3].

Table 1. Phytochemical analysis of black rice bran.

Parameter		Fermented Black Rice Bran	Non-fermented Black Rice Bran
Flavonoid		Positive	Positive
Alkaloid	Wagner	Negative	Negative
	Mayer	Negative	Negative
	Dragendorf	Negative	Negative
Tannin		Negative	Positive
Saponin		Positive	Negative
Quinon		Negative	Negative
Steroid		Negative	Positive
Triterpenoid		Positive	Negative

Phytochemical compounds play an important role in maintaining health. These compounds complement each other in the mechanism of action that occurs in the body. Phytochemical compounds have antioxidant activity, detoxification by enzymes, stimulation of the immune system, hormone metabolism and antibacterial and antiviral. Several studies have shown that phytochemicals are contained in fruits, vegetables and nuts. These bioactive components can inhibit the process of premature aging and reduce the risk of various diseases, such as cancer, liver disease, stroke, high blood pressure, cataracts, osteoporosis and gastrointestinal infections [4].

3.2. Antioxidant Activity

The antioxidant activity of the black rice bran was evaluated based on [1] method. Primary antioxidant or "chain-breaking antioxidant" acts as a free radical scavenger through termination of a chain oxidation reaction by donating H atoms to free radical compounds to produce more stable products[2].

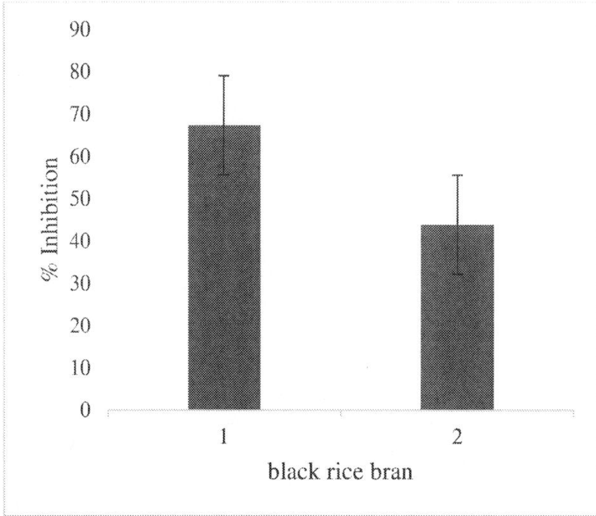

Figure 1. Inhibition by antioxidant of black rice bran

Antioxidant activity of fermented black rice bran (1) was more effective to inhibit free radicals than non fermented black rice bran (2). During fermentation, degradation of structural components of the cell wall is proven to be able to free phenolic bound and increase free phenolic components. The high antioxidant capacity in fermented black rice bran (67,37%) is thought to be a consequence of the increase in free phenolic components such as ferulic acid, oryzanol, and anthocyanin. Phenolic components such as ferulic acid (the main phenolic compound in black rice bran) have been reported to act as radical scavengers against radical DPPH [6,9].

4. Conclusion

Fermented black rice bran contained flavonoids, saponins and triterpenoid while non-fermented black rice bran sample contained flavonoids, tannins and steroids. Antioxidant activity of fermented black rice bran was more effective to inhibit free radicals than non fermented black rice bran.

References

[1] Bobo-Garcia G, Daidov-Pardo G, Arroqui C, Virseda P, Marin-Arroyo M R and Navarro M 2014 Intra-laboratory validation of microplate methods for total phenolic content and antioxidant activity on polyphenolic extracts, and comparison with conventional spectrophotometric methods *Journal Science Food Agriculture* **95(1)** 204-209

[2] Craft B C, Kerrihard A L, Amarowicz R and Pegg R B 2012 Phenol-based antioxidants and the in vitro methods used for their assessment *Compr Rev Food Sci Food Saf* **11** 148-173

[3] Cristobal M and Donald R 2000 *Aktivitas Antioksidan Flavonoid* (Oregon: Oregon State University Pr.)

[4] Hamburger M and Hostettmaun K 1991 Bioactivity in plants: the link between phytochemistry and medicine *Phytochemical* **30(12)** 3864-3874.

[5] Harborne J B 1987 *Metode Fitokimia* (Bandung: Institut Teknologi Bandung)

[6] Jun H I, Shin J W, Song G S and Kim Y S 2015 Isolation and identification of phenolic antioxidants in black rice bran *Journal of Food Science* **80(2)** C262-C268

[7] Markham K R 1988 *Techniques of Flavonoid Identification* (London: Academic Pr.)

[8] Oliveira M S, Kupski L, Feddern V, Cipolatti E, Badiale-Furlong E, and Souza-Soares L A 2010 Physico-chemical characterization of fermented rice bran biomass CyTa - *Journal of Food* **8** 236–269

[9] Razak D L A, Rashid N Y A R, Jamaluddin A, Sharifudin S A and Long K 2015 Enhancement of phenolic acid content and antioxidant activity of rice bran fermented with *Rhizopus oligosporus* and *Monascus purpureus Biocatal Agric Biotechnol* **4** 33-38.
[10] Schmidt C G, Goncalves L M, Prietto L, Hackbart H S and Furlong E B 2014 Antioxidant activity and enzyme inhibition of phenolic acids from fermented rice bran with fungus *Rizhopus oryzae Food Chemistry* **146** 371-377
[11] Schmidt C G, and Furlong E B 2012 Effect of particle size and ammonium sulfate concentration on rice bran fermentation with the fungus Rhizopus oryzae *Bioresource Technology* **123** 36-41.
[12] Zulfafamy K E, Ardiansyah and Budijanto S 2018 Antioxidative properties and cytotoxic activity against colon cancer cell widr of *Rhizopus Oryzae* and *Rhizopus Oligosporus*-fermented black rice bran extract *Curr Res Nutr Food Sci* **6(1)**

Acknowledgments
Author thank to Direktorat Jenderal Pendidikan Tinggi RI for financial support.

IOP Conf. Series: Earth and Environmental Science **335** (2019) 012020 doi:10.1088/1755-1315/335/1/012020

Tracer study of CO_2 enrichment for shallot growth in mini plant factory with CFD approach

A G Niam[1], S Widodo[1], M Solahudin[1], L Sucahyo[1] and S Nurjanah[2]

[1]Department of Mechanical and Biosystem Engineering, Bogor Agricultural University, West Java, Indonesia, email:ghautsun@apps.ipb.ac.id
[2]Student of a bachelor degree in Agricultural and Biosystem Engineering program, Bogor Agricultural University

Abstract. It has been proven that the effect of CO_2 enrichment leads to faster growth and higher yields. Using environmental parameters controlling based on needs, responses, and performance of crops, CO_2 level from 800 to 1500 ppm were able to accelerate the growth of shallot in mini plan factory. Therefore, CFD simulation was conducted to simulate and trace the distribution of CO_2 with varying airflow inlet. Airflow inlet is two wall air conditioners (ACs) with the 2 pk specification. Both of ACs was attached at opposite walls. There were two combinations of ACs position, which are top-top (both of ACs were positioned approximately 10 cm below the ceiling) and top-center (an AC positioned approximately 10 cm below the ceiling and another positioned 60 cm below the ceiling). In the simulation, CO_2 gas was injected into the inlet from 0.5 $l\ min^{-1}$ to 2 $l\ min^{-1}$. The simulation was used for tracing CO_2 injection and position that has the highest uniformity coefficient at the CO_2 levels of around 800 ppm. It is proven through a numerical approach, the best ACs position is a top-top combination, with CO_2 injected, is 2 $l\ min^{-1}$.

Keywords: CO_2 level, combination of ACs position, simulation, uniformity coefficient.

1. Introduction

The value of CO_2 gas concentration has an important role in understanding the behavior of plant growth, where the value of CO_2 gas concentration can be used as a key parameter in the photosynthesis so that the physiology of leaves can be monitored periodically [1]. The positive effects of the increase of CO_2 gas in the Earth's atmosphere are the increase of plant biomass production [2]. Therefore, it can be said that the phenomenon of CO_2 gas exchange on leaves can describe a variety of parameters on the ecophysiological of the plant [3]–[7], both at the molecular level, biochemistry, leaf tissue and canopy [7]–[9].

The effectiveness of CO_2 gas exchange on the leaves is very influential on plant photosynthesis rate which is a key factor of plant growth activity [10], so the effort to increase crop productivity can be done by enriching CO_2 gas in plant canopy [11]. By enriching the carbon dioxide gas concentration of about 1000 ppm, the cumulative biomass of tomato plants can increase by about 30-60% per plant [3]. But when the light intake conditions are small (like during the winter season), the recommended CO_2 concentration for tomato plants is around 700-900 ppm [4], [12], while for *Chrysanthemum* and *Saintpaulia ionantha* flowers, CO_2 enrichment is 900 ppm apparently able to reduce flowering time and increase the number of flowers significantly [12]. This proves that environmental control grows in the form of enrichment of CO_2 is capable of increasing plant growth, both in its vegetative and generative stages. It shows that this method has an important role in environmental engineering efforts to plant growth.

The shallot seed production system in Indonesia, in general, is still dependent on climate and topographic conditions [13], so the alternative production with environmental control efforts on the growth of shallots is a promising expansion of production sources. The alternative recommended in this article is the development of a mini plant factory technology package for shallots production in a controlled environment. One important parameter in the design of the technology package is the distribution model of CO_2 concentration in the growing environment. This parameter is necessary to ensure the supply of CO_2 to cultivated plants. A design tool that can describe and visualize the distribution of CO_2 gas concentrations in 3D is by CFD simulation approach.

CFD simulation approach proved to be able to visualize some physical parameters of plant growth environment such as temperature distribution in the root area [14], [15] to design the root-zone cooling system, the distribution of temperature and relative humidity on the leaf surface [16], air flow through the canopy a plant [17], [18], assessment of the application of natural ventilation [19], [20][21], and mechanical ventilation [22]–[24], the study of quality and concentration of air in a room [25]–[28], even assessment for sprayer applications on plants [29], [30]. Therefore, this article focuses on the tracer study of CO_2 enrichment for shallot growth in a mini plant factory with CFD approach.

2. Description of Mini Plant Factory

The factory with artificial lighting (PFAL) is defined as a closed crop production system with a building structure equipped with thermal insulators that are not translucent, minimal ventilation and a light source for plant growth using artificial light [31], [32]. The concept is based on the concept of a closed plant production system (CPPS) which consists of six important elements [33], namely: 1) impermeable wall structure, 2) multitier system, generally 4-8 layers with a height of about 40 cm between layers, 3) cooling device and dehumidifier as well as air circulator, 4) CO_2 enrichment units, 5) fertilizer system and 6) fertilizer controller.

The developed mini plant factory was made in a 40 ft. container consisting of a control room, chamber room (cultivation room) and nutrient room. The simulated room is only a chamber with a size of 672 x 212 x 261 cm. The room is equipped with 6 planting media shelves in the form of a substrate with a drip irrigation system. Each layer of the plant shelf is equipped with 4 units of lights as a lighting source for the photosynthesis process. The cooling source in the cultivation room comes from 2 AC units with a capacity of 2 pk. The position of the AC is placed on the front and rear walls with two different combinations, namely the top-top and top-center. The top position of the air conditioner is about 10 cm from the ceiling right in the middle of the wall, while the center position is around 60 cm from the ceiling. CO_2 gas with a variety of flow rates is exhaled through the blowing of air from both air conditioners to the chamber using a water pass pipe so that the flow of CO_2 gas will follow the air blowing from the air conditioner. Variations in CO_2 flow rates from 0.5 liters/minute to 2 liters/minute.

IOP Conf. Series: Earth and Environmental Science **335** (2019) 012020 doi:10.1088/1755-1315/335/1/012020

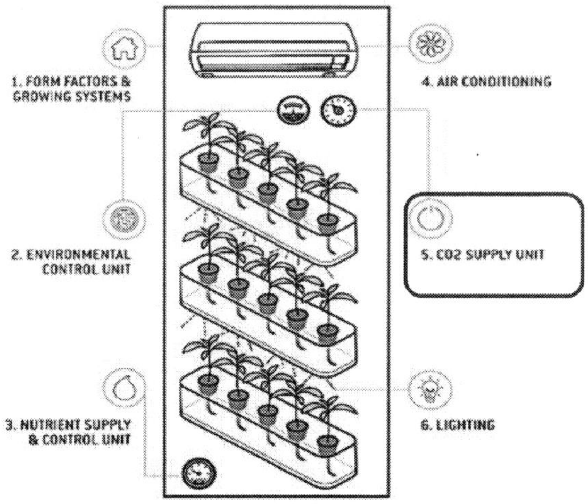

Figure 1. Six important elements of a closed plant production system [1]

3. Numerical simulation

CFD simulation has been done on Flow Simulation Solidworks software (Solidworks inc., 2016). The commonly used mathematical model is the Reynolds Averaged Navier-Stokes (RANS) equation which includes energy conservation, mass and momentum conservation with the finite volume method approach.

The chambers for plants in containers are impermeable, so that the air flow in the room can be laminar or turbulent. Therefore, the Flow Simulation Solidworks software considers laminar and turbulent flow, and to predict turbulent flow, Solidworks uses the Favre-Average Navier-Stokes (FANS) approach. However, for mixing gas concentrations inside the chamber that comes from injection of carbon dioxide gas, the approach taken is diffusion by involving a Lewis number, so that it can describe the concentration of the carrier gas and its essential gas, in this case carbon dioxide gas.

$$\frac{\partial \rho y}{\partial t} + \frac{\partial}{\partial x_i}\left[\rho y u_i - \frac{\rho R T}{p m}\left(\frac{\mu}{Pr\,Le} + \frac{\mu}{Pr_t\,Le_t}\right)\frac{\partial y}{\partial x_i}\right] = \frac{\partial}{\partial x_i}\frac{m_1 m_2}{m^2}\left[\frac{\rho y v_1 - y}{p}\left(\frac{\mu}{Pr\,Le} + \frac{\mu}{Pr_t\,Le_t}\right)\frac{\partial p}{\partial x_i}\right] \quad (1)$$

where, is the carrier gas and the essential mixture density; p is the carrier gas static pressure, R is universal gas constant, m is carrier gas and the essential mixture molar mass, m_1 is the essential gas molar mass, m_2 is the carrier gas molar mass, v_1 is the specific volume of the essential gas, and $_t$ are laminar and turbulent viscosity, Pr and Pr_t are laminar and turbulent Prandtl numbers, Le and Le_t are laminar and turbulent Lewis numbers, x_i is the i-th component of coordinate system, and u_i is velocity at the i-th component.

4. CFD model setup

The air inlet only comes from the AC that is carried with the CO_2 gas inlet. The mass flow rate of air from AC based on its specifications is around 17.7 $m^3 min^{-1}$, where by default the CO_2 content in the air is defined as 380 ppm. Then, pure CO_2 gas is injected into both ACs which are applied with the same mass flow rate, starting from 0.5 $l\ min^{-1}$ to 2 $l\ min^{-1}$. The combination and input scenario in this CFD simulation can be seen in table 1.

Table 1. The input scenario combination of simulation.

AC position	Mass flow rate CO_2 ($l\ min^{-1}$)
top-top	0.5, 1, 1.5, 2,
top – center	0.5, 1, 1.5, 2,

Accuracy in the CFD simulations strongly influenced by the appropriate set of boundary condition exerts. The surface material in the chamber (including the surface of rack materials, side walls, floor, and ceiling) was defined as impermeable no-slip walls or real walls.

Simplifying the model in the simulation is something that cannot be avoided, so that the problems in the iteration can be solved easily. Some of the assumptions in this simulation are: 1), incompressible, 2) steady state, 3) singe phase or no chemical reaction between essential gas and carrier gas, 4) CO_2 absorption by plants ignored, and 5) wall temperatures are constant.

5. Result and Discussion

Calculations in this simulation converge at 1.7 million mesh cells. Convergence is obtained by doing a mesh independency study to verify numerical calculation results or in other words reduce the error value of the iteration results and the parameter values obtained have shown a consistent number even though there is an increase in the number of cells. The main parameter of the calculation is the value of the concentration of carbon dioxide. The local domain that is considered is the space where the plant grows or the canopy zone of plant, so that in this case the calculated data that is noticed is in the zone of about 10 cm from the surface of the planting media found on each plant shelf. Plant shelf consists of 3 layers.

Levels of mass fraction of carbon dioxide gradients are presented in 15 levels with a range of values from 400 ppm to 1000 ppm. This is considered to represent the CO_2 enrichment target, ie until the optimum rate of photosynthesis is around 800 ppm until 1000 ppm. In addition to the concentration of CO_2, other parameters also contribute to the process of photosynthesis and respiration, such as temperature and relative humidity. The distribution of CO_2 concentration is presented in figure 2 with a parameter of mass fraction of carbon dioxide with ppm units. The image is a side view with a difference of 2 AC height locations; top-center and top-top.

Figure 2 shows that in general the concentration of CO_2 gas spreads fairly evenly between the plant shelves and moves towards the surface of the floor when the CO_2 gas is injected from the AC unit. This proves that the gravity factor calculated in the simulation works well. In addition, the spread of CO_2 gas to various corners of the room also proves that the fluid moves diffusively in addition to being affected by air blowing from the air conditioner.

The concentration of CO_2 for 0.5 $l\ min^{-1}$ input on top-top and top-center layouts is approximately 680 ppm to 742 ppm, and 572 ppm until 618 ppm. In scenario 2 (where CO_2 is injected 1 $l\ min^{-1}$) the CO_2 concentration is around 740 ppm until 786 ppm (top-top), and 732 ppm until 780 ppm (top-center). The concentration increase is close to the target value of the optimum CO_2 concentration. In general observations, the distribution of CO_2 that can be categorized as reaching the target value of optimum growing environmental requirements is found in scenarios 3 and 4 (1.5 $l\ min^{-1}$ and 2 $l\ min^{-1}$) which is around 800 ppm until 910 ppm. But on the other hand, it should also be considered the uniformity factor to ensure that the distribution of CO_2 gas can be accessed by all cultivated plants, so that uniformity of plant growth can be achieved. In general, the coefficient of CO_2 concentration uniformity of the simulation results is very good, which is above 98%. But the best value of uniformity is found at the value of the flow rate of 2 $l\ min^{-1}$ of carbon dioxide, where the uniformity of the top-top is almost the same as the top-center which is 99.78% and 99.82% respectively. The coefficient of uniformity for each flow rate can be seen in figure 3.

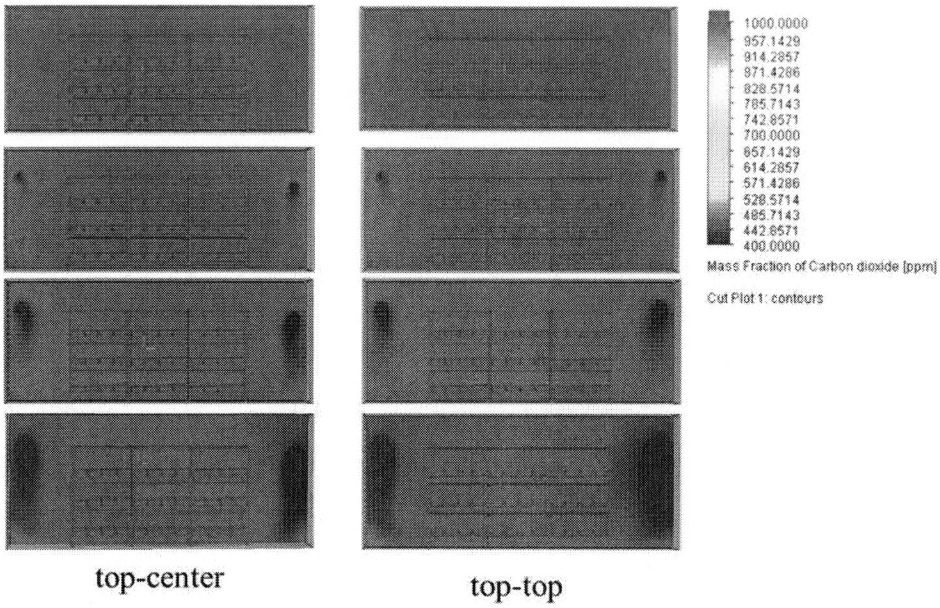

Figure 2 The distribution of carbon dioxide.

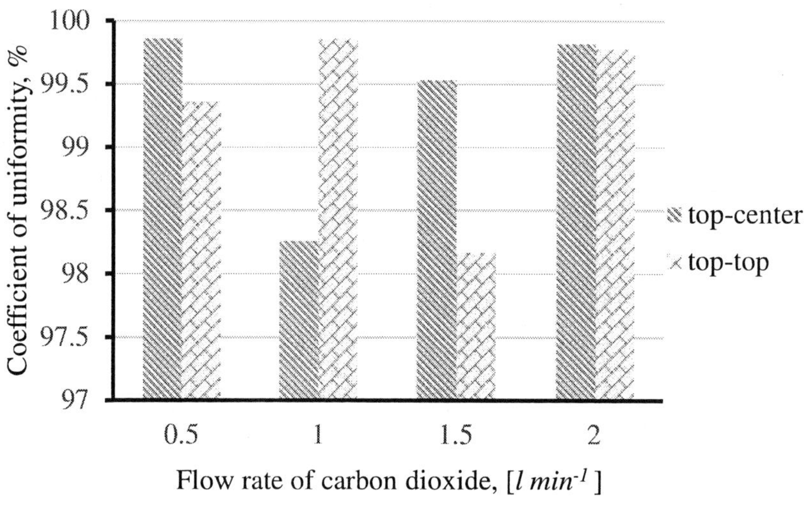

Figure 3 Coefficient of uniformity for variation of carbon dioxide flow rate.

6. Conclusion
The best scenario in designing the ACs position and carbon dioxide discharge in the mini plant factory respectively is the top-top position and 2 $l\ min^{-1}$.

References
[1] T. D. Sharkey, C. J. Bernacchi, G. D. Farquhar, and E. L. Singsaas, "Fitting photosynthetic carbon dioxide response curves for C3 leaves," *Plant, Cell Environ.*, vol. 30, no. 9, pp. 1035–1040, 2007
[2] J. S. Amthor, "Perspective on the relative insignificance of increasing atmospheric CO_2 concentration to crop yield," *F. Crop. Res.*, vol. 58, no. 2, pp. 109–127, 1998
[3] S. Nilsen, K. Hovland, C. Dons, and S. P. Sletten, "Effect of CO_2 enrichment on photosynthesis, growth, and yield of tomato," *Sci. Hortic. (Amsterdam).*, vol. 20, no. 1, pp. 1–14, 1983
[4] L. M. Mortensen, "Review: CO_2 enrichment in greenhouses. Crop responses," *Sci. Hortic. (Amsterdam).*, vol. 33, no. 1–2, pp. 1–25, 1987
[5] J. LI, J.-M. ZHOU, Z.-Q. DUAN, C.-W. DU, and H.-Y. WANG, "Effect of CO_2 Enrichment on the Growth and Nutrient Uptake of Tomato Seedlings," *Pedosphere*, vol. 17, no. 3, pp. 343–351, 2007
[6] E. Mateos-Naranjo, S. Redondo-Gómez, L. Andrades-Moreno, and A. J. Davy, "Growth and photosynthetic responses of the cordgrass Spartina maritima to CO_2 enrichment and salinity," *Chemosphere*, vol. 81, no. 6, pp. 725–731, 2010
[7] E. Nederhoff, "Carbon Dioxide Enrichment," *Hydroponics and Greenhouses*, no. 76, pp. 50–59, 2004
[8] M. C. Sánchez-Guerrero, P. Lorenzo, E. Medrano, A. Baille, and N. Castilla, "Effects of EC-based irrigation scheduling and CO_2 enrichment on water use efficiency of a greenhouse cucumber crop," *Agric. Water Manag.*, vol. 96, no. 3, pp. 429–436, 2009
[9] M. C. Sánchez-Guerrero, P. Lorenzo, E. Medrano, N. Castilla, T. Soriano, and A. Baille, "Effect of variable CO_2 enrichment on greenhouse production in mild winter climates," *Agric. For. Meteorol.*, vol. 132, no. 3–4, pp. 244–252, 2005
[10] Y. Kitaya, T. Shibuya, M. Yoshida, and M. Kiyota, "Effects of air velocity on photosynthesis of plant canopies under elevated CO_2 levels in a plant culture system," *Adv. Sp. Res.*, vol. 34, no. 7 SPEC. ISS., pp. 1466–1469, 2004
[11] T. Shibuya, J. Tsuruyama, Y. Kitaya, and M. Kiyota, "Enhancement of photosynthesis and growth of tomato seedlings by forced ventilation within the canopy," *Sci. Hortic. (Amsterdam).*, vol. 109, no. 3, pp. 218–222, 2006
[12] L. M. Mortensen, "Effect of intermittent as compared to continuous CO_2 enrichment on growth and flowering of Chrysanthemum × morifolium Ramat. and Saintpaulia ionantha H. Wendl," *Sci. Hortic. (Amsterdam).*, vol. 29, no. 3, pp. 283–289, 1986
[13] Suwandi, *Shallot Cultivation Outside the Season: Leading Technology Anticipating the Impact of Climate Change.* IAARD Press, 2014
[14] A. Niam, H. Suhardiyanto, K. Seminar, and A. Maddu, "Simulation of Temperature Distribution on the Planting Hole of Floating Hydroponic for Shallot Production in Tropical Lowland," *JTEP*, vol. 5, no. 3, pp. 235–244, 2017
[15] A. Niam, H. Suhardiyanto, K. Seminar, and A. Maddu, "CFD Simulation of Cooling Pipes Distance in The Growing Medium for Hydroponic Substrate in Tropical Lowland CFD Simulation of Cooling Pipes Distance in The Growing Medium for Hydroponic Substrate in Tropical Lowland," *IJERD*, vol. 15, no. February, pp. 56–63, 2017
[16] J. C. Roy, C. Vidal, J. Fargues, and T. Boulard, "CFD based determination of temperature and humidity at leaf surface," *Comput. Electron. Agric.*, vol. 61, no. 2, pp. 201–212, 2008
[17] A. M. Endalew, M. Hertog, M. A. Delele, K. Baetens, T. Persoons, M. Baelmans, H. Ramon, B. M. Nicolaï, and P. Verboven, "CFD modeling and wind tunnel validation of airflow through plant

canopies using 3D canopy architecture," *Int. J. Heat Fluid Flow*, vol. 30, no. 2, pp. 356–368, 2009

[18] H. Fatnassi, C. Leyronas, T. Boulard, M. Bardin, and P. Nicot, "Dependence of greenhouse tunnel ventilation on wind direction and crop height," *Biosyst. Eng.*, vol. 103, no. 3, pp. 338–343, 2009

[19] F. D. Molina-Aiz, H. Fatnassi, T. Boulard, J. C. Roy, and D. L. Valera, "Comparison of finite element and finite volume methods for simulation of natural ventilation in greenhouses," *Comput. Electron. Agric.*, vol. 72, no. 2, pp. 69–86, 2010

[20] F. D. Molina-Aiz, D. L. Valera, A. A. Peña, J. A. Gil, and A. López, "A study of natural ventilation in an Almería-type greenhouse with insect screens by means of tri-sonic anemometry," *Biosyst. Eng.*, vol. 104, no. 2, pp. 224–242, 2009

[21] W. A. Y. Mousa, W. Lang, T. Auer, and W. A. Yousef, "A pattern recognition approach for modeling the air change rates in naturally ventilated buildings from limited steady-state CFD simulations," *Energy Build.*, vol. 155, pp. 54–65, 2017

[22] M. J. Ferrua and R. P. Singh, "Improving the design and efficiency of the forced-air cooling process of fresh strawberries using computational modeling," *Procedia Food Sci.*, vol. 1, no. Icef 11, pp. 1239–1246, 2011

[23] M. Teitel, M. Atias, and M. Barak, "Gradients of temperature, humidity and CO_2 along a fan-ventilated greenhouse," *Biosyst. Eng.*, vol. 106, no. 2, pp. 166–174, 2010

[24] D. H. Willits, "Cooling fan-ventilated greenhouses: A modeling study," *Biosyst. Eng.*, vol. 84, no. 3, pp. 315–329, 2003

[25] Y. Dai, C. M. Mak, Z. Ai, and J. Hang, "Evaluation of computational and physical parameters influencing CFD simulations of pollutant dispersion in building arrays," *Build. Environ.*, vol. 137, pp. 90–107, 2018

[26] L. Rong, P. V. Nielsen, B. Bjerg, and G. Zhang, "Summary of best guidelines and validation of CFD modeling in livestock buildings to ensure prediction quality," *Comput. Electron. Agric.*, vol. 121, pp. 180–190, 2016

[27] L. Dong, H. Zuo, L. Hu, B. Yang, L. Li, and L. Wu, "Simulation of heavy gas dispersion in a large indoor space using CFD model," *J. Loss Prev. Process Ind.*, vol. 46, pp. 1–12, 2017

[28] G. Xu, E. C. Jong, K. D. Luxbacher, S. A. Ragab, and M. E. Karmis, "Remote characterization of ventilation systems using tracer gas and CFD in an underground mine," *Saf. Sci.*, vol. 74, pp. 140–149, 2015

[29] A. M. Endalew, C. Debaer, N. Rutten, J. Vercammen, M. A. Delele, H. Ramon, B. M. Nicolaï, and P. Verboven, "A new integrated CFD modeling approach towards air-assisted orchard spraying. Part I. Model development and effect of wind speed and direction on sprayer airflow," *Comput. Electron. Agric.*, vol. 71, no. 2, pp. 128–136, 2010

[30] A. M. Endalew, C. Debaer, N. Rutten, J. Vercammen, M. A. Delele, H. Ramon, B. M. Nicolaï, and P. Verboven, "A new integrated CFD modeling approach towards air-assisted orchard spraying-Part II: Validation for different sprayer types," *Comput. Electron. Agric.*, vol. 71, no. 2, pp. 137–147, 2010

[31] T. Kozai, "Propagation, grafting and transplant production in closed systems with artificial lighting for commercialization in Japan," *Propag. Ornam. Plants*, vol. 7, no. 3, pp. 145–149, 2007.

[32] T. KOZAI, "Resource use efficiency of closed plant production system with artificial light: Concept, estimation, and application to plant factory," *Proc. Japan Acad. Ser. B*, vol. 89, no. 10, pp. 447–461, 2013

[33] T. Kozai, G. Niu, and M. Takagaki, *Plant Factory an indoor vertical farming system for efficient quality food production*. Elsevier Inc., 2016

[34] A. Sobachkin and G. Dumnov, "Numerical Basis of CAD-Embedded CFD," *NAFEMS World Congr. 2013*, no. February, pp. 1–20, 2013

[35] W. Malalasekera and H. K. Versteeg, *An Introduction to Computational Fluid Dynamics - The Finite Volume Method*, vol. 44, no. 2. 2006

Acknowledgments

This work was funded by Applied Research Grant from the Ministry of Research, Technology and Higher Education, Indonesia, under contract no: 129/SP2H/PTNBH/DRPM/2018, February 01, 2018. We would like to extend our thanks to the Mechanical and Biosystem Engineering Department, FATETA-IPB, that allowed us to use their facilities.

Development of KMS to support quality shoes production

N R Feta[1], K B Seminar[1] and I Hermadi[1]

[1] Department of Computer Science, Bogor Agricultural University, Indonesia. e-mail: nenengrachmaliafeta@gmail.com

Abstract. PT. PAI is a manufacturing company of Tomkins brand shoes. The company produces four types of shoes such as a child, junior, women, and men, from every kind of shoe there are different shoe names and models. The information or knowledge that each employee has about how to produce high-quality shoes has not been appropriately managed and stored, plus the employee's entry and frequent employee mutations due to shoe production in one of the stages of production, which results in a loss of knowledge. A lot of defective shoe production and reduce production targets. Therefore, this study aims to create a knowledge management system in the production section. The method in this study was Awad & Ghaziri using five main stages of Knowledge Management Life Cycle. Based on the research that has been conducted, a web-based knowledge management system has been successfully created. The system developed is in the form of knowledge sharing and knowledge transfer applications. On the main login page, there are three levels of access rights, namely experts, admin and employee. Based on a series of tests, this system concerning functionality has been by the needs of users in the company and has been built correctly.

Keywords: shoes production, knowledge management system, life cycle

1. Introduction

PT. PAI is a Tomkins brand shoes manufacturing company. This company has a total of 1,190 employees for now. The company produces 4 types of shoes such as *children, junior (teenagers), women (men), men*, from each type of shoe there are different shoes names and models. There is a lot of knowledge used in the production process, such as material process in the warehouse, *cutting* techniques (cutting material), *sewing* (stitching on the shoes), *press* outsole (shoes bottom), *assembling* process (unification of the shoes top and bottom part), quality control criteria, *finishing* and *packing* process. Knowledge of human resources (HR) in producing shoes is very important, if the human resources have high quality, then the shoes production will have high quality as well [1].

Based on the employee's transfer data from company HRD until the end of 2017, 45 employees were transferred from the cutting process, 73 employees were transferred from the sewing process, 58 employees were transferred from the press outsole process, 80 employees were transferred from the assembling process and the other 51 employees were transferred from the laboratory process. The amount of employee's transfer from one part to another causes loss of knowledge from experienced employees before the knowledge was stored and documented properly.

New employees also have a difficulty in learning the used process in the production, because the employees who have the knowledge are often busy so they don't have time to perform the transfer knowledge. The employees knowledge and experience are not documented in any form, but some excel documents are only stored on the local computer, and often only recorded and placed in a box on the archive table, this state causes difficulties when other parts of different buildings look seek the document. This causes frequent errors due to lack of understanding of the production process so that many shoes are defective. Currently, there are no learning process facilities and problem handling facilities available to the employees.

Therefore, this research aims to form a KMS arrangement at PT. PAI, that is how to identify both tacit (unexplored knowledge) and explicit (knowledge that has been documented, stored or recorded) [2], and storing the knowledge so that it can be used as useful information and references for the employees. Web-based applications use a combination of KMS and technology to make data, information, and knowledge become accurately, accessible, and quickly accessed [4].

Previous research related to KMS Development concerning the management solution knowledge for plastic industry, and using identification methods designed by Fernandez and using throwaway prototyping as a methodology for developing KMS prototypes [5]. Another research conducted aims to: propose mass utilization of medical record information system and propose interconnection access for medical records of patients among health institutions throughout Indonesia by developing knowledge management system [6].

Based on the existing problems, researchers will make a Knowledge Management System (KMS) design for the shoes industry using Knowledge Management System Life Cycle (KMSLC) methodology. So that it can produce a web-based system that is appropriate to the conditions in the Company. With the application of KMS, it is expected that each employee can communicate and exchange knowledge which can ultimately improve the company's performance so that the company's quality is maintained.

2. Research Methods

This study chosen the KMSLC concept developed by [7], but it is only performed until the verification and validation stage of the knowledge management system, was presented in Figure 1.

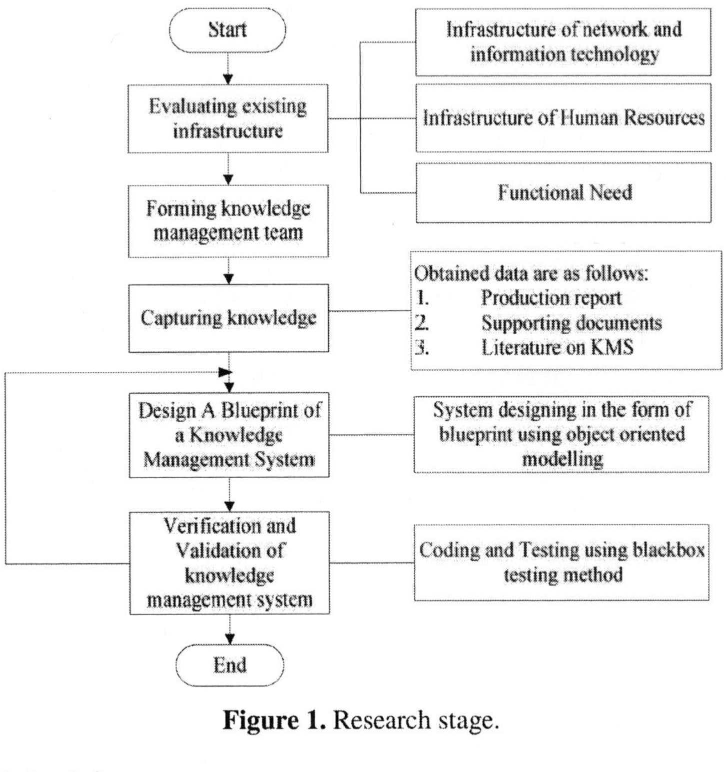

Figure 1. Research stage.

2.1. Evaluating existing infrastructure

Evaluating hardware infrastructure, software infrastructure, network infrastructure, human resources infrastructure and knowledge infrastructure and analyzing the process of knowledge culture that runs in the company [8].

IOP Conf. Series: Earth and Environmental Science **335** (2019) 012021 doi:10.1088/1755-1315/335/1/012021

2.2. Forming a knowledge management team

Forming a knowledge management team that would work together to develop a knowledge management system so it is produced an appropriate KMS design with the company's condition.

2.3. Capturing knowledge
Capturing knowledge by conducting literature studies from knowledge source documents regarding to shoe production. Observation by observing, recording, and translating the process in solving the problems related to the production of defective shoes. And conduct interviews to capture knowledge from the minds of the expert with regard to how the stages in the production process of shoes.

2.4. Design a blueprint of a Knowledge Management System
Creating a design of knowledge management system application.

2.5. Verification and validity of Knowledge Management System
The design conversion of Knowledge Management System that has been designed into the program code. The program will use a codeigniter framework and uses programming language PHP and database MySQL. This stage also verifies whether the knowledge will be implemented and associated with a design is correct with that required by the user. While to know the system is valid or not, then do the test against the functionality of the system by using black box testing methods.

3. Results and Discussions

3.1. Evaluate existing infrastructure

3.1.1. Infrastructure hardware. Based on the results of interviews have been carried out to your IT staff, every production building have the facilities of 2 personal computer (PC) and 3 administrative staff manually assigned to manage the report production of shoes every day. Of 2 PC, 1 PC used by the head of department and 1 PCS used by administration staff to support its performance. To help in the work are also provided internet connection for 24 hours and the photo copy machine. Obtained also some information about the needs of the technology infrastructure needed by companies, which needed a server, application and data base.

3.1.2. Infrastructure software. Based on the results of the interview, the software in the company is as shown in Table 1.

Table 1. Existing software infrastructures.

No.	Software	The Name Of The Server/Domain	Description
1.	Sales Information System	http://www.tomkins.id/	Information systems are used for product promotion and sales of products.
2.	SAP	Desktop Application	*System Application and Product in data processing* are *Enterprise Resources Planning* (ERP) are in use for data *entry* type-the type of raw material in the production of shoes.

3.1.3. Infrastructure of the Human Resources (HR). Infrastructure HUMAN resources related with production of shoes as shown in the Table 2.

IOP Conf. Series: Earth and Environmental Science **335** (2019) 012021 doi:10.1088/1755-1315/335/1/012021

Table 2. Existing human resources (HR)

No.	HR	Department	Expertise
1.	Nana Sujana	Raw Materials Warehouse	Expert material in the Raw Materials Warehouse
2.	Hamid Bintan	*Cutting*	*Cutting* Expert (Cutting material)
3.	Pipin Aripin	*Laboratory*	Lab, *Emboss* and *Printing* Expert
4.	Agus Concerned	*Sewing*	Expert Sewing (Suturing material)
5.	Muhammad Adam	*Assembling*	Assembling Expert (*Upper* and *outsole Attachment*)
6.	Abdul Hafid	*Press/Rubber* Warehouse Material So	Press a rubber *Outsole* Expert and Fine Good Warehouse Expert
7.	Subani	*Stockfit*	*Stockfit* Expert
8.	Dodi Suparyadi	*Packing*	*Packing* Expert
9.	Eva Yuli Saefudin	Quality Control	Experts checking the quality of the production
10.		Administrative Staff	- Report the results of the production - Reports of the shoes A, B, and C *grade* - Job report Daily (Planning of *size*, color and shoes model that will be product)

3.1.4. Network infrastructure

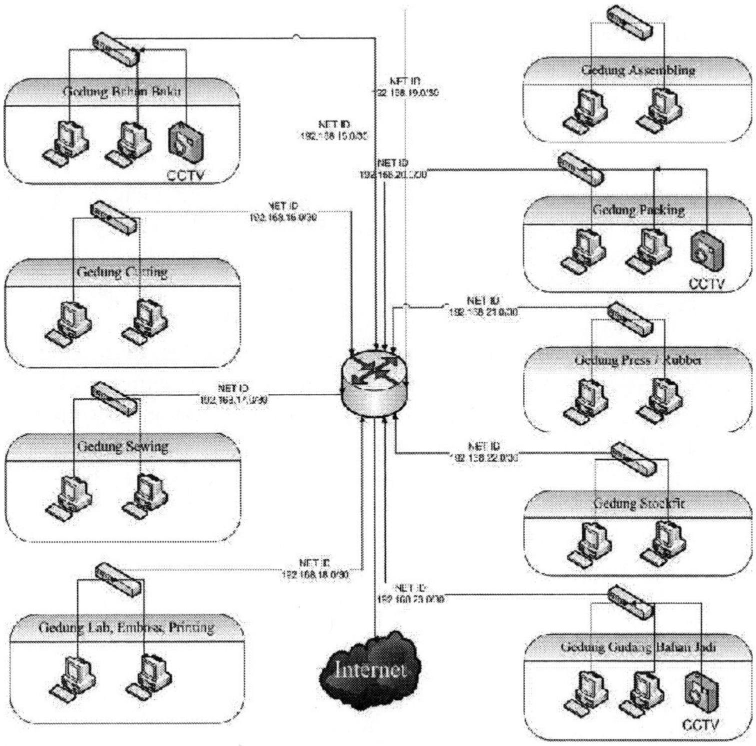

Figure 2. Network infrastructure production division.

3.1.5. Knowledge infrastructure. Based on the observation result on the production division, the existing knowledge which is related to the production of shoes seen in table 3.

Table 3. Existing knowledge.

No	Knowledge	*Explicit*		*Tacit*	Description
		Hardcopy	*Softcopy*		
1.	*Order Status* Report	✓	✓		Employees
2.	Reports Of Production Results	✓			Employees
3.	Shoes A, B and C *grade* Report	✓	✓		Employees
4.	*Stock Packing* Data	✓			Employees
5.	*Delivery Order* Report	✓	✓		Employees
6.	*Form Test Result*	✓			Employees
7.	*Production Specification*	✓			Employees
8.	*Per order* material Requirements Standard	✓	✓		Employees
9.	The Minutes Of The Meeting	✓			Employees
10.	Experience participating in SOP training in shoe production			✓	Experts/ Employees
11.	Experience leadership in dealing with teamwork issues			✓	Expert
12.	Innovation in developing SOP creatively in the field			✓	Experts/ Employees
13.	Expertise in applying various strategies, methods and techniques for shoe production processes for employees			✓	Expert

3.2. Forming a team of Knowledge Management

The formation of KM team is conducted to identify the resources or required stakeholders. Organizing team members is very important for the efficient Knowledge Management (KM), which is defined as the process to manage continuously all knowledge assets for organization profit [9]. KM team has a detail specification as follows:

1. Project manager : responsible for the manufacturing system as well as being the coordinator of the
2. Knowledge Engineer : Gave advice in the search for knowledge, taking the knowledge required of a knowledge source, as well as ensuring that the overall system can run
3. Knowledge Designer : Design as well as designing the display in accordance with the needs of the system in the development of a KMS into a blue print
4. Knowledge Sources : Experts who become a source of knowledge, they consist of head of the department, IT staff, administrative staff of each building.
5. Programmer and Database Admin : Help in creation of programs, database admin, namely those that help in making the data base of knowledge originating from knowledge sources.
6. Software tester : Helps to test the system whether it is in accordance with the requirements or not

3.3. Capturing knowledge

In this stage, the process is conducted by observing, recording, and translating process in the shoes production and related problems in the shoes production.

From the interview results, the company does not have a system or application that can help capture the running knowledge. The process is conducted still manually and not structured. The existing knowledge also spreads in each production building and employees. Based on the results of observation and interview the source of knowledge are : Order Status Report, Production Report, Shoes A, B and C grade Report , Stock Packing Data , Delivery Order Report, Form Test Result, Production Specification and per order Material Requirements Standard.

Experience, innovation, expertise obtained from experts in the production division includes raw materials department experts, cutting department, department of laboratory, sewing, assembling, press/rubber, stockfit, packing, fine goods and quality control. Based on the knowledge source above then the knowledge is grouped into:

1. *Explicit* knowledge is captured from various media related to the shoe production process include *order status reports*, production reports, A, B and C *grade* reports, *stock packing* data, *delivery order* report, *form test result*, *production specification* and *per order* material requirement standard material.
2. The *tacit* knowledge captured by experts related to the shoe production process includes knowledge of shoe production as the result of ideas, experiences, expertise from the experts. Knowledge map in the process of shoe production can be seen in figure 3.

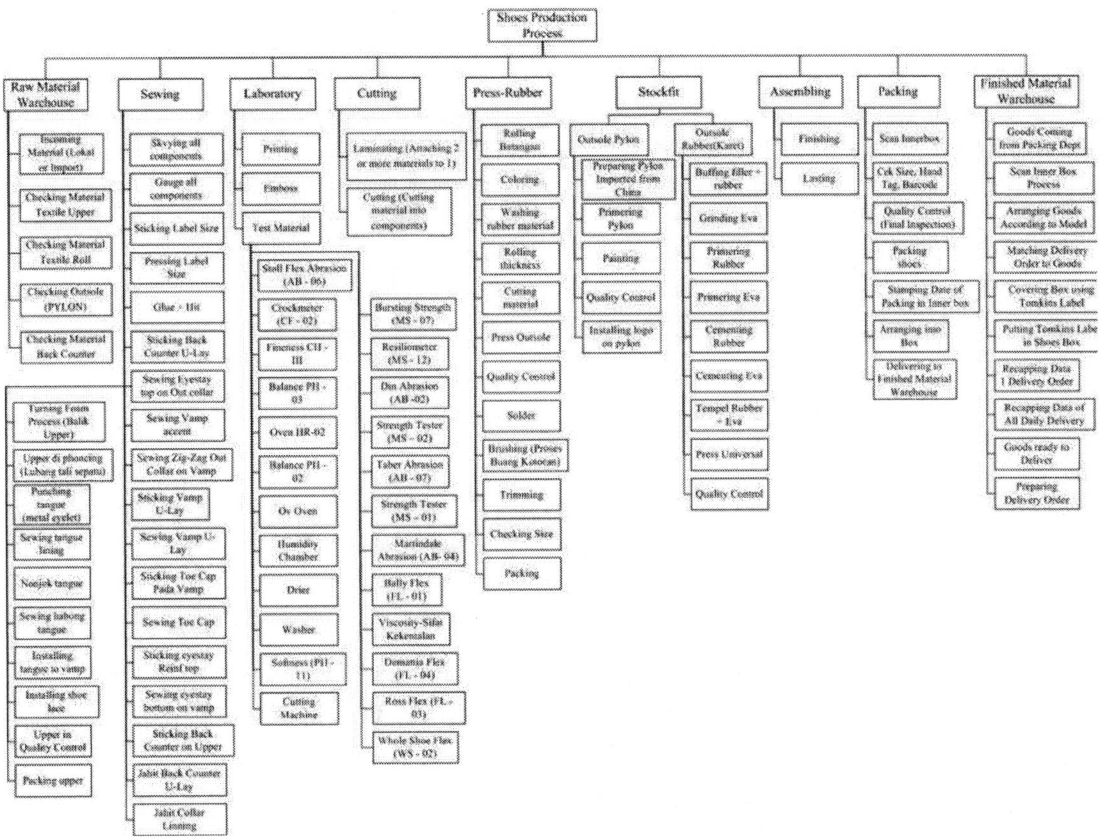

Figure 3. Knowledge mapping process of shoes production.

3.4. Design a blueprint of a Knowledge Management System

In this stage, the use case diagram serves to illustrate the behavior of to be established system, referring to [10]. Use case diagram can be seen in Figure 4 and entity relationship diagram as data modeling can be seen in Figure 5.

IOP Conf. Series: Earth and Environmental Science **335** (2019) 012021 doi:10.1088/1755-1315/335/1/012021

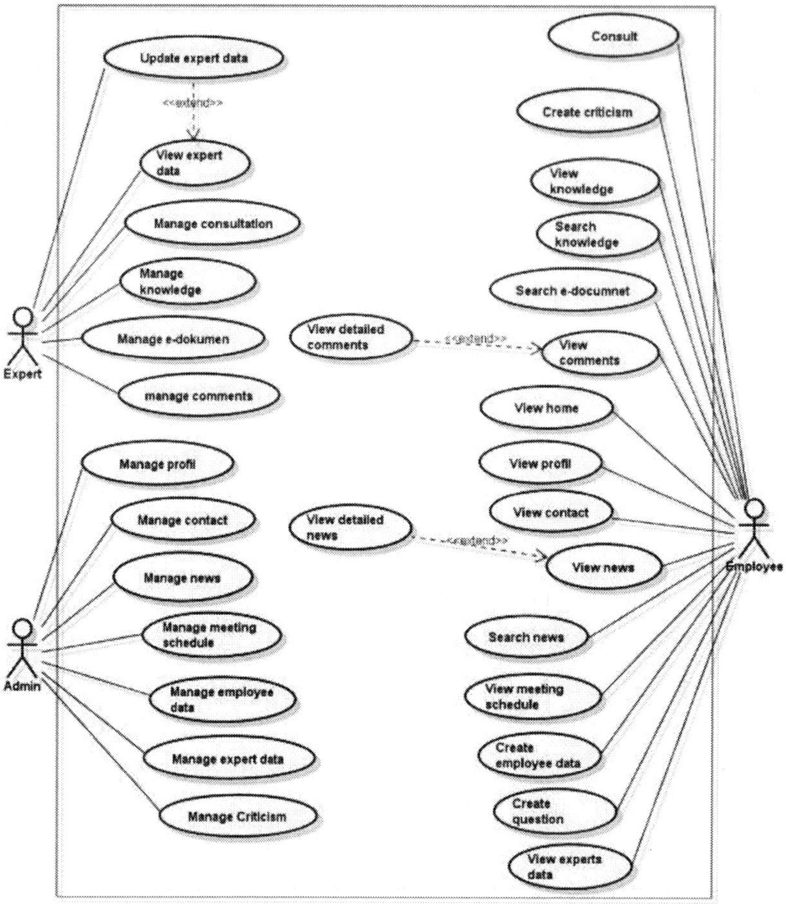

Figure 4. Use case diagram.

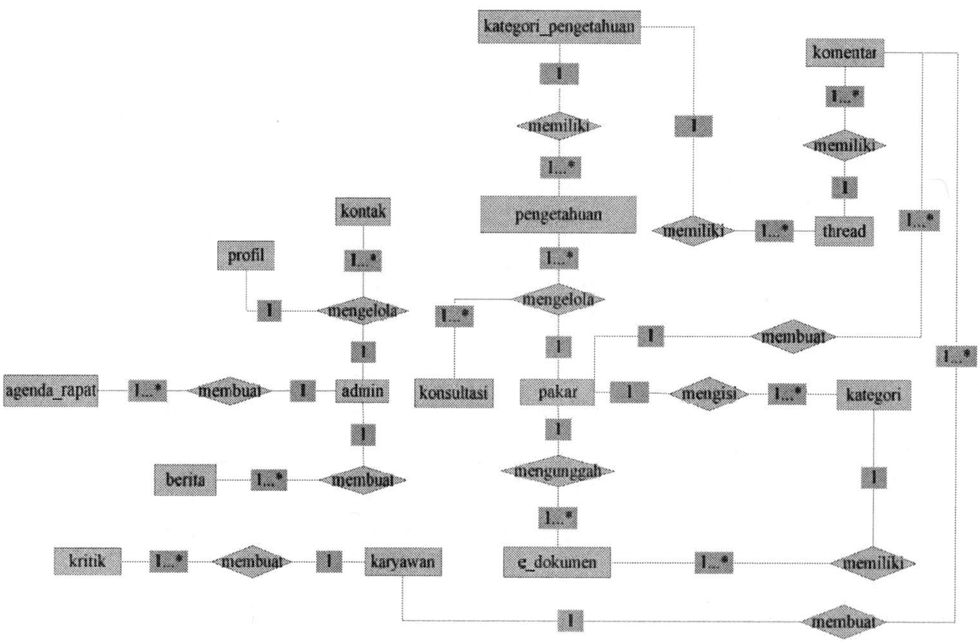

Figure 5. Entity relationship diagram.

On the ERD there are 15 entities, those are: admin, expert, categories, agenda_rapat, profile, contacts, knowledge, e_dokumen, thread, kategori_pengetahuan, comment, news, consultation, criticism and employees, referring to [11].

3.5. Knowledge Management System verification and validity

This stage is performed to evaluate whether KMS meets the criteria or not [12]. Testing is conducted to determine and ensure that shoes production KMS is correct and can function properly with the required specifications. Testing conducted by *blackbox testing* using test cases that try all the functions with the correct and incorrect case, then discover the error in the incorrect or missing function, the error *interface*, the error on the input value and functional validity. Here is a display of created KMSPS system in figure 6 and 7.

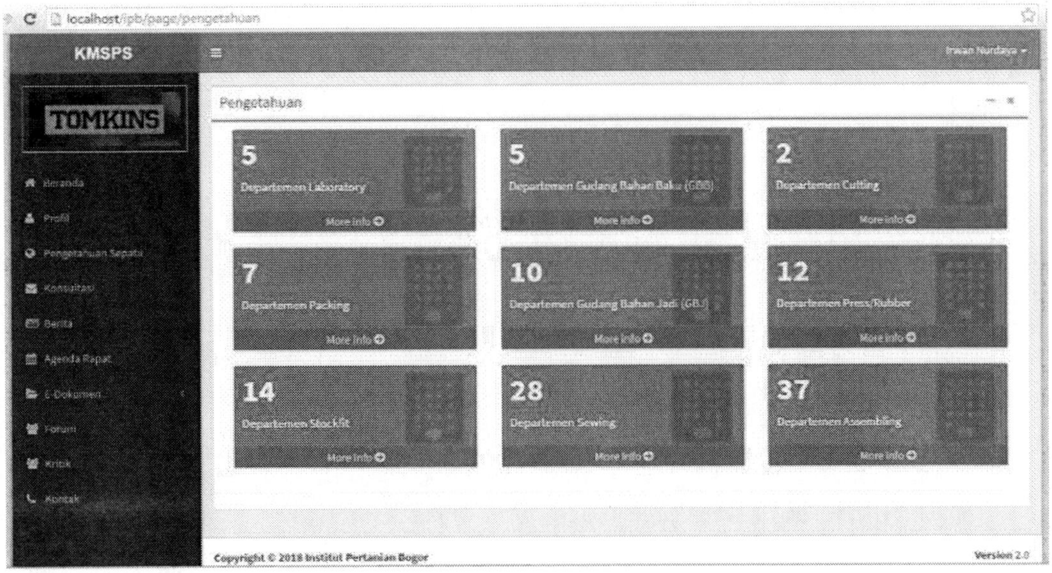

Figure 6. Page knowledge of user employees.

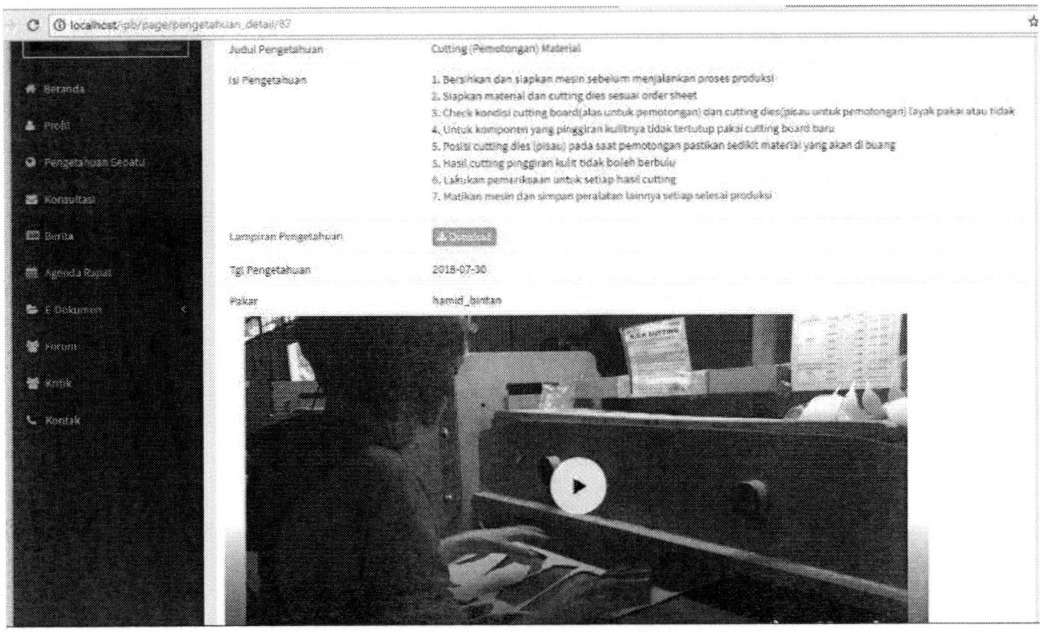

Figure 7. Page detail knowledge of user employees.

4. Conclusions

KMSPS (Knowledge management system of shoes production) is developed as a means to share knowledge about the ways and processes of shoe production that can be used by employees and experts. KMSPS is succeed uses KMSLC approaching with framework codeigniter and PHP and MySQL database. KMPS test result indicates that all user functions are fulfilled and properly functioning.

The stored knowledge has 10 categories include raw material department, cutting department, laboratory department, sewing department, assembling department, press / rubber department, stockfit department, packing department, fine goods and quality control department.

KMSPS has also successfully developed a Knowledge Sharing System on the forum menu and Knowledge Transfer on the consultation menu which is a type of Knowledge Management System. KMSPS (Knowledge management system of shoes production) of PT. PAI requires support from the management to realize and contribute actively in managing and implementing this system.

References

[1] Suprianto 2014 *Perancangan knowledge management system: Studi Kasus Badan Kepegawaian Daerah Pemerintah Provinsi DKI Jakarta* (Jakarta: Universitas Indonesia)
[2] Fernandez B 2010 *Knowledge Management System and Process* (United States of America: ME Sharpe, Inc) pp 8-9
[3] Nasution AR 2014 *Pengembangan Sistem Manajemen Pengetahuan Zakat* (Bogor: Institut Pertanian Bogor)
[4] Rus I and Lindvall M 2002 *Introduction: Knowledge Management In Software Engineering* IEEE Software vol 19(3) pp 26-38
[5] Sari MAP 2013 *Perancangan Knowledge Management Solution untuk Industri Plastik: Studi Kasus PT. ABC* (Jakarta: Universitas Indonesia)
[6] Halim E, Phoebe PH and Hebrard M 2017 *Proc. Int. Conf. on Information Management and Technology (ICIMTech) Indonesia Medical Knowledge Management System* (Jakarta) pp 1-6
[7] Awad and Ghaziri 2010 *Knowledge Management Upper Saddle River* (NJ: Pearson Prentice Hall)
[8] Wardani M, Hariyanto B and Ardiyanti A 2011 *Proc. Int. Conf. on Electrical Engineering and Informatics Implementation of Knowledge Management System In PT Intimas Wisesa.* (Bandung) pp 1-6
[9] Masmoudi A and Mezghani E 2017 *Proc. Int. Conf. on Enabling Technologies: Infrastructure for Collaborative Enterprises A Web-based Knowledge Management System for Scientific Research Team* (Polandia) pp 1-6
[10] Rosa and Shalahuddin 2013 *Rekayasa Perangkat Lunak (Terstruktur dan Berorientasi Objek)* (Bandung: Informatika)
[11] Heryanto I and Raharjo B 2009 *Menguasai Oracle, SQL, dan PL/SQL* (Bandung: Informatika)
[12] Rahmatia D and Surendro K 2013 *Proc. Int. Conf. on ICT for Smart Society (ICISS) Knowledge Management System Design Model for Smart Enterprise* (Jakarta) pp 1-6

Acknowledgements

The author gratefully acknowledges support from PT. PAI has given permission and guided me for the implementation of this research.

RiTx; Integrating among Field Monitoring System (FMS), Internet of Things (IOT) and agriculture for precision agriculture

B D A Nugroho[1,2] and H K Aliwarga[1]

[1] Department of Agriculture and Biosystem Engineering, Universitas Gadjah Mada, Indonesia, e-mail: bayu.tep@ugm.ac.id

Abstract. As developing country with industrial revolution 4.0, agro-technology, Internet of Things (IoT), E-Commerce and logistics technology is an important point in Indonesia. Although there has been many agro-knowledge and information systems for famers, but a little information in integrating among agriculture, IoT and field monitoring systems (FMS) technology for farmers in Indonesia. RiTx or Agricultural Technology is one of application for farmers with integrating among agriculture, IoT and field monitoring system technology. RiTx provides real-time, localized soil and weather data and field-level agronomic intelligence, which support farmers to decide their decisions during planting until harvesting, so failure during planting season can be eliminated. The province of East Nusa Tenggara (NTT) is cluster of islands located in the south – eastern part of Indonesia. The monsoon season stretches from October to March each year. Outside of the monsoon season, the climate of NTT is relatively dry. This study focuses on applying FMS technology with IoT for agricultural precision. For this study, field monitoring system was set up in the field since 26 July 2017 in Kupang, NTT Province that is consisted of three main components, i.e., FieldRouter, Datalogger and the sensors. Here, there are several sensors that have been installed in the field, e.g., solar radiation, rain-gauge and soil moisture. As the results, the IT field monitoring system showed good performance and reliable for precision agriculture. The actual field conditions were monitored well in term of image, numeric, and graphical data acquisition. Based on monitored data, plant growth can be well monitored. In addition, dynamic changes of environmental parameters can be monitored as well. Finally, it can help farmers in the future, not only in Indonesia or South-East Asia but also in the world, so prosperity for farmers can be granted.

Keywords: field monitoring systems, internet of things, precision agriculture

1. Introduction

East Nusa Tenggara (NTT) is a province with typical monsoon climate with low annual rainfall and relative dry. Increasing population and reducing arable area have increased the challenges to raise rice productivity. Furthermore, climate change has affected on any sectors including agriculture, forestry and fishing (Bayu, 2016). During El Nino years, the start of the monsoon is often delayed and the amount of rainfall reduced, exacerbating the vulnerability of communities in the province (IRI-Columbia, 2013). The increasing variability of weather, attributed to climate change, has also affected farming in NTT, in particular affecting the start and duration of wet seasons (Montgomery et.al, 2010). This uncertainty contributes to the food security households. In contrast, La Nina cycles often lead to high rainfall and flooding, also contributing to crop failures. These climatic factors are the major causes of food insecurity in NTT, along with other factors such as pests and plant diseases (Muslimatun and Fanggidae, 2009).

Sustainability of rice farming is depending on water resource. When its scarcity became increasing due climate changes, water saving technology for rice farming is essential. In the other hand, industrial revolution 4.0 in agricultural sector is increasing in recent years (Gardjito *et al.* 2008). Many companies have been developing agricultural application based on android platform. But, the agricultural application still in specific feature; e-commerce, agro-knowledge and market place, not yet any integrating feature in one application, especially in agricultural technology application. RiTx or Agricultural Technology is one of application for farmers with integrating among agriculture, IoT and field monitoring system technology. RiTx provides real-time, localized soil and weather data and field-level agronomic intelligence, which support farmers to decide their decisions during planting until harvesting, so failure during planting season can be eliminated.

However, to test the FMS and IoT, we have set up FMS in the field in Kupang, East Nusa Tenggara Province. To provide for this investigation, we needed data on rice field environment such as soil and meteorological data through continuous measurements. Soil moisture data are important to identify water availability in the field. Meanwhile, information on meteorological data such as precipitation, air temperature, relative humidity, solar radiation and wind speed are required to consider natural environmental effect in determining water availability in the field.

Aims of this research is application of Information Technology (IT) Field Monitoring System in providing precise environmental data to support precision agriculture in Indonesia and can be monitored with RiTx Apps. The output of the application is expected to be used by farmers and increase the land and water productivities under the climate change situation.

2. Materials and methods

2.1. Study area

This study is located in 8° – 12° S dan 118° – 125° E, East Nusa Tenggara Province (Figure 1). The province is divided into 20 district and one municipality, Kupang, which had the highest population density in the province in 2012 (around 2,496 people per km^2) (BPS, 2010). Outside of the monsoon season, the climate of NTT is relatively dry. Inter-annual climate variability is caused by El Nino Southern Oscillation (ENSO) that affects the start and end of the monsoon as well as the amount of rainfall.

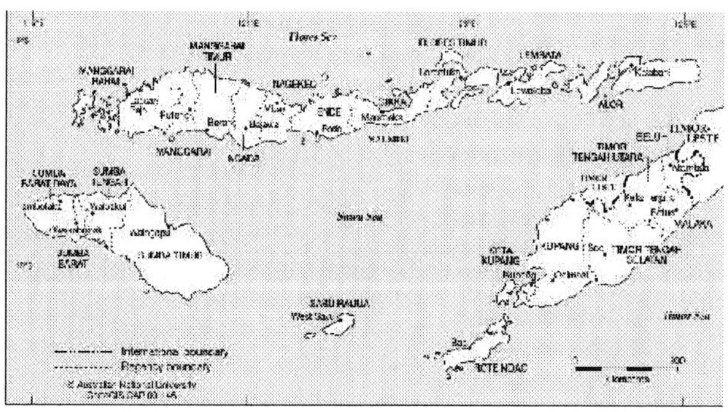

Figure 1. The newest regencies of East Nusa Tenggara showing borders, capital cities and regency names (ANU, 2015).

2.2. Methods

The field monitoring system was set up under natural environment in Tarus sub district, Kupang, NTT since 26 July 2016. Rice plant was planted on 29 July and harvested on 22 November 2017. The schema of IT field monitoring system was figured out in figure 2.

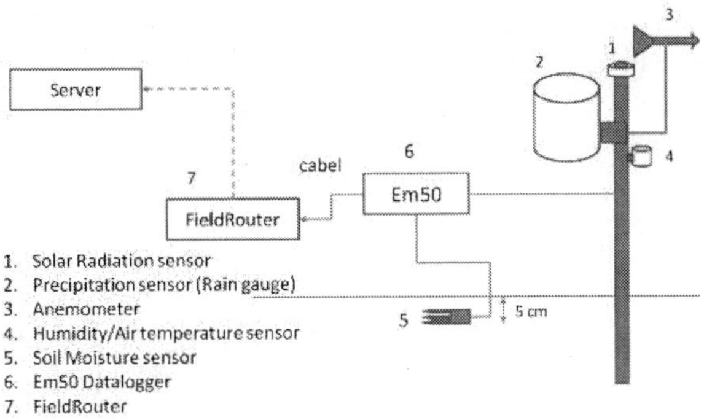

Figure 2. Scheme of Field Monitoring System (FMS).

There were 5 sensors that have been installed in the field, i.e., solar radiation, rain gauge, anemometer humidity and air temperature, and soil moisture sensors. All sensors were connected by cable to data logger. All measurement parameters were stored in that logger every 5 minutes. Then, FieldRouter will accumulate the data and send them using internet connection to server. The server realtime will received the picture of field condition. User can access the data through the RiTx Apps publish the data in term of numerical and graphical data.

3. Results and Discussions

3.1. Developing Field Monitoring System (FMS)

Developed field monitoring system, all monitored data including numerical, graphical and images can access data through RiTx apps in real-time data every 5 minutes update. For this monitoring, RiTx apps can be downloaded, and was prepared to present 5-minutes update monitored data. The interface of the RiTx Apps can be seen on figure 3.1) In that app, all data were updated real-time, and early warning system for field condition also available. So, if any low or high weather and soil condition in the field, the farmer will get notification and they can anticipate the condition based on recommendation from the apps.

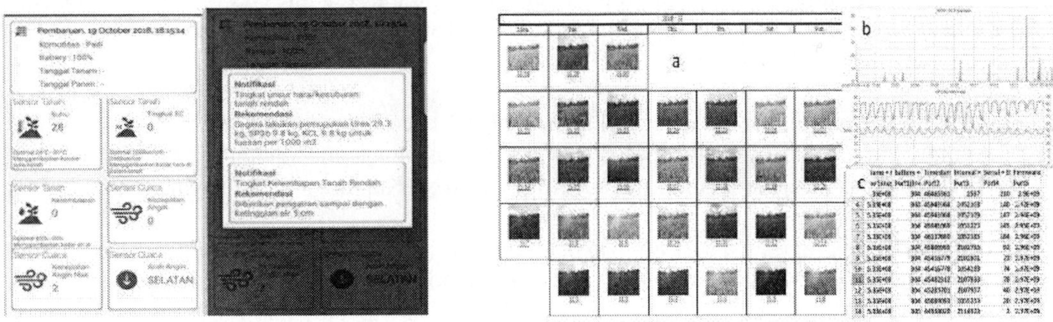

(a) (b)

Figure 3. (a) Interface of RiTx Apps based weather and soil data and recommendation; (b) data presented in website based system information: a) daily image during one month, b) graphical data, c) numerical data.

In addition, there are two more types data, i.e., numerical and graphical data. Numerical data represented environmental data measured by the sensors every 5 minutes update. Graphical data represented the dynamic changes of each environmental parameter. The sample both numerical and graphical data can be seen in figure 3.2). Based on this data, farmers can be estimated next conditions in the field, so failure of planting and harvesting can be avoid.

3.1.1. Dynamic changes of environmental parameters. On the RiTx Apps, dynamic changes of environmental parameters can be monitored. Figure 4a shows dynamic changes of daily solar radiation. Solar radiation was fluctuated. During August to September, the higher solar radiation was occurred during August 28 to September 7. The lowest solar radiation occurred on October when their values reached 60 MJ/m2/day.

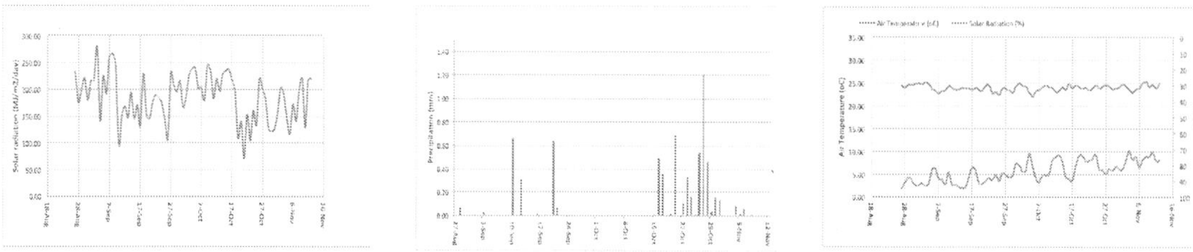

(a) (b) (c)

Figure 4. (a) dynamic changes of daily average solar radiation; (b) dynamic changes of daily precipitation; and (c) air temperature and dynamic change of daily average solar radiation.

Figure 4b shows dynamic changes of precipitation during planting season. Precipitation had fluctuated trend with in the beginning of planting (August 27-September 3), precipitation was less, and then the trend was increased from September 4 to 24. But, during September 24 to October 7, the trend was decreased again. The high precipitation occurred on October 27.

Figure 4c shows dynamic changes of solar temperature and air temperature. Dynamic change in solar radiation and air temperature was fluctuated, which is air temperature range around 25^0C during planting season.

4. Conclusions

In this study, the IT field monitoring system showed good performance and reliable for monitoring in the field. The soil and weather data can be monitored in real-time 5 minutes update. The real field conditions are well monitored, in term of images, numerical, and graphical data. Based on monitored data, dynamic changes of environmental parameters can be monitored as well. These results can be proved that agro-technology can be applied to support precision agriculture in Indonesia. In the future, it will be useful for farmers to avoid failure in planting and harvesting due to climate change conditions.

References

[1] Australia National University. 2015. Maps Online ANU college of Asia & The Pacific. Available online: http://asiapacific.anu.edu.au/mapsonline/base-maps/nusa-tenggara-timur-0

[2] BPS 2010 Nusa Tenggara Timur dakam Angka 2010. (Kupang : BPS Provinsi Nusa Tenggara Timur)

[3] Bayu Dwi Apri Nugroho 2016 *Fenomena Iklim Global, Perubahan Iklim dan Dampaknya di Indonesia* (Yogyakarta : Universitas Gadjah Mada Press)

[4] IRI - Columbia International Researh Institute for Climate and Society Earth Institute Columbia University 2013 *Climate Risk Management in Indonesia and the Philippines.* (In: Columbia University, New York available online : http:// iri.columbia.edu/wp-content/uploads/2013/07/ Indonesia2.pdf.)

[5] Gardjito, Rudiyanto, Agustina H, Setiawan BI, Mizoguchi M, Ito T, Priess J 2008 *Environmental Monitoring with Field Server via Internet.* (In: LIPI and STORMA Workshop, Jakarta, 20-21 February 2008)

[6] Montgomery, R., Rohandi and Kinseng, R. 2010 *Farm risk reduction assessment for Nusa Tenggara Inslands.* (In: Asia Research Centre Working Paper 31, Asia Research Centre, London School of Economics and Political Science, London)

[7] Muslimatun S, Fanggidae S 2009 *A brief review on the persistent of food insecurity and*

Acknowledgements
This research was part of a research project entitled *"Climate Projection and Adaptation Strategy of System of Rice Intensification (SRI) Cultivation against Regional Climate Change by integrated climate-crop-soil-water model approach in East Nusa Tenggara*" funded by Indonesian Climate Change Trust Fund (ICCTF) during March 2016 to March 2018 (2 years).

Land priority determination for paddy field extensification in Subang Regency - Indonesia

P B K Santoso[1], Widiatmaka[2], S Sabiham[3], Machfud[4], and I W Rusastra[5]

[1]Natural Resources and Environment Management Program, Graduate School, Bogor Agricultural University, Campus IPB Baranangsiang Bogor 16144, Indonesia e-mail: pbksantoso@gmail.com
[2]Department of Soil Science and Land Resources, Faculty of Agriculture, Bogor Agricultural University, Campus IPB Darmaga, Bogor 16680, Indonesia
[3]Department of Soil Science and Land Resources, Faculty of Agriculture, Bogor Agricultural University, Campus IPB Darmaga, Bogor 16680, Indonesia
[4]Department of Agricultural Industrial Technology, Faculty of Agricultural Technology, Bogor Agricultural University, Campus IPB Darmaga, Bogor 16680, Indonesia
[5]Center for Socioeconomic and Agricultural Policy, Cimanggu Research Campus, Tentara Pelajar Street Kav 3, Bogor 16111, Indonesia

Abstract. Extensification efforts in agriculture to increase paddy production can be done through the selection of land for the development of paddy fields from available land. Every inch of land has different characteristics, which can be used for the selection of land suitable for a use, including for paddy fields. This study aims to delineate available land and set priorities for the development of paddy fields in Subang Regency. The main methodology used is a multicriteria-based land evaluation. The land attribute is weighted using the AHP technique from the assessment of 8 experts. Land availability was obtained from a combination of Weighted Linear Combination method and Geographic Information System. The results of the study found suitable land for paddy field extensification. Prioritization results showed that 76,115 hectares of land had high priority, 7,165 hectares had moderate priority, 55,280 hectares had low priority, and an area of 19,561 hectares is land that is not recommended for use. The land available for extensification of paddy fields from suitable land is 9,885 hectares with a composition of 73.24% (7,240 hectares) high priority, 21.20% (2,096 hectares) moderate priority, 3.36% (530 hectares) low priority, and 0.19% (19 hectares) are not recommended for paddy field extensification.

Keywords: GIS, Multicriteria, Priority of paddy fields

1. Introduction

Indonesia needs an increase in paddy production to offset rapid population growth. Increased production can be done by supplying inputs of specific location-based. The suitability of land is evaluated in the land mapping unit in Subang Regency, West Java Province, at a scale of 1: 25,000 [1]. The results of the study recommend local specific inputs for paddy fields in Subang Regency. The potential of the land in Subang Regency allows for the extensification of identified paddy fields outside of the actual paddy fields, which have the potential to develop paddy fields. As in the case of existing fields, such fields require local specific inputs according to characteristics that differ from place to place and can be spatially described.

The shape of the Subang Regency extends from the coast to the hilly areas, which is considered to be more dynamic in the paddy field-alternation process [2]. Alternative land available has the hope to maintain the area of paddy fields in order to maintain the amount of paddy production needed. The increasing of paddy production can be done by intensification or extensification. Intensification is an effort to maximize available paddy fields by improving paddy production inputs. Extensification is an effort to expand paddy fields in other locations. This study tries to select the land for extensification that requires references for decision making in terms of determining the potential of paddy fields [1],[3],[4]. Decision making requires several alternatives to be chosen for subsequent implementation [3] and is an objective tool for the allocation of land use [5] including for the allocation of paddy fields.

The conceptualization of problems in the land suitability evaluation was the application of the MCDM (Multiple Criteria Decision Making) methodology which was chosen for unidimensional evaluation [6]. The MCDM method assists decision making processes involving policy priorities, trade-offs, and uncertainties [7]. The choice of location or allocation of land use for specific needs requires decision makers to involve several alternatives that are suitable to be evaluated based on multiple criteria [7],[8],[9]. All identified criteria have different levels of importance, so the decision-making process must consider the weight according to the relative importance of the results of alternative evaluations [8],[10].

Some alternative land allocation decisions require criteria from not only biophysical but also social and economic factors [2],[11],[12],[13]. The best alternative for paddy field LULC (land use/land cover) involves decision analysis that refers to several criteria (Multicriteria Decision Analysis) to determine subjective preferences between different criteria [10],[14]. Different weighted criteria have weight obtained from expert justification with the AHP (Analytical Hierarchy Process) technique. Priority of land is generated from spatial weighting using WLC (Weighted Linear Combination) in GIS (Geographic Information System).

Some researchers have previously integrated AHP with GIS for agricultural land suitability analysis [10],[15],[16],[17]. The study uses multi-criteria analysis for biophysical and socio-economic parameters. Multi-criteria techniques are also used for land suitability analysis [18], vegetation [17], agriculture [16], land evaluation [19]. Some researchers use GIS and multicriteria for the purpose of determining a location, for example: [20] determine suitable locations for gas power plants at Natanz City in Iran, [13] chose land suitability for pasture use and agriculture in northeastern Iran, [21] choose the location of the city park in Inskadar Malaysia.

The availability of land that can be converted into paddy fields is land that has a lower landrent value than paddy fields [22],[23], namely forest LULC, moor/fields, and bushes. This study uses the distance criteria for the land to-be to a place or object. Determining the alternative of weight of each land to-be utilizes criteria based on factors of land availability, land suitability, ease of work, and affordability. Each factor is used and based on the factors used. Specifically this research has the purpose of: (1) seeking priority land for paddy fields, (2) determining the availability of land for paddy fields. The results of this study provide information on the spatial location of the LULC paddy fields provided for land use planning, especially for LULC wetland purposes.

2. Materials and methods

2.1. Study Area

Subang Regency is located in a geographical position 107°31 '- 107°54' East Longitude and 6°11'-6°49 'South Latitude (**figure 1**). Based on the topography, the Subang Regency region has mountainous regions in the South with altitudes between 500 – 1,500 meters above sea level (ASL), hilly areas and plains in the Middle with an altitude of 50-500 meters above sea level, and lowland areas in the North with height of 0-50 meters above sea level [24]. Most of the Subang Regency (80%) has a slope of 0 ° - 17 °, 11% of the area has a slope of 18 ° - 45 °, and 9% of the area has a slope of more than 45 °. In general, the Subang Regency region has a tropical climate with an average rainfall of 2,352 mm per year with the number of rainy days of 100 days [24].

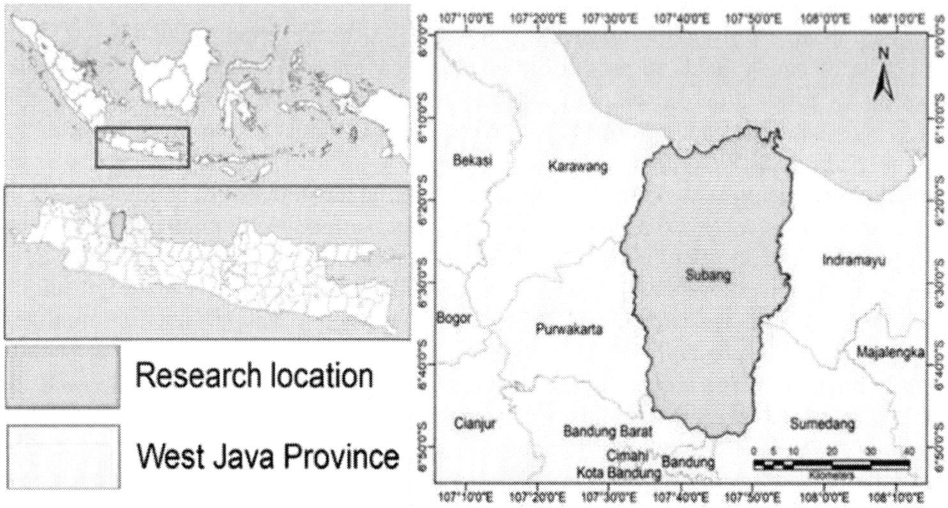

Figure 1. Research location to alocate land for paddy field, Subang Regency, West Java Province.

2.2. Requirements data

Attributes in the AHP hierarchy require spatial data generated from various existing data sources (**table 1**). This study uses GIS to produce all the attributes needed.

Table 1. Spatial data needs to determine the criteria weight of each factor, criteria, and alternative

Spatial Attributes	Data	Resolution / Scale / Acurration	Source	Note
Land availability	Spot imagery 7	1.5 meter	[25]	Forest, bushland, moor / fields
Land suitability	Land suitability map	1:50,000	[1]	S1,S2,S3
Ease of work	Land Unit Map	1:50,000	[26]	Clay content <15%
Affordability	Road map	1:50,000	[27]	Collector road, path, farm road
Street	Road map 50k	1:50,000	[27]	Toll road, national road, provincial road, district road
Settlement	Spot imagery 7	1.5 meter	[25]	Settlement / place of activity
Paddy fields	Spot imagery 7	1.5 meter	[25]	Paddy Field
Water sources	Topographic Map	1:25,000	[26]	Water Body
Market	Topographic Map	1:25,000	[28]	The coordinates of the paddy market location
	Paddy Mill	1.5 meter	[29]	The coordinates of the paddy mill location
Industry	Regional Spatial Plan (RSP) 2011-2031	1:50,000	[30]	Industrial zone

2.3. Hierarchy

The hierarchy of land priority decision models for wetlands has a factor, criteria and alternative structure (**figure 2**). Biophysical attributes consider the biophysical availability of land, land suitability, ease of work, and affordability [3] which are used as factors in the multicriteria decision hierarchy. The criteria hierarchy selects the location of the land that takes into account the distance from the road, the distance from the settlement, the distance from the existing paddy field, the distance from the water source, the distance from the market, and the distance from the industry. Alternatives in the multicriteria hierarchy choose LULC forests, fields / fields, and bushes. The third LULC is an alternative because the landrent value is lower than the LULC paddy field [22],[23].

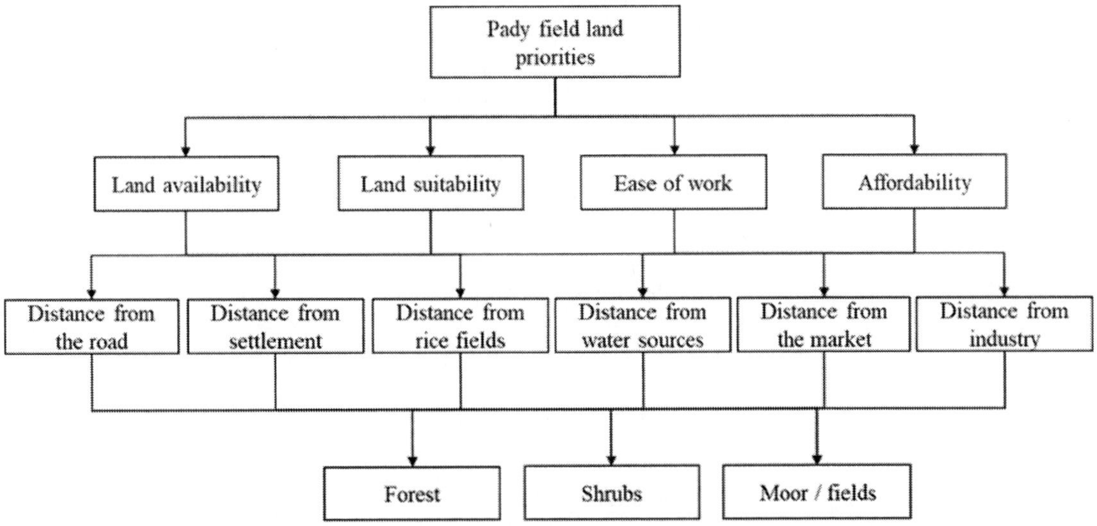

Figure 2. The hierarchy of land priority decision models for wetlands has a factor, criteria and alternative structure.

2.4. Weighting attribute criteria

Weighting each attribute utilizes the calculation results from the AHP method which involves 8 experts as the resource person. The AHP method is used because it is one of the representative multicriteria decision analysis methods used in this study. The AHP method have three basic principles: decomposition, comparative assessment, and priority synthesis [31],[32],[33]. Decision problems are decomposed into hierarchies that capture important elements of the problem. Comparative assessment requires the assessment of pairwise comparisons of elements at the hierarchical structure level. Priority syntheses take each of the priority scale ratios derived at various levels of the hierarchy and build a set of priorities for the elements at the lowest level of hierarchy, the alternative [32].

Problem solving is formed in a hierarchical framework with a level of factors, criteria and alternatives [33]. The AHP procedure involves three main steps: (i) developing the AHP hierarchy, (ii) assigning important weights to each hierarchical structural element using pairwise comparison methods, and (iii) establishing overall priority rankings [32],[33],[34],[35]. This study involved eight experts with expertise specifications of planning experts, environmental experts, land morphologists, agronomists, economists, and socioeconomic experts.

2.5. Pairwise Comparison

The number of attributes in the hierarchy is 13 (4 attributes of the factor hierarchy, 6 attributes of the criteria hierarchy, and 3 attributes of the alternative hierarchy). Each attribute is carried out by pairwise comparisons based on the hierarchy and carried out 29 paired matrices. The factor hierarchy has 1 comparison matrix, the criteria hierarchy has 4 comparison matrices, and the alternative hierarchy has 24 comparison matrices.

If there are m attributes that are compared in each comparison, a $\boldsymbol{A}_{mxm}$ matrix will be formed with elements a_{ij} (*i,j*=1,2,..., m). Element a_{ij} shows the level of importance the *i* attribute relatively to the *j* attribute. If $a_{ij} > 1$ then said that i attribute is more important than j attribute, and if $a_{ij} < 1$ then said that *i* attribute is less important than *j* attribute. If $a_{ij} = 1$ then *i* attribute is as important as *j* attribute. For all *i=j* then the value of a_{ij}=$a_{jj} = 1$, and for each elements a_{ij} and a_{ji} fullfil the equation $a_{ij}.\, a_{ji} =$

IOP Conf. Series: Earth and Environmental Science **335** (2019) 012023 doi:10.1088/1755-1315/335/1/012023

1, with *i,j* = 1,2,3,......*m*. The relative importance between the two criteria is expressed by a numerical scale from 1 to 9 (**table 2**).

Table 2. Relative values for elements in paired matrices

Score Value (a_{ij})	Description
1	Equally important
3	A little more important
5	More important
7	Very important
9	Absolute very important
2,4,6,8	Middle value between two values

After the paired matrix $\boldsymbol{A}_{mxm}$ is formed, the next step to complete the $\boldsymbol{A}_{mxm}$ pairwise comparison matrix becomes normal, which makes the number of each column element equal to 1 using the equation:

$$\bar{a}_{ij} = \frac{a_{ij}}{\sum_{k=1}^{m} a_{kj}} \tag{1}$$

The average element values in each row of the matrix $\boldsymbol{A}_{normal}$ represent the weight vector $\bar{w}$ obtained from equation (3).

$$\bar{w}_i = \frac{\sum_{k=1}^{m} \bar{a}_{ik}}{m} \tag{2}$$

There is a $\boldsymbol{S}_{mxm}$ matrix with each s_{ij} element value showing the *i*-selected value in the *jth* attribute. After the weight vector $\bar{w}$ and the *S*-score matrix have been obtained, a vector $\bar{v}$ that is the total value is needed by multiplying the matrix ***S*** by vector $\bar{w}$, namely:

$$\bar{v} = \boldsymbol{S}_{mxm} . \bar{w}_{mx1} \tag{3}$$

The *i* value in $\bar{v}(v_i)$ represents the overall value in the *i*-choice (*i*=1,2, ... , *m*). The choice rating is a descending order of the overall score. For paired matrices performed by several experts, each item for the last answer is used a geometric multiplication formula [36], with the formula:

$$\bar{b}_{ij} = \sqrt[e]{\prod_{g=1}^{e} (\bar{a}_{ij})_g} \tag{4}$$

With

$\bar{b}_{ij}$	:	geometric mean
e	:	the number of experts
$(\bar{a}_{ij})_g$	:	assesment by g^{th} expert
i,j	:	1,2, ... , m
g	:	1,2, ... , e
$\prod$	:	multiplication

2.6. Calculate Consistency

The Consistency Index (CI) is obtained by calculating the scalar value λ as the average j^{th} vector element which is the ratio of the jth element of the $A.\bar{w}$ vector to the corresponding vector element $\bar{w}$. So:

$$CI = \frac{\lambda - m}{m - 1} \tag{5}$$

$$CR = \frac{CI}{RI} \tag{6}$$

RI can be obtained from the consistency table (**table 3**) proposed by [31]. Consistent decision makers produce $CI = 0$, however the inconsistency value less than 10% (CR <0.1) can be accepted.

Table 3. Random Index (*RI*) Values

m	2	3	4	5	6	7	8	9	10
RI	0	0.58	0.90	1.12	1.24	1.32	1.41	1.45	1.51

2.7. Determination of Land Availability for Paddy Field Extensification

The determination of priority attributes is adjusted to the hierarchy of multicriteria decisions that have been formed. Spatial attribute priority calculation uses WLC technique which has two components, attribute weight (w_j) and value function ($f(r_{ij})$) [32]. In accordance with the hierarchy in AHP, the weight of each attribute in each hierarchy utilizes the results of pairwise comparisons in the corresponding AHP method. Value functions use spatial attributes that correspond to the decision hierarchy (**figure 1**). The relationship between value functions and attribute weights is expressed in equation (7) [37],[32].

$$V(\boldsymbol{x}_i) = \sum_{j}^{m} w_j v_j(\boldsymbol{x}_i) = \sum_{j}^{m} w_j r_{ij} \tag{7}$$

Where w_j is normal weight and $\sum_j^m w_j = 1$. Whereas $v_j(x_i)$ is a value function for the value of *j*-th attribute. Then $\mathbf{x}_i$ is $\mathbf{x}_{i1}, \mathbf{x}_{i2}, \dots, \mathbf{x}_{im}$ are spatial value and $\mathbf{r}_{ij}$ are spatial attributes that are transformed into comparable scales.

Distance analysis (Euclidean Distance) utilizes the center distance attribute to the outer boundaries of the district administration which results in spatial distance attributes. Furthermore, normalized spatial distance attributes use a linear scale transformation method that generates standard attribute values with comparable units [37]. The function of nonlinear values standardizes criteria values ranging from 0 to 1, with 0 representing the lowest value and 1 represents the highest value (**table 4**) [7]. Standardization of values use the following equations (8) and (9) [38],[39].

$$v(x_{ij}) = \frac{x_{ij} - min_j(x_{ij})}{max_j - min_j}, \text{ maximize the value of j atribute} \tag{8}$$

$$v(x_{ij}) = \frac{max_j(x_{ij}) - x_{ij}}{max_j - min_j}, \text{ minimize the value of j atribute} \tag{9}$$

Table 4. Value of standardization of attributes at a comparable scale

Attribute	Value	Description
Land availability	1 – 0	Value 1 is suitability and 0 is not suitability
Land suitability	1 – 0	Value 1 is suitability and 0 is not suitability
Ease of work	1 – 0	Value 1 is suitability and 0 is not suitability
Affordability	1 – 0	Value 1 is suitability and 0 is not suitability
Distance from the road	1 – 0	Value 1 is not suitability and 0 is suitability
Distance from settlement	1 – 0	Value 1 is not suitability and 0 is suitability
Distance from paddy fields	1 – 0	Value 1 is suitability and 0 is not suitability
Distance from water sources	1 – 0	Value 1 is suitability and 0 is not suitability
Distance from the market	1 – 0	Value 1 is suitability and 0 is not suitability
Distance from industry	1 – 0	Value 1 is not suitability and 0 is suitability
Forest	1 – 0	Value 1 is suitability and 0 is not suitability
Shrubs	1 – 0	Value 1 is suitability and 0 is not suitability
Moor / fields	1 – 0	Value 1 is suitability and 0 is not suitability

The empirical conceptual model of WLC is used as the basis for the development and validation of the attributes of spatial data on land suitability for paddy field extensification. The involvement of researchers who use the WLC method shows its strength to achieve acceptable results without reducing scientific standards [40]. For the purposes of technical spatial data analysis, all attributes of spatial data involved are converted to raster format with a cell size of 30mx30m with the system coordinate UTM 48S. Completion of the WLC technique uses ArcgGis 10.31 software.

2.8. Stages to Obtain Land Availability

WLC method produces land suitability which is a linear combination between each attribute weight and value function. The suitability of land must be adjusted to the spatial pattern of Subang Regency to obtain land availability for the extension of paddy fields. The steps to obtain land availability for the extension of paddy fields are presented in **figure 3**.

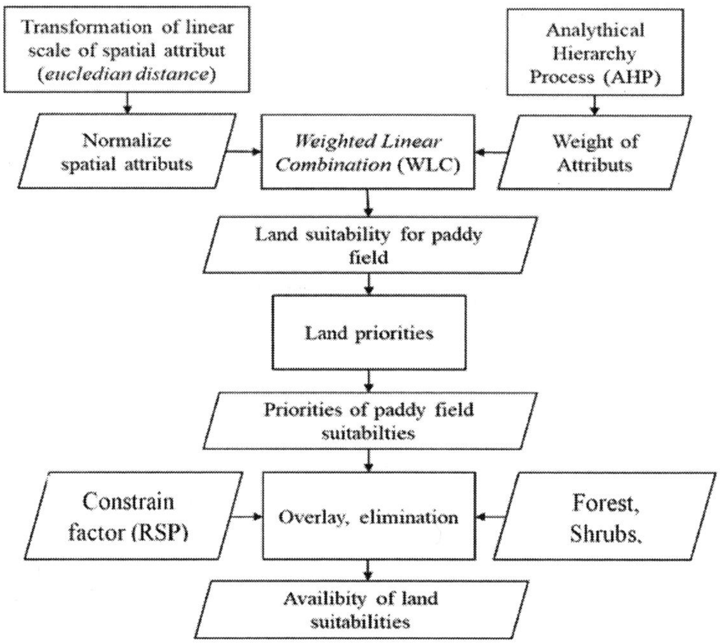

Figure 3. Stages to obtain land availability for extensification of paddy fields.

3. Results and discussion

3.1. Weight and standardization attributes

Factor, criteria and alternative assessments involved eigth experts with their respective expertise in land morphology, environmentalists, planning experts, economists, geodesists, social experts, agricultural socio-economic experts, and agronomists. Each element of the combined pairing matrix is produced from equation (4) which is presented in **table 5** for factor pairwise comparisons, **table 6** for pairwise criteria comparisons, and **table 7** for alternative paired comparisons.

Table 5. Pairwise comparisons in the factor hierarchy involving 8 experts using the AHP method in Subang Regency

Factor	Land availability	Land suitability	Ease of work	Affordability
Land availability	1.00	2.00	6.00	3.00
Land suitability	0.50	1.00	2.00	2.00
Ease of work	0.17	0.50	1.00	1.00
Affordability	0.33	0.50	1.00	1.00

Table 6. Pairwise comparisons in the criteria hierarchy involving 8 experts using the AHP method in Subang Regency

Factor Criteria	DFTR	DFS	DFRF	DFWS	DFTM	DFI
Land Availability						
Distance from the road (DFTR)	1.00	3.00	1.00	1.00	3.00	4.00
Distance from settlement (DFS)	0.33	1.00	1.00	1.00	1.00	2.00
Distance from rice fields (DFRF)	1.00	1.00	1.00	1.00	4.00	4.00
Distance from water sources (DFWS)	1.00	1.00	1.00	1.00	4.00	6.00
Distance from the market (DFTM)	0.33	1.00	0.25	0.25	1.00	2.00
Distance from industry (DFI)	0.25	0.50	0.25	0.17	0.50	1.00
Land suitability						
Distance from the road (DFTR)	1.00	3.00	1.00	1.00	3.00	4.00
Distance from settlement (DFS)	0.33	1.00	1.00	1.00	1.00	2.00
Distance from rice fields (DFRF)	1.00	1.00	1.00	1.00	3.00	3.00
Distance from water sources (DFWS)	1.00	1.00	1.00	1.00	5.00	6.00
Distance from the market (DFTM)	0.33	1.00	0.33	0.33	1.00	3.00
Distance from industry (DFI)	0.25	0.50	0.33	0.17	0.33	1.00
Ease of work						
Distance from the road (DFTR)	1.00	1.00	1.00	1.00	3.00	5.00
Distance from settlement (DFS)	1.00	1.00	1.00	1.00	2.00	4.00
Distance from rice fields (DFRF)	1.00	1.00	1.00	1.00	3.00	4.00
Distance from water sources (DFWS)	1.00	1.00	1.00	1.00	4.00	6.00
Distance from the market (DFTM)	0.33	0.50	0.33	0.33	1.00	2.00
Distance from industry (DFI)	0.20	0.25	0.25	0.17	0.50	.00
Affordability						
Distance from the road (DFTR)	1.00	2.00	2.00	1.00	3.00	4.00
Distance from settlement (DFS)	0.50	1.00	1.00	1.00	2.00	2.00
Distance from rice fields (DFRF)	0.50	0.50	1.00	1.00	2.00	3.00
Distance from water sources (DFWS)	1.00	1.00	1.00	1.00	3.00	3.00
Distance from the market (DFTM)	0.33	0.50	0.50	0.50	1.00	2.00
Distance from industry (DFI)	0.25	0.50	0.33	0.33	0.50	1.00

The AHP method approach produces the weight of each attribute on the hierarchy of factors, criteria, and alternatives. The weight of each attribute is obtained from pairwise comparisons of geometric mean result matrices from 8 experts. The results of pairwise comparisons in the factor hierarchy produce the value of the weight of land availability, land suitability, ease of work, and affordability. The attribute weight in the criteria hierarchy is the multiplication of paired comparisons in the criteria hierarchy with the factor weight that produces the overall value on the criteria attribute. The overall value of the attributes in the alternative hierarchy is obtained from doubling the weight of the alternative attribute with the weight of the criteria; the end result of the overall value of the alternative hierarchy is the average of the overall value of each alternative produced. The final value of each alternative is the average of each weight value from the contribution of each criterion. All paired comparisons yield a consistency value of less than 10% which indicates that the decision was not given by chance [31]. All results of calculations based on hierarchy are presented in **table 8**.

The result of weighting attributes shows that in the hierarchy of factors, "land availability" has the highest role in determining the priority of land for extensification of paddy fields. The "land suitability" factor has a second weight after "land availability". This shows that after the land is available, it must be considered the suitability of the land for paddy commodities. Furthermore the affordability factor has a third weight in the factor hierarchy. This shows that after land is available and has land suitability for wetland; planners must pay attention to how easy the priority location of the land can be reached. The lowest in the hierarchy of factors, the "ease of work" factor has the lowest weight, this shows that in determining the priority of land for extensification of paddy fields, the ease of working factor is the last factor in considering the priority location of land for extensification of paddy fields.

Table 7. Pairwise comparisons in the alternative hierarchy involving 8 experts using the AHP method in Subang Regency

Land availability									
	a. DFTR			b. DFS			c. DFRF		
	Forest	Shrubs	Moor	Forest	Shrubs	Moor	Forest	Shrubs	Moor
Forest	1.00	0.74	0.72	1.00	0.48	0.45	1.00	0.70	0.43
Shrubs	1.35	1.00	0.94	2.09	1.00	1.17	1.42	1.00	0.89
Moor	1.39	1.06	1.00	2.22	0.85	1.00	2.31	1.13	1.00
	d. DFWS			e. DFTM			f. DFI		
	Forest	Shrubs	Moor	Forest	Shrubs	Moor	Forest	Shrubs	Moor
Forest	1.00	0.63	0.42	1.00	0.66	0.47	1.00	0.73	0.98
Shrubs	1.59	1.00	0.94	1.51	1.00	0.92	1.38	1.00	1.77
Moor	2.37	1.06	1.00	2.15	1.08	1.00	1.02	0.57	1.00
Land suitability									
	a. DFTR			b. DFS			c. DFRF		
	Forest	Shrubs	Moor	Forest	Shrubs	Moor	Forest	Shrubs	Moor
Forest	1.00	0.74	0.69	1.00	0.76	0.66	1.00	0.76	0.72
Shrubs	1.35	1.00	1.35	1.32	1.00	1.22	1.32	1.00	1.29
Moor	1.45	0.74	1.00	1.52	0.82	1.00	1.38	0.77	1.00
	d. DFWS			e. DFTM			f. DFI		
	Forest	Shrubs	Moor	Forest	Shrubs	Moor	Forest	Shrubs	Moor
Forest	1.00	0.74	0.69	1.00	0.72	0.69	1.00	1.05	0.98
Shrubs	1.36	1.00	1.29	1.39	1.00	1.35	0.95	1.00	1.35
Moor	1.44	0.77	1.00	1.45	0.74	1.00	1.03	0.74	1.00
Ease of work									
	a. DFTR			b. DFS			c. DFRF		
	Forest	Shrubs	Moor	Forest	Shrubs	Moor	Forest	Shrubs	Moor
Forest	1.00	0.77	0.88	1.00	0.81	0.84	1.00	0.76	0.76
Shrubs	1.29	1.00	1.49	1.23	1.00	1.35	1.32	1.00	1.22
Moor	1.13	0.67	1.00	1.19	0.74	1.00	1.32	0.82	1.00
	d. DFWS			e. DFTM			f. DFI		
	Forest	Shrubs	Moor	Forest	Shrubs	Moor	Forest	Shrubs	Moor
Forest	1.00	0.72	0.62	1.00	0.76	0.84	1.00	0.76	0.84
Shrubs	1.39	1.00	1.00	1.32	1.00	1.29	1.32	1.00	1.29
Moor	1.63	1.00	1.00	1.19	0.77	1.00	1.19	0.77	1.00
Affordability									
	a. DFTR			b. DFS			c. DFRF		
	Forest	Shrubs	Moor	Forest	Shrubs	Moor	Forest	Shrubs	Moor
Forest	0.31	0.77	0.65	1.00	0.95	0.84	1.00	0.76	0.67
Shrubs	0.36	1.00	1.10	1.05	1.00	1.22	1.32	1.00	1.22
Moor	0.33	0.91	1.00	1.19	0.82	1.00	1.49	0.82	1.00
	d. DFWS			e. DFTM			f. DFI		
	Forest	Shrubs	Moor	Forest	Shrubs	Moor	Forest	Shrubs	Moor
Forest	1.00	0.72	0.62	1.00	0.98	1.01	1.00	1.08	1.01
Shrubs	1.39	1.00	1.10	1.02	1.00	1.22	0.92	1.00	1.35
Moor	1.63	0.91	1.00	0.99	0.82	1.00	0.99	0.74	1.00

Table 8. The value of the weight of each attribute from the calculation aggregation results involving 8 experts using the AHP method in Subang Regency

Hierarchy	Attribute	Weight
Aim	Priority of land for paddy fields	1
Factor	Land availability	0.46
	Land suitability	0.25
	Ease of work	0.11
	Affordability	0.18
Criteria	Distance from the road	0.17
	Distance from settlement	0.09
	Distance from rice fields	0.19
	Distance from water sources	0.39
	Distance from the market	0.09
	Distance from industry	0.07
Alternative	Forest	0.28
	Shrubs	0.36
	Moor/fields	0.36

Attribute weight in the criteria hierarchy, the attribute weight of the proximity of the water source has the highest weight. This is very reasonable considering that water is the main input in paddy production. The second weight after criteria for proximity to water sources, proximity to existing paddy fields provides convenience in the extension of paddy fields in terms of infrastructure such as irrigation networks and road access to locations. Next is weighting criteria for distance from the road, priority of land for extensification of paddy fields adjacent to the road (Toll, National, Province, and Regency) vulnerable to conversion [1]. In planning, the distance from the road criteria has a low weight. It is followed by weighting criteria for distance from residential and market. This fact shows that the priority of land approaching settlements is also vulnerable to conversion even though the rate of conversion of paddy fields to settlements tends to be slow. The market chosen in this study is the paddy market and paddy mills. The criteria for proximity to this market are used to consider the costs incurred due to transportation from paddy fields to the market. Based on alternative land that can be converted into paddy fields, LULC shrubs and field moorings have the same weight, while forest LULC has the lowest weight of the three alternatives.

The standardization of each spatial attribute has an ideal limit in the preparation of land suitability for paddy field extensification (**figure 4**). These standard attributes are basic data for determining priorities of land suitability for paddy field extensification. Conformity limits of each attribute are divided into 4 priorities, namely: priority 1, priority 2, priority 3, and priority 4. Determination of the limits of each priority using quartile values that produce 4 parts (**table 9**). Suitability levels were also divided into 4 priorities based on the equal intervals by [41].

Table 9. Land priority classification based on quartile

Priorities	Description	Interval values
P1	Land has a high priority for paddy field extensification	0.75 - 1
P2	Land has a moderate priority for paddy field extensification	0.50 - 0.75
P3	Land has a low priority for paddy field extensification	0.25 - 0.5
P4	Available land is not recommended for paddy field extensification	0 - 0.25

Normalization of spatial attribute values using equation 8 and equation 9 using GIS tools. Normal spatial attributes are classified according to four priority classes (**table 7**) which produce priority spatial attributes (**figure 4**). The spatial factor attribute has four formers, namely land availability, land suitability, easyness, and affordability.

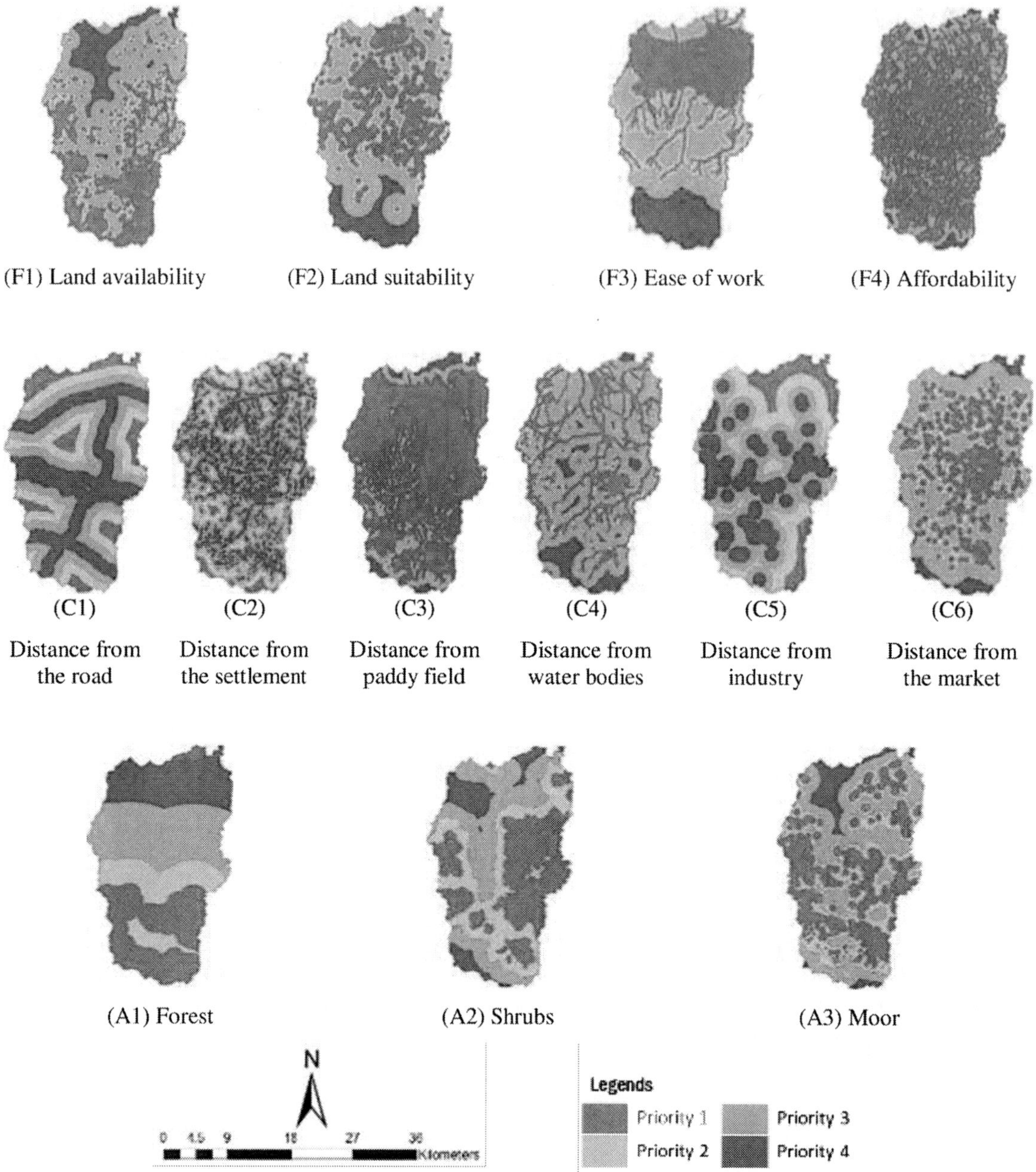

Figure 4. The suitability of the location of each attribute for paddy field extension.

P1 is a land that is prioritized for the development of paddy fields that have high land suitability. P2 is an area of land that can be used for the development of paddy fields that have several constraints or distance limiting factors. P3 is the location of land that can be cultivated for paddy fields with the consequence that the land has a larger limiting factor. P4 is a land that has a very large limiting factor, so it is not recommended to develop paddy fields.

3.2. Availability of land for extensification of paddy fields

WLC aggregation on each spatial attribute with the weight of the AHP calculation results in spatial distribution for paddy field extensification. The results of the calculation resulted in

IOP Conf. Series: Earth and Environmental Science **335** (2019) 012023 doi:10.1088/1755-1315/335/1/012023

34.90% (76,115 hectares) of areas with high priority for paddy field extensification, 30.79% (67,165 hectares) of moderate priority for paddy field extensification, 25.34% (55,280 hectares) had low priority for paddy field extensification, and 8.97% (19,561 hectares) are not recommended for paddy field extensification. This shows that Subang Regency has potential land for paddy fields which are not yet used for paddy fields [1]. Visually, delineation of land suitability for paddy field extensification is presented in figure 5a.

Land suitability (figure 5a) is a land location that has not considered the Urban Planning in Subang Regency and the type of land use can be converted into paddy fields. Therefore, the availability of existing land is chosen in the direction of the pattern of space for wet agriculture in the RTRW (regional spatial plan) and selecting land from forest use (production that can be converted, limited production), shrubs, and moor/fields (figure 5b).

The fact shows that from the 76,115 hectares of land suitability which has a high priority is only 9.51% (7,240 hectares) which can be used for paddy field extensification. Available land with moderate priority is obtained from land suitability with moderate priority of 3.12% (2,096 hectares) of 67,165 hectares. Land suitability of 55,280 hectares with low priority can be obtained 0.96% or 530 hectares which is the availability of land with low priority and the availability of land that is not recommended only 0.10% (19 hectares) of 19,561 hectares of land suitability that is not recommended. The total available land for extensification of paddy fields is 9,885 hectares, with a composition of 73.24% (7,240 hectares) of available land with high priority, 21.20% (2,096 hectares) of available land with moderate priority, 3.36% (530 hectares) land is available with low priority, and 0.19% (19 hectares) of land that is not recommended for paddy field extensification.

The availability of land obtained shows the results of the calculation of the normalized WLC method for determining land suitability priorities. Land availability shows the entire area of Subang Regency that has not considered the Spatial Plan and land cover that can be converted into paddy fields, namely forests, shrubs, and fields. Multicriteria evaluation results based on GIS are visualized in figures 5(a) and figure 5(b). Figure 5(a) shows the availability of land suitable for paddy fields. Figure 5(b) shows the availability of land suitable for paddy fields that choose LULC (land use/land cover) in the forest, shrubs, and moor / fields.

(a) Suitability for paddy fields

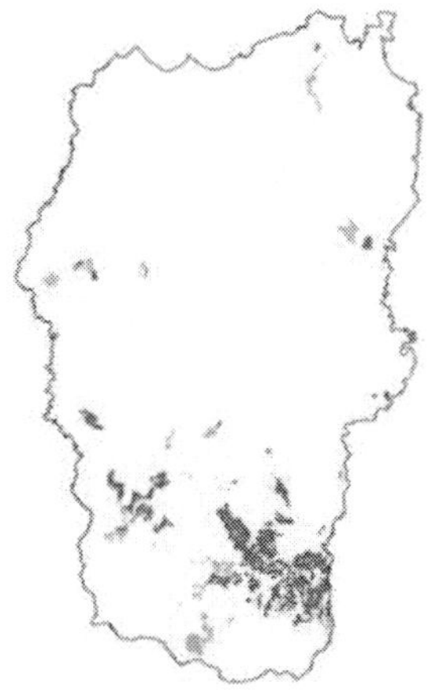

(b) Availability of land for paddy fields

Figure 5. Land suitability and availability of land that can be used for the extension of paddy fields.

4. Conclusion

Land suitability evaluation for paddy fields in the Subang regency found 158,121 hectares of land suitable for paddy fields. The land area suitable for paddy fields is divided into 4 priorities, namely 76,115 hectares of high priority land (48.14%), moderate priority 7,165 hectares (4.53%), low priority 55,280 hectares (34.96%), and an area of 19,561 hectares of land (12.37%) which is not recommended. The available land is then stacked with the Urban Planning. The results of elimination using the Spatial Plan show that the land suitable for paddy fields covering 9,885 hectares, with a composition of 73.24% (7,240 hectares) of available land with high priority, 21.20% (2,096 hectares) of available land with moderate priority, 3.36% (530 hectares) of land available with low priority, and 0.19% (19 hectares) of land that is not recommended for paddy field extensification. There is an area of 9,885 hectares which is not the actual paddy field that has the potential to be developed into paddy fields. From the available land, an area of 9,/336 hectares is high and moderate priority land. Subang Regency has the potential to increase paddy production from the addition of paddy fields from the availability of land acquired.

The results of land suitability analysis for paddy fields reflect the potential for expanding agricultural land use. In general, land evaluation based on multicriteria and GIS shows high potential to use in land use planning.

References

[1] Widiatmaka, W. Ambarwulan, P.B.K. Santoso, S. Sabiham, Machfud, M. Hikmat, "Remote sensing and land suitability analysis to establish local specific inputs for paddy fields in Subang, West Java," *Procedia Environmental Sciences,* vol. 33, pp. 94–107, 2016.

[2] Widiatmaka, W. Ambarwulan, R.P. Tambunan, Y.A. Nugroho, Suprajaka, Nurwadjedi, P.B.K. Santoso, "Land use planning of paddy field using geographic information system and land evaluation in West Lombok, Indonesia," *Indonesian Journal of Geography*, vol. 45, no. 1, pp. 89-98. 2014.

[3] S. Baja, "*Perencanaan Tata Guna Lahan dalam Pengembangan Wilayah Pendekatan Spasial & Aplikasinya,*" Yogyakarta (ID), Andi Offset, 2012.

[4] J. Langenmeyer, E. Gomez-Baggethum, D. Haase, S. Scheuer, T. Elmqvist, "Bridging the gap between ecosystem service assesments and land-use planning through Multi-Criteria Decision Analysis (MCDA)," *Environmental Science & Policy*, vol. 6, Pp. 45-56, 2015.

[5] M. Nyeko, "GIS and Multi-Criteria Decision Analysis for Land Use Resource Planning," *Journal of Geographic Information System,* vol. 4, pp. 341-348, 2012

[6] J.M. Pereira, L. Duckstein L, "A multiple criteria decision-making approach to GIS-based land suitability evaluation," *International Journal Geographical Information Systems*, vol. 7, no. 5, pp. 407-424, 1993.

[7] P. Jankowski, "Integrating geographical information systems and multiple criteria decision-making methods," *International Journal Geographical Information Systems*. vol. 9, no. 3, pp. 251-273, 1995.

[8] O. Marioni, "Implementation of the analytical hierarchy process with VBA in ArcGIS," *Computers & Geosciences*, vol. 30, pp. 637–646, 2004.

[9] J. Malczewski, "A GIS-based approach to multiple criteria group decision-making," *International Journal Geographical Information Systems*, vol. 10, no. 8, pp. 955-971, 1996.

[10] J. Malczewski, "GIS-based multicriteria decision analysis: a survey of the literature," *International Journal of Geographical Information Science*, vol. 20, no. 7, pp. 703–726, 2006.

[11] A.N. Dang, A. Kawasaki, "Integrating biophysical and socio-economic factors for land-use andland-cover change projection in agricultural economic regions," *Ecological Modelling*, vol. 344, pp. 29–37, 2017.

[12] Food and Agriculture Organization, "A framework for land evaluation," *Soils Bulletin*, no. 32, 1976.

[13] E. Memarbashi, H. Azadi, A.A. Barati, F. Mohajeri, S. Van Passel, F. Witlox, "Land-Use Suitability in Northeast Iran: Application of AHP-GIS Hybrid Model," *International Journal of Geo-Information,* vol. 6, no. 396, pp. 1-15, 2017.

[14] M. Kordi, S.A. Brandt, "Effects of increasing fuzziness on analytic hierarchy process for spatial multicriteria decision analysis," *Computers, Environment and Urban Systems*, vol. 36, no. 1, pp. 43–53, 2012.

[15] H. Zabihi, A. Ahmad, I. Vogeler, M.N. Said, M. Golmohammadi, B. Golein, M. Nilashi, "Land suitability procedure for sustainable citrus planning using the application of the analytical network process approach and GIS," *Computers and Electronics in Agriculture*, vol. 117, pp. 114-126, 2015.

[16] H. Akıncı, A.Y. Özalp, B. Turgut, "Agricultural land use suitability analysis using GIS and AHP technique," Computers and Electronics in Agriculture, vol. 97, pp. 71–82, 2013

[17] R.B. Zolekar and V.S. Bhagat, "Multi-criteria land suitability analysis for agriculture in hilly zone: Remote sensing and GIS approach," *Computers and Electronics in Agriculture*, vol. 118, pp. 300-321, 2015.

[18] S. Owusu, M.L. Mul, B. Ghansah, P.K. Osei-Owusu, V. Awotwe-Pratt, D. Kadyampakeni, "Assessing land suitabilityfor aquifer storage and rechargein Northern Ghana using remote sensing and GIS multi-criteriadecision analysis technique model," *Earth Syst. Environ*, pp. 1-11, 2017.

[19] S. Bandyopadhyay, R.K. Jaiswal, V.S. Hegde, V. Jayaraman, "Assessment of land suitability potentials for agriculture using a remote sensing and GIS based approach," *Int. J. Rem. Sens*, vol. 30, no. 4, pp. 879–895, 2009.

[20] F.S. Alavipoor, S. Karimi, J. Balist, A.H. Khakian, "A geographic information system for gas power plant location using analytical hierarchy process and fuzzy logic," *Global J. Environ. Sci. Manage*, vol. 2, no. 2, pp. 197- 207, (2016).

[21] F.N. Rasli, K.D. Kanniah, C. Muthuveerappan, C.S. Ho, "An Integreted Approach of Analitical Hierarchy Process and GIS for Site Selection of Urban Parks in Iskandar Malaysia," *International Journal of Geoinformatics,* vol. 12, no. 2, pp. 67-77, 2016.

[22] R. Barlowe, "*Land Resource Economics: The Economics of Real Estate*," New York (US), Prentice-Hall Inc., 1986.

[23] E. Rustiadi, S. Saefulhakim, D.R. Panuju, "*Perencanaan dan Pengembangan Wilayah*," Jakarta(ID), Yayasan Pustaka Obor Indonesia, 2011.

[24] Center Bearo Statistic Subang District, "*Subang Dalam Angka,*" Subang (ID), BPS Subang, 2015.

[25] Indonesian National Institute of Aeronautics and Space, "*Citra SPOT 6 ecw format-Archief*," Jakarta(ID), LAPAN, 2014.

[26] Indonesian Center for Agricultural Land Research and Development, "Land suitable map," Bogor(ID), BBSDLP, 2016.

[27] Indonesian Ministry of Public Works, "*Road map in Java*," Jakarta(ID), Kementerian Pekerjaan Umum, 2016.

[28] Geospatial Information Agency, "*Peta Rupa Bumi Indonesia*," Bogor(ID), BIG, 2016.

[29] P.B.K. Santoso, "GPS Survey," primary data, 2016.

[30] Development Planning Agency at Sub-National Level, "*RTRW Kabupaten Subang 2011-2031*," Subang(ID), Bapeda, 2012.

[31] T.L. Saaty, "Decision making with the analytic hierarchy process," *International Journal of Services Sciences*, vol. 1, no. 1, pp. 83-98, 2008.

[32] J. Malczewski and C. Rinner C, "*Multicriteria Decision Analysis in Geographic Information Science,*" New York (US), Springer Science, 2015.

[33] R. Vachnadze, "Prioritization of performance measures using analytic hierarchy process," *International Journal of the Analytic Hierarchy Process*, vol. 8, no. 3, pp. 490-501, 2016.

[34] E. Ekmekci, P. Aragonés-Beltrán, R.R. de Sanabria Gil, Y.I. Topcu, "A decision model for determining public locations suitable for automated external defibrillator (AED) deployment: a

case study in the City of Valencia," *International Journal of the Analytic Hierarchy Process*, vol. 9, no. 2, pp. 151-166, 2017.
[35] H. Akinci, A.Y. Özalp, M. Özalap, "Investigating impacts of large dams on agricultural lands and determining alternative arable areas using GIS and AHP in Artvin, Turkey," *SUJEST*, vol. 5, no. 1, pp. 83-95, 2017.
[36] Marimin, "*Pengambilan Keputusan Kriteria Majemuk*," Jakarta(ID), PT Gramedia Widiasarana Indonesia, 2014.
[37] J. Malczewski, "*On the Use of Weighted Linear Combination Method in GIS: Common and Best Practice Approaches,*" Oxford (UK), Blackwell Publishers, 2000.
[38] P. Jankowski, "*Multicriteria decision-making. The International Encyclopedia of Geography*," San Diego (US), San Diego State University, 2017.
[39] G. Romano, P.D. Sasso, G.T. Liuzzi, F. Gentile, "Multi-criteria decision analysis for land suitability mapping in a rural area of Southern Italy," *Land Use Policy*, vol. 48, pp. 131–143, 2015.
[40] R.C. Estoque, Y. Murayama, "Suitability Analysis for Beekeeping Sites in La Union, Philippines, Using GIS and Multi-Criteria Evaluation Techniques," *Research Journal of Applied Sciences*, vol. 5, no. 3, pp. 242-253, 2010.
[41] Y. Meng, J. Malczewski, S. Boroushaki, "A GIS-Based Multicriteria Decision Analysis Approach for Mapping Accessibility Patterns of Housing Development Sites: A Case Study in Canmore, Alberta," *Journal of Geographic Information System*, vol. 3, pp. 50-61, 2011.

Acknowledgements
Thank you to the Regional Government of Subang Regency, West Java Province for allowing the author to conduct research in the administrative area of Subang Regency from 2016-2017.

Lean management in the rice industry; case study at Widasari, Indramayu district

T Perdana[1,2], A H Sadeli[1,2], F R Hermiatin[2] and T Ginanjar[2]

[1]Department Agro Socio-Economic, Faculty of Agriculture Universitas Padjadjaran, Sumedang, West Java 45363, Indonesia e-mail: tomy.perdana@unpad.ac.id
[2]Agricultural Logistics and Supply Chain System (AGRILOGICS) Research Group, Faculty of Agriculture, Universitas Padjadjaran, Sumedang, West Java 45363, Indonesia

Abstract. Rice is a strategic product in Indonesia development because it becomes an essential aspect of economic growth. The inefficiency of food agribusiness system becomes a crucial aspect of developing the sustainable industry. Those affected by the lack of using technology that impacts the high amount of waste and losses, the inefficiency of time, etc. It could impact the competitiveness, equity of value-added, unstable price and poverty. Some amount of rice grade materials rejected from the production line due to the unsatisfactory quality. This article aimed to analyze efficiency in the food industry as a preparation in the development of agricultural digital supply chain. The analysis shows that inefficiency in the rice industry due to a lot of manual activities that impact the time and expense efficiency (labor intensive rather than the use of technologies). Improving efficiency in the food industry is not only about time and cost management, but it becomes complex issues which are involved in several barrier factors. It should be anticipated with appropriate technology, improving skills, and also develop an integrated logistics system. Importantly, improving efficiency in rice industry needs to develop soft skill as the key success factors to improve efficiency in the digital supply chain for the rice industry.

Keywords: rice industry, lean management, efficiency

1. Introduction

Understanding how agricultural business growth is a key endeavor in this digitalization eras. The economy will continue to grow and develop and today, we face the digitalization era as a revolution economic worldwide (hereinafter called the fourth industry revolution). The agricultural sector is one of the strategical sectors in the developing country likely in Indonesia. Agriculture is seen as a potential economic sector as well as challenges sector include with tons of risk from the production process until the products delivered to the end customers. Perishable, short of the live-time period, socio barrier and lack of technologicy adoption are an image of the agricultural sector in developing country [1]. Moreover, logistics and supply chain management become one of the major challenges due to the several factors such as product characteristics, handling process, distribution, market as well as socio characteristics. Those factors could impact to the availability and continuity of products ([13]; [14]; [15]; [19]).

Rice is a staple food and automatically as an essential and strategic commodity in Indonesian economic growth. The average of rice consumption expenditure in the fifth years (2013-2017) is about IDR 58,248 per capita per month or around 29.39 percent of total food expenditure of Indonesian is allocated for rice [15]. Food expenditure can be illustrated how the government develops food sovereignty to develop economic independence is the control of local resources and major strategic products as an effort to develop community wealth (hereinafter called Nawa Cita program). Meanwhile, the food industry has a complex and dynamic logistics system. The inefficiency of production becomes one of the barriers to develop the value of product [6]. This day, efficiency in logistics is an important aspect to develop food economics growth and advantage of products in the global competitiveness. Efficiency and effectiveness of production system have continued serious addresses in the food industry due to the competitiveness, consumer awareness, and economic revolution. Recently, a whole food supply chain actors have witnessed the dynamic form of competitiveness in the agricultural markets. It should have an understanding of this situation as an effort to develop competitiveness and find alternative food networks [17].

Misconception and high fragmented in the food industry could affect in several aspects of food logistics. The food industry in Indonesia dominated by small-scale farmers that witnessed the emerging problems such as uncertainty of production volume, the inadequacy of hard and soft technology, nothingness of production planning and conceptual [4]. Those issues caused by several factors such as food actor is dominated by smallholders which are not business oriented, lack of knowledge because they only doing business for generations, and also lack funding due to it has no network and information access coupled with no conceptual alignment with the banking system. Ironically, farmers are not a price maker anymore but become price taker because farmers do not have power and competitiveness in the food industry, thus resulting price fluctuation and limited information access about consumers need ([11]; [8]).

West Java Province is the second main production of rice in Indonesia after East Java Province with a total contribution of around 16% of total production with the consumption need is an amount of 98.39 Kg per capita per year [16]. Indramayu district is one of the main production centers in West Java Province with a total contribution of around 12.66% of total production in West Java Province. Even though Indramayu district is the point of the production area, the actors are witnessed competitiveness and helpless to increase the advantage. In addition, rice has a tendency to influence inflation both the regional and national economic situation. That is how rice becomes a sensitive commodity in the economic growth this day. Furthermore, the inefficiency of rice logistics system becomes a crucial aspect in the global competitiveness. Improper post-harvest handling, the inadequacy of plan production, technology, food logistics infrastructure (as an example less of grain storage warehouse, transportation mode, the absence of rural logistics service, etc.), and limited of farmers knowledge are the obstacle to improve price competitiveness. Price stability is an issue to increase farmer wealthy outright competitiveness. Price stability also contributed to inflation as an indicator of economic stability. Effectiveness in rice logistics could develop business process outright develop advantage and competitiveness and value of products with the result that producer could gain stable price [11]. Efficiency in the whole logistics process could reduce waste (both from production until the post-harvest process) and any kind of factors that contributed to the production cost [19]. Other problems in the rice logistics are about continuity of production, lack of funding becomes an obstacle for farmers to adopt new technology. In order to fill the current issues, it needs an appropriate approach to develop efficiency and effectiveness of rice logistics and supply chain management.

Lean production (known as lean manufacturing) is a stimulus to understanding a study problem about efficiency. Lean production is a key tool to find a new foundation for the improvement of the rice production process and had several impacts in the globalization competitiveness. A comprehensive design of industrial production that stressed on continuing improvement, system efficiency, and value-adding action had truly based on lean production concept [24]. Designing the production process as an effort to minimize waste in the whole supply chain process and gain the value of products is the aim of lean production. Improving competitiveness and value of products in the industrial process need the integral concept of the rice manufacturing landscape ([25]; [4]). In order to develop rice optimization production with the efficiency concept, lean production could be a starting point to develop the model. Different principles in lean production theory could create detailed of the conceptual model and it is a

IOP Conf. Series: Earth and Environmental Science **335** (2019) 012024 doi:10.1088/1755-1315/335/1/012024

suitable strategy for rice logistics process [23]. Lean production concern in the development of production, procurement and process improvement in the distribution process [9]. The ultimate aimed of implementation of lean production for rice logistics and supply chain industry is to enhance the productivity of products, improve quality of products, shorten lead times, reduce waste of time and cost, etc. In line with the issues, lean manufacturing had been developing food chain as an effort to delivering beneficiaries along food logistics actor in the United Kingdom [4].

In this study, value stream mapping was used to analyze the general idea of the production process, improving efficiency and reducing waste in the production system. Lean production has used an effort to enhance the capacity of production, develop a model with the robust, effective and efficient concept that could implement for the whole actors. Value stream mapping could illustrate the logic of production process, post-harvest flow and how the actors delivering the products to the end consumers. Improving efficiency for rice industry is considering involve some element in the lean production such as bottleneck removing, cellular manufacturing, competitive benchmarking, cycle time reductions, cross-functional workforce, pull system, planning and scheduling strategy, self-directed work programs, process capacity measurement, quality management, and operation management system [12]. Lean production has a key success factor that could gain value of products and advantageous to develop a concept based on production improvement, institutional framework, service provider and logistics actors. A multi-dimensional combination approach is a foundation concept to merging individual practice in a multiplicative function ([6]; [7]).

Based on the explanation above, the goals of this study is to explore the potential factor in the rice industry with using lean production concept as an effort to reduce waste, develop efficiency and effectiveness of rice industry. The novelty of this study is development efficiency of lean production management for rice industry involving the whole actors in the rice logistics and supply chain in Widasari sub-district, Indramayu, West Java. Our role in this study is a chance to analyze the conceptual lean production for rice commodity and introduction a new concept for the rice industry as a preventive study to tackling the fourth industrial revolution.

2. Material and Method

The study is based on rice supply chain cases at Widasari, Indramayu District, West Java Province, Indonesia. In this paper, value stream mapping was used as an instrument to develop the model of efficiency with the concept of lean production management. The case study was applied as a research method to gain the information and data as well as exploring the complex issues of rice logistics and supply chain system. Moreover, an in-depth interview, field observation, literature studies, and document analysis were used to gain in-depth understanding of the rice industry in Indramayu District. About 35 informants were an interview to obtain the actual data of the rice industry.

Value stream mapping is a tool to develop production planning with exploring production operations planning, materials management and control, process improvement, database/system design and development and develop project activity in support of the industrial process, reducing waste all along the process of the factory [18]. The value stream is a set of action involve value added and non-value added that are required to take a product through the main flows, starting the procurement of agro-inputs, production process, post-harvest process until the products deliver to the end user [19; [20]]. The concept of understanding considers the flow of both information and materials within the overall supply chain.

This model concern about production optimation process, harvest and post-harvest process and distribution process in rice supply chain and also identify all of the kind the of waste in the value stream and take an action how to eliminate these. All of the conceptualization is linking and visualizing the actual material and information entire rice supply chain as a means to develop value stream [10]. Value stream mapping is a predefined toolset of standardized icons and divides into a two-step, the first step is drawing current state map consist of current data and information on the shop floor. The data and information is basic information to develop ideal future state map. Data and information should recognize in the current step mapping is customer demand, cycle time (C/T), process time (P/T), change over time (C/O), number of operations (Op.), capital (cap.), available time, uptime or downtime, quality or defect rate (Q), number of product variations, batch size and inventory level.

Furthermore, the second step is drawing future state ideas as a point of key step to deliver the information about analysis. Future state map is redesigning the process of current state map based on lean principles [20]. Future state map should observe several principles i.e., rebalancing process, move toward continuous flow, changing the batch size, leveling out production, reducing changeover time, improve quality and linking all the stage in the illustration map as a visual of future state analysis ([18; [19]].

3. Result and Discussion

3.1. Rice Supply Chain

Understanding the rice supply chain is an important aspect to analyze current state as well as future state map. The study is underpinning the current process all over the rice supply chain as an understanding of the manufacturing process of a rice commodity. Figure 1 illustrates the process mapping of the rice supply chain in Indramayu district.

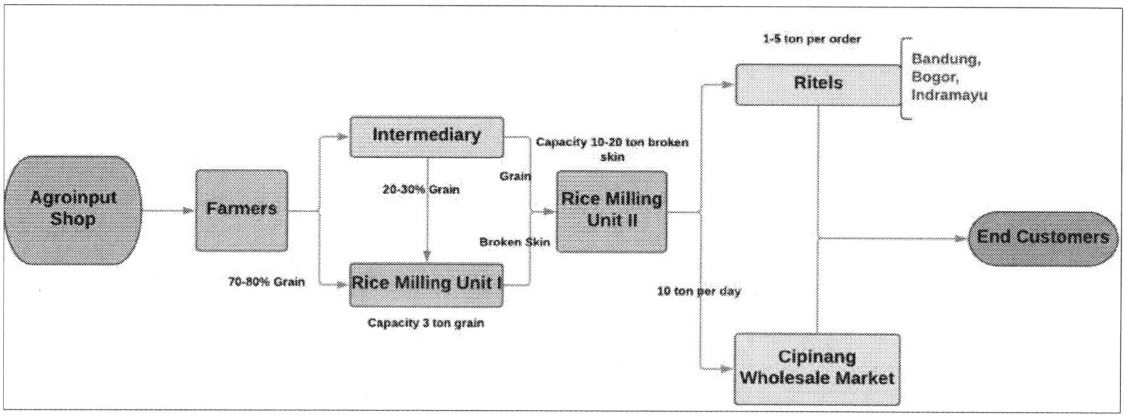

Figure 1. Current rice supply chain map at Widasari, Indramayu district.

Rice supply chain in Widasari, Indramayu was involved multiple actors, i.e., farmers as a producer, intermediaries (known as bandar) as an intermediaries between farmers and RMU (rice milling unit), and in this picture, RMU is divided into two type, i.e., RMU I is only do processing process from grain into broken skin and RMU II is an actor who does post-harvest from broken skin until marketing rice. In this figure shows that the rice supply chain develops based on distribution roles to produce rice. It is mean that rice supply chain has been linkage and it is an opportunity to develop efficiency more effective to develop the system. Meanwhile, 80% of grain is supplied directly by farmers to RMU I without going through a dealer or other intermediary. This indicates that farmers have an alternative market to sell grain produced to get a better price and competitive bargaining position. Meanwhile, intermediaries contribute to supply grain around 20% and in general, the intermediary is the partner of each RMU I or RMU II. However, it does not influence the price fluctuation and farmers still do not have a bargaining position. All actors in Indramayu (farmers, intermediaries, and RMU) are still as price taker and market act as price maker.

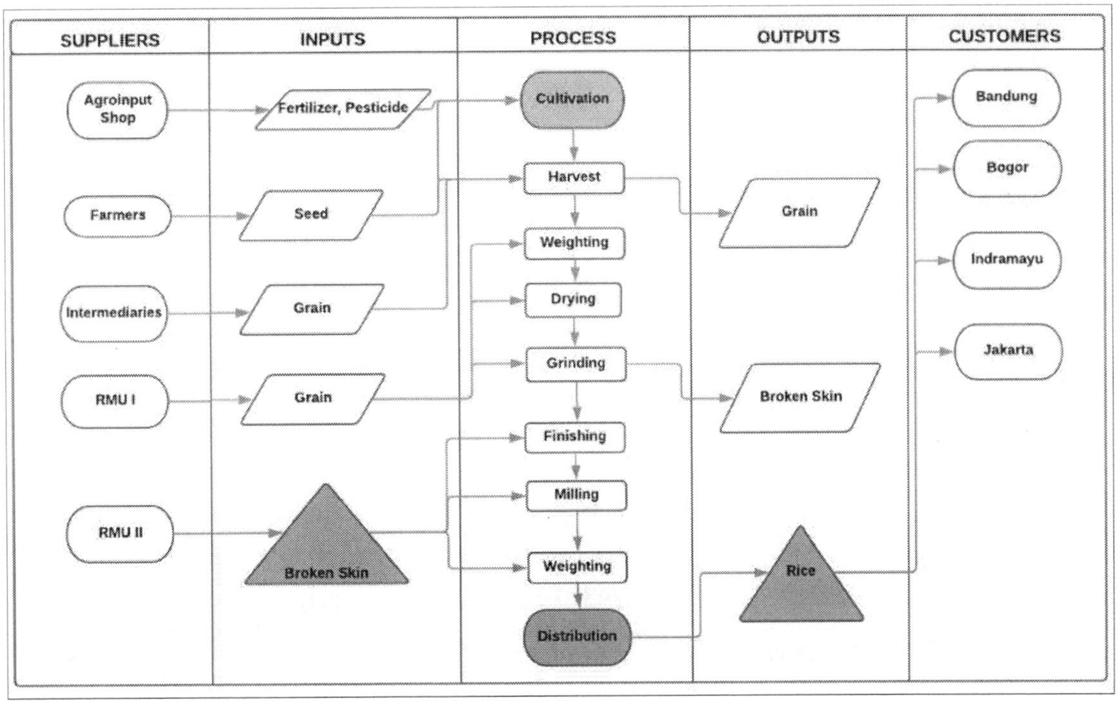

Figure 2. SIPOC mapping on production process of the rice industry.

Figure 2 illustrates the current situation of rice production flow. The SIPOC mapping consists of the supplier (means actors entire rice supply chain), input, process (starting from procurement until distribution), output and customers (means shop or market). Supplier consists of agro-input shop, farmers, intermediaries, and RMU. Meanwhile, local government acts as a policymaker and does several supports for the production process. The output of this supply chain is rice that divided into two grade; medium and premium quality. Meanwhile, RMU I is only produce broken rice to fulfill RMU 2 needs. Business process of rice value stream is described in the schematic picture.

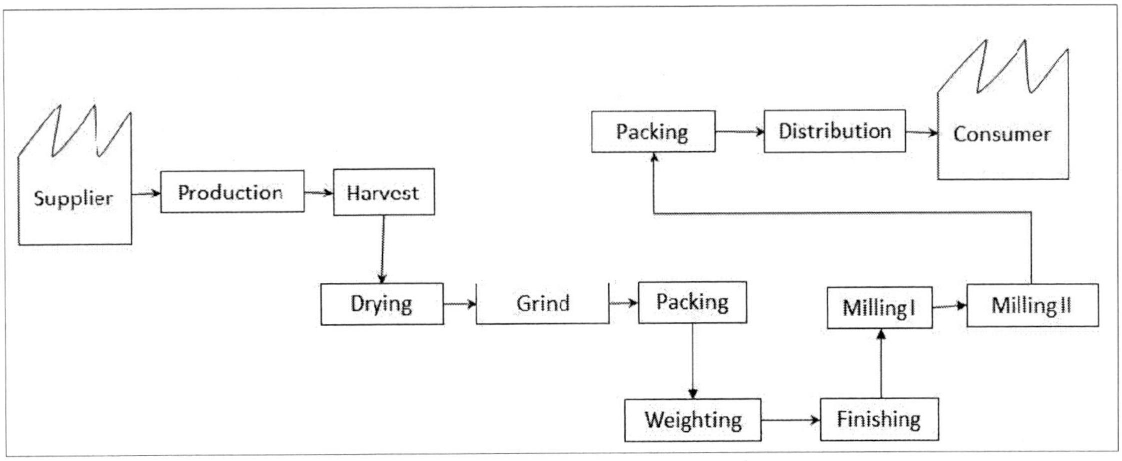

Figure 3. Schematic picture of rice industry.

According to figure 3, the business process of the rice supply chain is having some process from production until the distribution process. Farmers do production process and after harvest, product delivered to the intermediary and/or directly distributes to RMU I. It seen that farmers do not have the

IOP Conf. Series: Earth and Environmental Science **335** (2019) 012024 doi:10.1088/1755-1315/335/1/012024

competitiveness and less value-adding process. RMU II do a lot of value adding and have a chance to get more some value of products as a strategy to build the value of products. According to the capacity of production, from 1 Ha field could produce 7-8 ton grain with the conversion from grain to rice is about 62.77%. RMU I only can process 3 ton grain per day and RMU II can be handling 10 ton broken skin per day.

3.2. Current State Mapping

Following table 1, all the activities depend on estimations process. PERT (Program Evaluation and Review Technique) analysis is used to describe the spend time on each process activities. Rice value stream is seen that the production (cultivation) process is lack of technology. It was difficult to maintain the quality of products. Meanwhile, the drying process is still using the traditional method rather than using new technology. Value stream should show non-value adding activities and maintain a pull-off process as well as lead time. It is seen that rice value stream mapping has a lot of losses and uncertainty of quality and time effectiveness.

Table 1. PERT analysis in the rice value stream process.

No	Activities	Activity Type	Expected Time (t_e) (Hour)	Activity Value
1	Agro-input Procurement	Operation	2.00	NVA
2	Land Processing	Operation	20.33	NNVA
3	Basic Fertilization	Inspection	2.67	VA
4	Planting	Operation	5.33	VA
5	Cultivation	Operation	19,920	VA
6	Pest and Diseases Protection	Inspection	96.67	VA
7	Harvest	Operation	4.67	VA
8	Drying	Operation	48.00	VA
9	Grinding	Operation	3.00	VA
10	Packing	Operation	1.00	NNVA
11	Weighting	Operation	1.51	VA
12	Finishing	Operation	3.58	VA
13	Milling I	Operation	4.42	VA
14	Milling II	Operation	4.08	VA
15	Packing	Operation	1.14	VA
16	Distribution	Transportation	9.33	NNVA

Regarding information flow in figure 4, the current state mapping or an overview of the state is the performance of the rice supply chain in Indramayu District. In the picture, it is clear that in each post-harvest process of rice there is a waiting time before the rice moves from process one to the next process. This is an obstacle to waste management, especially waste at the time of processing. The length of time wasted on each process will indirectly affect the quality of rice produced, the speed performance of the production time to affect the costs used during the post-harvest process [3].

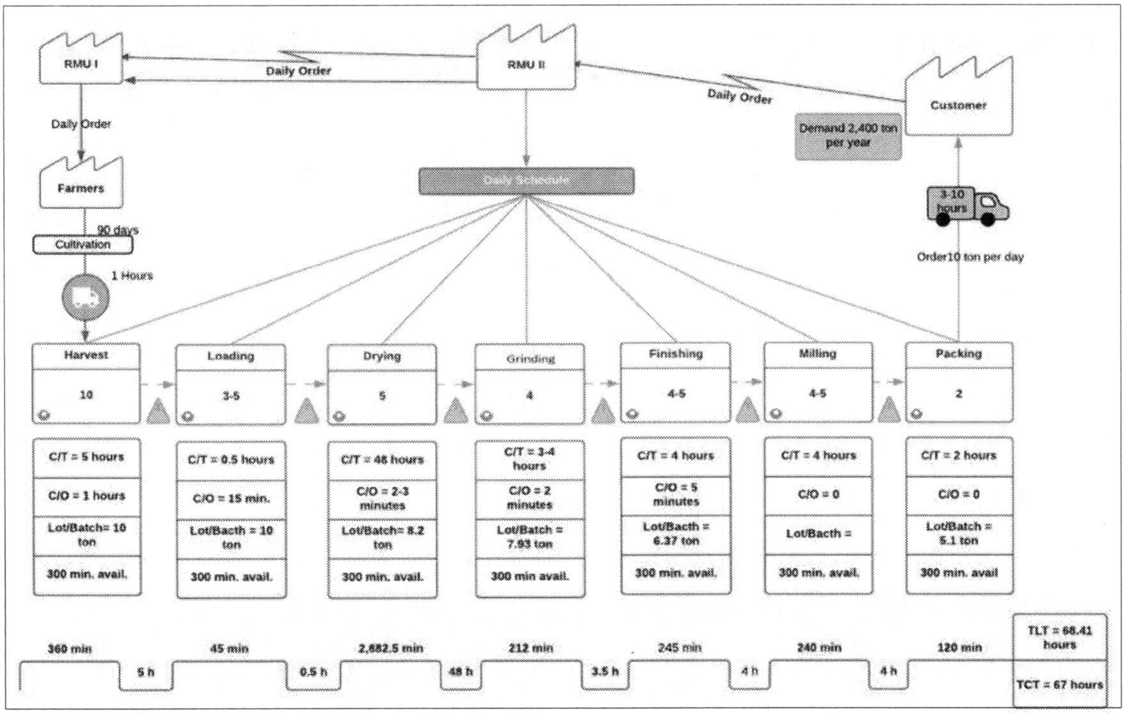

Figure 4. Current state mapping of rice industry.

The communication in the business process is divided into a 2-way system, first direct and indirect communication using the telephone. This communication is about negotiation, determination of quality and price. The drying of rice is still done manually, which is still using solar energy so that the process of drying rice is still relatively long and affects the subsequent post-harvest process. The average farmer can produce 10 tons of grain, with an estimated conversion of grain to rice by 62.74%. RMU is an important factor in marketing and plays a role in increasing the value of rice produced by farmers. TCT in the current state around 67 hours, with TLT long 68 hours and 41 minutes which means that there is a more time-consuming activity. Therefore, time management is important to improve the performance of the rice supply chain in Indramayu District [9].

Table 2. Classification analysis of activities and average time of current state rice production.

Activities	Time (hour)	Percentage
VA	20,095	99.84%
NVA	2.00	0.01%
NNVA	30.66	0.15%
TOTAL	**20,127.73**	**100.00%**

Most of the activities in the rice production are adding the value of any activities that add value to the consumer. Value adding process in rice production is an effort to provide the best customer service. Meanwhile, any kind of activities which do not add value is waste. That would decrease TCT of the production to set approaches for tracking the progress in the entire supply chain.

3.3. *Future State Mapping*

Future state map is a step to create an idea as an effort to improve and plan to achieve the efficiency of the rice production process. The implementation of rice daily production is reducing waste. TAKT time of rice supply chain is about 120.7 minutes per Kg. It is mean that every 120.7 minutes of rice has completed in whole processing. In the current state map sight that the roughing process consumes about 28 minutes more than TAKT time.

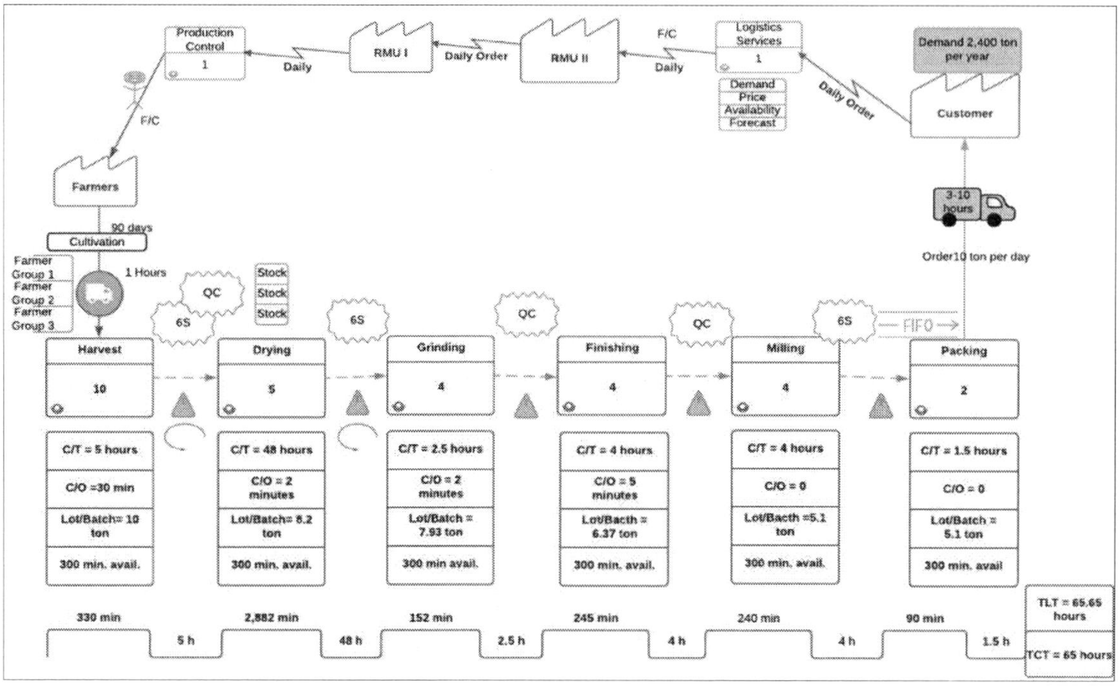

Figure 5. Future state map of rice industry.

According to current state case scenario, there is some waste in the post-harvest process and it should be reduced in order to improve the efficiency of the value chain, cost, processing time and some activities. Furthermore, rice value stream should develop a contract between RMU II, RMU I and farmers as well as RMU II and market. It is a guarantee to do forecasting (continuity of products) and get a stable price. Moreover, maintaining the relationship with market or customer, RMU II should collaborate with logistics services which are help RMU II to do forecasting, negotiation and other business administration process, as well as logistics services should play a role as the center of rural and city logistics and supply chain. In this scenario, most of the communication will be done by an electronic system which could decrease waste of time in order to fulfill the efficiency of customer services and production. Moreover, in the downstream part should develop production controls that play a role to control continuity of production, production services for farmers (sharing knowledge and actual information related production problems), improvement of cultivation scheduling, and social engineer. This collaboration is adopting distributed government in order to distribution process for each actor and for sharing value as an effort to improve wealthy. Furthermore, the production control services should visit farmers more often and all the information should obtain based on direct communication. It is an endeavor to apply socio engineering process. As the preventive way for continuity of production, basis production should build from several farmers or farmers group.

Regarding rice future state map, networking between farmers, RMU I and RMU II must be strengthened. Thus, the intermediary is not required; therefrom decrease Non Value Added activities. Decreasing Non-Value Added activities is reminding the parties to obtain the highest profit in global competitiveness. Furthermore, in the production process, a farmer who from an individual or are grouped is under the same contract, term and right. The production control unit also will be able to manage farmers need such as agro-input procurement (such as fertilizer, pesticide, and seed) and also funding as to support cultivation process. The goal of the production control unit is to maintain the quality and quantity of products before delivered to the RMU (Rice Milling Unit) I. Accordingly, the time spent at this process can be minimized. This scenario shows that rice production should calculate of stock as a strategy to maintain price, availability of product and controlling the quality of the product. Quality control must be applied to start from drying process until the packaging process. In this scenario, 6s (*seiri* (sort), *Seiton* (Set in Order), *Seiso* (Shine or sweeping), *Seiketsu* (Standardize), *Shitsuke* (Sustain), and Safety) system was used to eliminated waste within the post-harvest process. In addition, the packaging process is one of the important aspects to increase market strategy. Branding and packaging become an essential aspect in the global competitiveness. Branding and packaging is a part to increase the value of products and enhance agribusiness industry in the globalization era [20]. Lead time on post-harvest process was decreased gradually from 68.41 hours to 65.65 hours. Moreover, the estimation of the efficiency is about 99.87% of the total lead time is adding the value of production and only 0.13% of the time is wasted on the loading process.

4. Conclusion

This paper has studied the efficiency of the rice supply chain in Indramayu District through value stream mapping in order to analyze the performance of the rice supply chain. The result was illustrated into several processes, supply chain mapping, SIPOC diagram, Schematic Picture, and current state map was used to the understanding of actual information about rice industry. Moreover, future state map was illustrated the optimum scenario to improve the performance of rice supply chain. In order to improve efficiency, the rice industry should improve socio engineering between farmers, RMU I and RMU II and also involving logistics services and production control to maintain quality, continuity and market guarantee. Furthermore, quality control, stockiest system, and technology engineering should involve as an effort to increase the value of products. In addition, branding and packaging of products must be added as a strategy in the marketing process as an effort to increase the value of products as well as enhance the competitiveness of rice industry in the global competitiveness.

References

[1] Blackburn, Joseph and Gary S 2009 Supply chain strategies for perishable products: the case of fresh produce *Production and Operations Management* **18.2** 129-137

[2] Behrouzi F and Wong K Y 2011 An investigation and identification of lean supply chain performance measures in the automotive SMEs *Scientific research and essays* ***6*(24)** 5239-5252

[3] Christopher M and Gattorna J 2005 Supply chain cost management and value-based pricing *Industrial marketing management* ***34*(2)**115-121

[4] Department for environment, food, and rural affairs 2006 *Food Industry Sustainability Strategy* Available at; <https://www.gov.uk/government/uploads/system/uploads/attachment_data/file/69283/pb11649-fiss2006-060411.pdf> [accessed 30 December 2017]

[5] Dolan C Humphrey J and Harris-Pascal C 1999 Horticulture commodity chains: the impact of the UK market on the African fresh vegetable industry

[6] Flynn, Barbara B, Roger G S and Sadao S 1995 The impact of quality management practices on performance and competitive advantage *Decision sciences* **26.5** 659-691

[7] Kuo J C and Chen M C 2010 Developing an advanced multi-temperature joint distribution system for the food cold chain *Food control* ***21*(4)** 559-566

[8] LaLonde B J and Pohlen T L 1996 Issues in supply chain costing *The International Journal of Logistics Management* **7(1)** 1-12

[9] Lamming R 1993 *Beyond Partnership: Strategies for Innovation and Lean Supply* (Prentice Hall)

[10] Lange A, Siebert R and Barkmann T 2015 Sustainability in land management: an analysis of stakeholder perceptions in rural northern Germany *Sustainability* ***7*(1)** 683-704

[11] Lewis M A 2000 Lean production and sustainable competitive advantage *International Journal*

of Operations & Production Management ***20*(8)** 959-978

[10] McDonald T, Van Aken E M and Rentes A F 2002 Utilising simulation to enhance value stream mapping: a manufacturing case application *International Journal of Logistics* ***5*(2)** 213-232

[12] Opara L U 2002 Engineering and technological outlook on traceability of agricultural production and products *Agricultural Engineering International: CIGR Journal*

[13] Shah, Rachna, and Peter T W 2003 Lean manufacturing: context, practice bundles, and performance *Journal of operations management* **21.2** 129-149

[14] Simchi-Levi D, Kaminsky P and Simchi-Levi E 2000 *Designing and Managing the Supply Chain* (Singapore: Mc. Graw Hills Book Co.)

[15] Statistik B P 2018 *Statistik Indonesia* (Jakarta : Badan Pusat Statistik)

[16] Statistik B P 2018 Jawa Barat dalam angka (Jakarta : Badan Pusat Statistik Provinsi Jawa Barat)

[17] Stevens G C 1989 Integrating the supply chain *International Journal of Physical Distribution & Materials Management* ***19*(8)** 3-8

[18] Renting H, Marsden T K and Banks J 2003 Understanding alternative food networks: exploring the role of short food supply chains in rural development *Environment and planning A.* ***35*(3)** 393-411

[19] Russell R S and Taylor-Iii B W 2008 *Operations Management Along the Supply Chain* (John Wiley and Sons)

[20] Rother M and Shook J 2003 Learning to See: Value Stream Mapping to Add Value and Eliminate Muda (Lean Enterprise Institute)

[21] Tarapituxwong S, Tantranon N and Duangpatra C 2015 Production process improvement using value stream mapping: a case study of organic coffee firms in Thailand *FEU Academic Review* ***8*(2)**128

[22] Utami H N, Sadeli A H and Perdana T 2016 Customer value creation of fresh tomatoes through branding and packaging as customer perceived quality *International Society for Southeast Asian Agricultural Sciences–Philippines* ***22*** 123-136

[23] Van Donselaar K, Kokke K and Allessie M 1998 Performance measurement in the transportation and distribution sector *International Journal of Physical Distribution & Logistics Management* ***28*(6)** 434-450

[24] Womack J P and Jones D T 2002 11 from lean production to lean enterprise *Managing Innovation and Change* p.138

[25] Womack J P, Jones D T and Roos D 1990 *Machine That Changed The World* (Simon and Schuster)

Acknowledgment

We would like to acknowledge to Universitas Padjadjaran which funded Internal Grant (HIU).

Development of web-based for egg surveillance to maintain egg distribution

R Budiman[1], S P Thoriq[1], M A Suyuthi[1] and L Nurhikmat[1]

[1]Bogor Agricultural University, Bogor, Indonesia e-mail: rhavif_budiman@apps.ipb.ac.id

Abstract. Recently there was news that the price of eggs increasing, and one of the problems is caused by uneven distribution. A system for particular interest about the data and information for eggs distribution is urgently needed to solve this problem. The objective of this research is to design a system that is able to visualize egg distribution in Bogor through WEB GIS. Development of this information system is Web-*based Development Life Cycle* (WDLC). This method consists five steps, there are web development planning, engineering web requirements, web application design, web technical implementation, and usage web system. As the result, a prototype design of web GIS has been developed to be able to visualize eggs distribution in Bogor. This system can be used as a tool to determine egg distribution of a location either in Bogor or other location in Indonesia. It is recommended that the government ca use this system to support even eggs distribution in Indonesia.

Keywords: egg distribution, web-based development life cycle, web GIS

1. Introduction

Indonesia is a large country with a large number of inhabitant and high population growth rate. Because of that, the demand of food and one of them is egg. Eggs are one type of food that is widely known to have high nutrition and good for human body. Eggs are of particular interest from a functionality point of view, because they offer a moderate calorie source (about 150 kcal/100 g), a protein of excellent quality, great culinary versatility and low economic cost [1]. In Indonesia, recently there was news that the price of eggs increasing, and one of the problems is caused by uneven distribution. A system for particular interest about the data and information for eggs distribution is urgently needed to solve this problem. A system for particular interest about the data and information for eggs distribution is urgently needed to solve this problem.

Geographical Information Systems (GIS) is a system that have ability to integrate geospatial data from many varieties of sources. Obviously, maps play an important role in GIS. Web based GIS is a type of GIS that combines a geographical information system and web technology. The reach of the GIS system multiples manifold when it is implemented through the internet. Not only does it get a greater user set, it also means that the user need not invest in a resource intensive machine to access the Web based GIS system. This makes Web based GIS both cheap and more readily accessible [2]. With different theme or purposes and integrated with spatial data and maps, GIS is a solution to understand spatial data for laymen and experts [3]. while the Web GIS according to ESRI is is a type of distributed information system, comprising at least a server and a client, where the server is a GIS server and the client is a web browser, desktop application, or mobile application. Web GIS offers many advantages including ease of use, support for unified updates, diverse apps and more.

IOP Conf. Series: Earth and Environmental Science **335** (2019) 012025 doi:10.1088/1755-1315/335/1/012025

This paper will discuss about WEB GIS that builds display of information about the number of eggs in an area, the price of eggs in an area, and the area in the form of heat maps that are useful to facilitate the public and the government to obtain information about the distribution and number of eggs in an area.

2. Material and Method

The method used in this study is Web-based Development Life Cycle (WDLC) which presents in figure 1. This method consists of five stages, namely web development planning, engineering web requirements, web application design, technical web implementation, and usage web system [4].

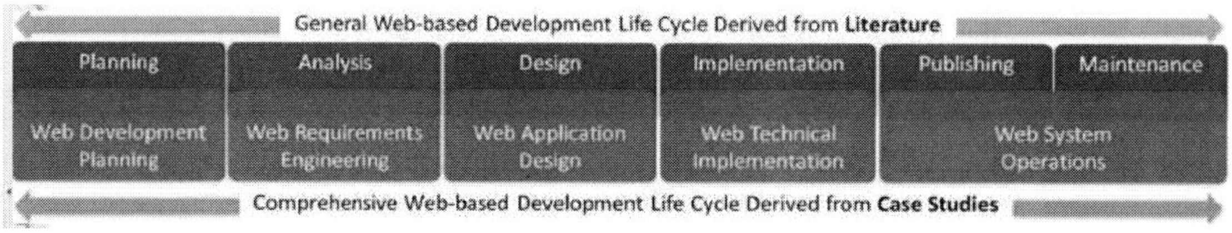

Figure 1. Web-based development life cycle [4].

2.1. Web Development Planning (WPA)

In web development planning step are to understand the schedule development systems, the strategy are schedule, clearer goal and work plan.

2.2. Web Requirement Engineering (WRE)

In engineering web requirement step, needs data that involve gathering and delivering needs from the user and needs analysis which will result requirements and model specifications. This requirement document needs to be validated for make sure the requirements specified are correct. In this step, involves, search egg distribution problem, and search book, involves eliciting responses and gathering requirements through direct communication, collecting relevant book or any related documents.

2.3. Web Application Design (WAD)

In web application design, all specifications are prepared and developed design needed. The prototype was made to show the feasibility of the design. This stage is done for produce a complete requirement before the construction of the system in fact.

2.4. Web Technical Implementation (WTI)

In web technical implementation step, technical implementation covers the development of information-based systems web, which is the creation of databases, building interfaces, and implementation code so that it can be used in various web browsers. In usage web system, Web-based information systems are distributed well on the web hosting or on the user's own site. Thus, optimization of search engine has been applied.

2.5. Web System Operation (WSO)

In usage web system, Web-based information systems are distributed well on the web hosting or on the user's own site.

3. Result and Discussion

3.1. Web Development Planning (WPA)

Web Development Planning is focus in planning the understanding user needs, characteristics, motivation, website goal and defining the user expectations and requirements of the new or not yet developed website [5]. In this step, WEB GIS developer planning development system. Plan that

IOP Conf. Series: Earth and Environmental Science **335** (2019) 012025 doi:10.1088/1755-1315/335/1/012025

developer doing is data processing, needs analysis, design, implementation code program, system testing, and system usage.

3.2. Web Requirement Engineering (WRE)
In this step, there are two main activities. First, requirement elicitation that involve articulating and gathering the client needs. Secondly, requirement analysis which at the end will produce model and specification requirements [6]. In engineering web requirement step, the activities that carried out is collecting system requirements and collecting data.

3.2.1. Data collection and analysis needs. In the initial stages of engineering web requirements is to collect egg consumption data in an area, because to make it easier, the area chosen is the Bogor area, based on [7] Badan Pusat Statistik (BPS) Jawa Barat, in average per capita consumption of egg is 2,199 kg per week in 2017. Because there is no data that mentions the number of eggs that vary in the community, it is assumed that the number of eggs in circulation is the same as the number of eggs consumed by the people of Bogor. Based on [8] BPS 2016, in figure 2, there are names of regions, area, population and estimated number of eggs in Bogor city.

Table 1. sample data for egg distribution.

Sub district	Area (ha)	Population (person)	Estimated egg (kg)
Cariu	8.555	46.186	101.563
Cibinong	4.611	326.519	718.015
Dramaga	2.632	100.679	221.393
Kelapa Nunggal	6.860	104.120	228.959
Jonggol	13.463	122.697	269.810
Bojong Gede	2.824	236.486	520.032

3.2.2. Data collection and analysis needs. This information system has the following functions: (1) displays the map of districts in Indonesia, (2) displays general information about Web GIS egg distribution, (3) displays legends and tables as egg distribution information that supports maps, and (4) data management carried out by the administrator. Map processing systems have the following functions: (1) display maps, tables, legends along with egg distribution data throughout Indonesia, and (2) map navigation includes zoom-in, zoom-out, grab tool and pan on the map.

System functional requirements that describe the entry and exit of data can be seen in figure 2.

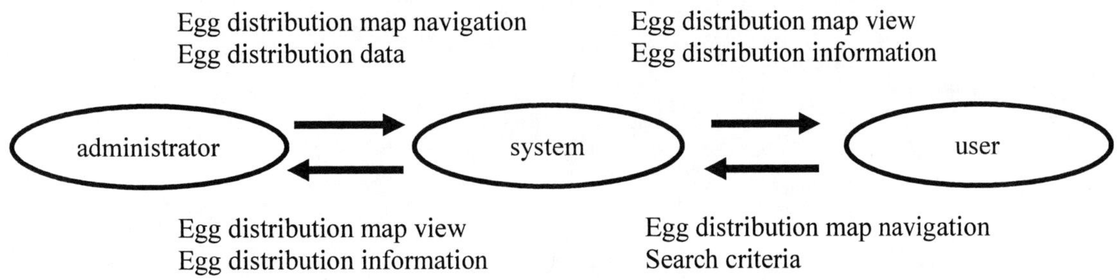

Figure 2. Data flow diagram.

3.2.3. Determining User. One of the critical steps is determining user, in determining user determine why user use or visit this website and identify any constraint they will experience during visiting the website [9]. Users of this system are classified into two, the first is general users, namely those who need information regarding the distribution of eggs in Indonesia. And second is administrator who has access to manage spatial data on egg distribution.

3.2.4. Research tools. The software used in the development of Web GIS and the development of egg distribution systems is ArcGIS 10.3. PostgreSQL 9.2 is used as software for database management, while GeoServer 2.2 Snapshot as a web map server for managing data in the form of maps. Open Street Map is used for map mapping applications, stylers for setting styles (SLD), and XAMPP 7.1 as web servers.

The hardware that will be used in the development of the egg distribution GIS Web is a personal computer with the following specifications: Intel (R) Core (TM) i5-7200U CPU @ 2.50GHz, 8192MB RAM, and 1 TB hard disk.

3.3. Web Application Design (WAD)

Web application design is represented in two ways, namely system interface design, and system architecture design.

3.3.1. Research tools. The design of the Web GIS interface can be seen in figure 3. The main page of the system consists of four main parts, namely:

1. Header, containing system titles, menu buttons, and buttons about the system.
2. The body that contains the map of Indonesia along with the map navigation menu includes zoom-in, zoom-out, grip tool and pan map.
3. Body that contains queries
4. Footer, containing information on the GIS web maker.

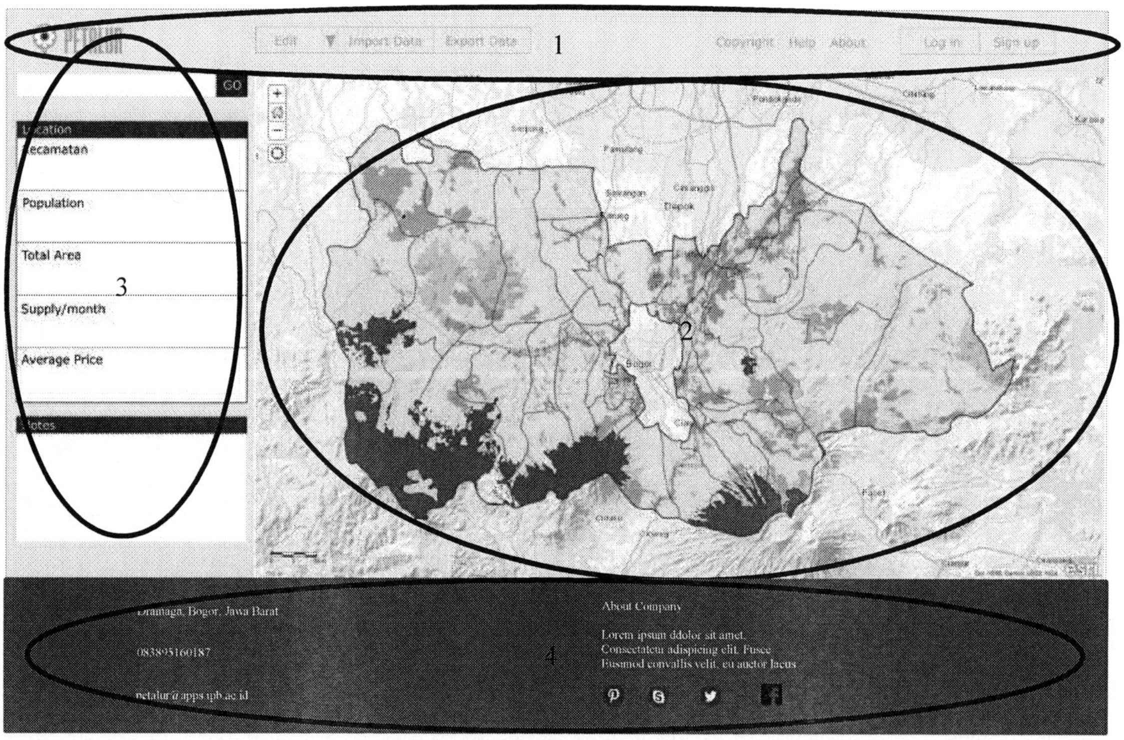

Figure 3. Website design.

IOP Conf. Series: Earth and Environmental Science **335** (2019) 012025 doi:10.1088/1755-1315/335/1/012025

3.3.2. Raised map button. The purpose of Raised map button is to make it easier for users to distinguish the destination map and other maps.

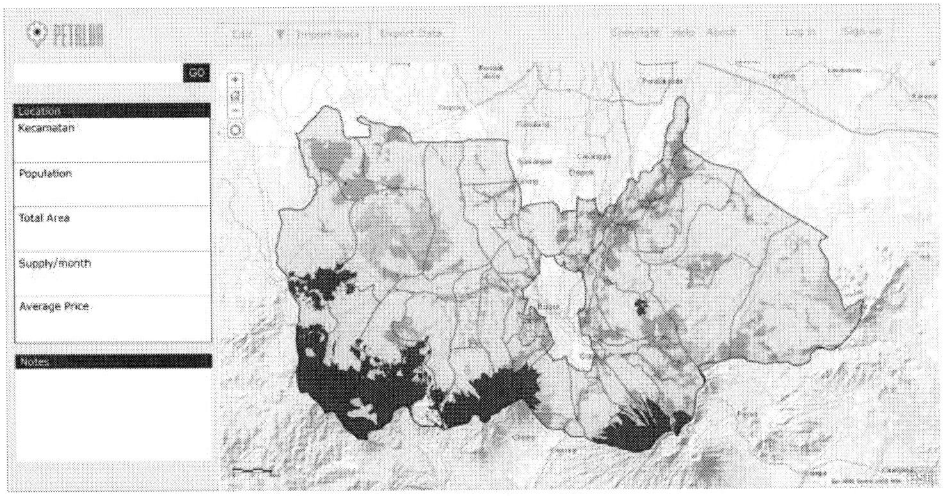

Figure 4. Display before click.

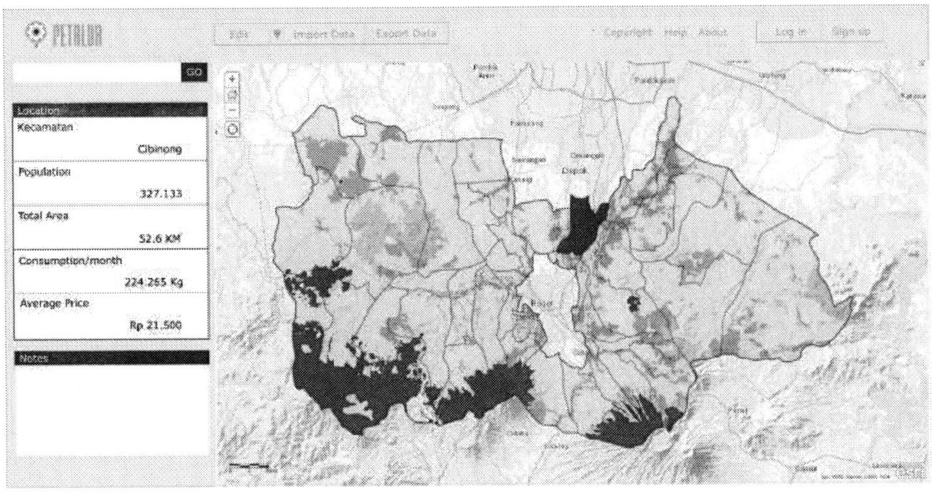

Figure 5. Display after click.

3.3.3. System arhitecture design. The design of this website architecture was developed using a client-server architecture. The client-server architecture at least has a user interface using bootstrap to build the interface on the client side and distributed data stored on the server side. The client will send the request via the web browser to the web server. The web server then receives requests from users and sends them through a web protocol like HTTP (Hyper Text Transfer Protocol). HTTP will generate the connection between client and server. Map requests sent by users through the web browser will be received by the web server. Furthermore, the web server will forward it to Geoserver which is a geospatial data server, and GeoServer will forward to the data server where the egg distribution database is stored. The website architecture used in system development can be seen in figure 4.

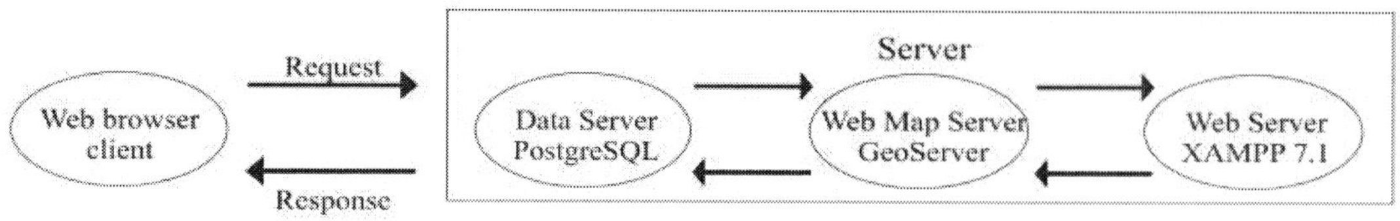

Figure 6. client-server architecture.

3.3.4. Future Implementation. The implementation for the development of web-based for egg surveillance is monitoring and evaluation. The monitoring and evaluation egg surveillance is important tool targeting two elements, that is effectiveness and sustainability. Monitoring and Evaluation should span the life cycle of the system to providing a continuing stream of data and feedback in the ideal situation [10]. This web GIS using monitoring and evaluation system to estimate the amount of the egg in the boundary area. the data that thatget on the monitoring can improve the prospect of sustaining successful outcome at evaluation process. the outcome of this web GIS is based by the transparency and accuracy of estimating the amount of eggs to maintain the distribution of egg at some area.

4. Conclusion

A web-based geographic information system for egg distribution data in Indonesia uses QGIS 2.18.21 which serves to display information names of regions, area, population and estimated number of eggs. This system was built with the Web-based Development Life Cycle (WDLC) research method and implemented on the Windows 10 operating system, ArcGIS 10.3, and PosgreSQL as database management, and integrated with Google Maps and Open Street Map. This GIS allow users to obtain information on names of regions, area, population and estimated number of eggs.

References

[1] Carrillo S, Rios V H, Calvo C, Carranco M E, Casas M, and Perez-Gil F 2012 *N*-3 fatty acid content in eggs laid by hens fed with marine algae and sardine oil and stored at different times and temperatures *Journal of Applied Phycology* **24** 593-599

[2] Chakraborty D, Sarkar D, Agarwal S, Duttal D, and Sharma JR 2015 Web based GIS application using open source software for sharing geospatial data *International Journal of Advanced Remote Sensing and GIS* **4** 1224-1228

[3] Krakk M 2004 The role of the map in a Web-GIS environment *Journal of Geographical Systems* **6** 83-93

[4] Abdul-Aziz A, Koronios A, Gao J, and Sulong M S 2012 A methodology for the development of web-based information system: web development team perspective. *Proceedings of the Eighteenth Americas Conference on Information System* (Seattle Washington: AIS Electronic Library) p 1-9

[5] Sniegula A and Glinka K 2017 Usability testing in web development life cycle *Entrepreneurship and Management* **18** 131-145

[6] Abdul-Aziz A, Koronios A, Gao J, and Sulong MS 2012 A methodology for the development of web-based information system: web development team perspective *Proceedings of the Eighteenth Americas Conference on Information System* (Seattle Washington: AIS Electronic Library) 1-9

[7] Badan Pusat Statistik [BPS] 2017 *Per Capita Weekly Average Consumption of Several Food Items* (Jakarta: Indonesia)

[8] [BPS] Statistics of Bogor regency 2017 *Bogor City in Figures* (Bogor: Indonesia)

[9] Kamatchi R, Iyer J, and Singh S 2013 Software engineering: web development life cycle *Internasional Journal if Engineering Research Technology* **2** 1-4

[10] Catherman R 2013 *Monitoring and Evaluating Programs and Project using GIS* (California: MEDRIX)

IOP Conf. Series: Earth and Environmental Science **335** (2019) 012025 doi:10.1088/1755-1315/335/1/012025

Acknowledgement
I am deeply indebted to my supervisor, Professor Kudang Boro Seminar, for warm support, inspiration and thoughtful guidance.

IOP Conf. Series: Earth and Environmental Science **335** (2019) 012026 doi:10.1088/1755-1315/335/1/012026

Efficient mapping for fodder cultivation with commercial small drone

N H Pandjaitan[1], M I Rau[1], Sutoyo[1], T U Karunia[1], R S B Waspodo[1], N Mustikasari[2]

[1] Department of Civil and Environmental Engineering, Faculty of Agricultural Technology, Bogor Agricultural University, Bogor 16680, Indonesia, email:norahp@apps.ipb.ac.id
[2] Indonesian Center for Agricultural Land Resources Research and Development (ICALRRD), Jl. Tentara Pelajar, Bogor 16000 Indonesia

Abstract. To optimise land use, an overview regarding its topography of the proposed land is essential to be portrayed, especially to depict the slope, elevation, calculation of cut and fill, as well as planning for water supply. However, in practice, high-resolution topographic data are hard to obtain and are highly priced. As an alternative solution, commercial small drones could be utilized to improve the input for such data. Commercial small drones could be utilized as a photogrammetric approach to obtain detailed digital surface model (DSM). This study aimed to get high-resolution of topography condition using commercial small drones and evaluate the result comparing to terrestrial surveying. The approach was done using Drone DJI Phantom 3 Professional, where the ground control points (GCPs) were undertaken with GPSMAP 64s SEA. The study area was ±23 ha proposed for fodder cultivation, located at Ngantru Village, Bojonegoro, East Java Province in Indonesia. Outcomes obtained from the analysis were orthophoto, digital models, and contours. The results yielded an efficient approach for mapping compared to field/terrestrial surveying. The drone approach only cost IDR ±4,000,000, whereas terrestrial mapping (using electronic total station) was worth up to IDR ±25,000,000. As the expense altered significantly, the process of collecting data varied as well. The overall data acquisition using drone only took 2 hours to complete, while field survey may require up to 2 days of acquisition. Although results accuracies were also an issue, the output reliability for drone with errors from 0.5 to 6 m was somewhat acceptable. In conclusion, small commercial drones were one way to undertake efficient mapping (in terms of cost and time) compared to field/terrestrial survey.

Keywords: Digital surface model, Topographic mapping

1. Introduction

To optimise land use, an overview regarding its topography of the proposed land is essential to be portrayed. In many perspectives, including fodder cultivation as well, topographic mapping is one of such main parameters to be put into an account when undertaking planning for agricultural production. In agriculture, especially in fodder cultivation, topographic mapping is essential to depict the slope, elevation, calculation of cut and fill, as well as planning for water supply. As a result, this will indeed affect the number of productivity at the end.

However, in practice, high-resolution topographic data are hard to obtain and are highly priced. As an alternative solution, commercial small drones could be utilized to improve the input for such data. Since its significant development for civil/commercial type of platforms, especially in 2012 and 2013 [1], drone has recently been very easy to utilize. Commercial small drones could be performed as a photogrammetric approach to obtain detailed digital surface model (DSM). Reconstructing original 3D models of space objects based on the two-dimensional images obtained from different points of sight has been a hot research area in the field of photogrammetry and computer vision [2].

This study aimed to get high-resolution of topography condition using commercial small drones and evaluate the result comparing to terrestrial surveying.

2. Materials and Methods

2.1. Study Location

The fodder cultivation study was located at Sentra/Sekolah Peternakan Rakyat (SPR) Mega Jaya, Ngantru Village, Bojonegoro Regency, East Java Province, Indonesia. SPR Mega Jaya was positioned approximately at 573003 mE and 9218285 mS (WGS 1984, UTM 49S). The situation surrounding SPR Mega Jaya is given in the Figure 1.

Figure 1. Study location at SPR Mega Jaya.

2.2. Tools and Software

Tools used for this study included commercial drone, i.e. DJI Phantom 3 Professional, and a consumer-grade receiver GPS, i.e. GPS Garmin GPSMAP 64s SEA (Figure 2). Software utilized for the study were ArcGIS 10.4.1, Agisoft Photoscan Professional, and Pix4D Capture (iOS).

IOP Conf. Series: Earth and Environmental Science **335** (2019) 012026 doi:10.1088/1755-1315/335/1/012026

(DJI Phantom 3 Professional) (Garmin GPSMAP 64s SE)

Figure 2. Drone and GPS.

3. Results and Discussion

3.1. Data Acquisition

The data acquisition was initially undertaken by building up both flight plan and ground control points (GCPs). As the flight mission was built using Pix4D Capture [3], information of the flight height, captured area, as well as time estimation of the data acquisition could be attained. The proposed area of interest of ±23 ha was captured in 85 m height which took approximately 14 minutes of acquisition. The depiction of the flight mission for SPR Mega Jaya in Pix4D Capture was showed in Figure 3.

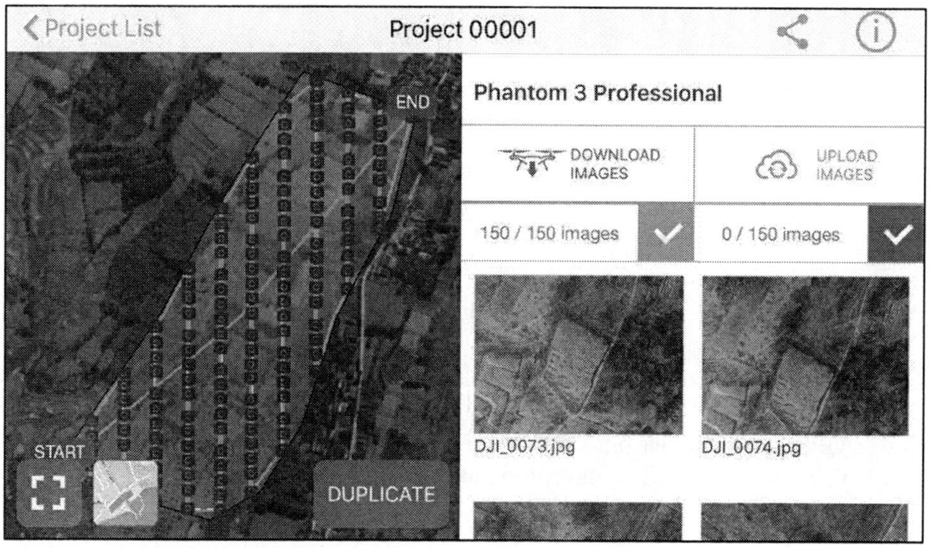

Figure 3. Flight mission for SPR Mega Jaya with Pix4D Capture.

As soon as the flight mission was established, GCPs were required to be well-planned as well. As for this study, GCPs were placed in sparse locations, yet represent the extent of the purposed area of interest. The GCPs were measured using a consumer-grade GPS which has been reported to have 5-10 m accuracies [4-6]. Number of GCPs used within the study was 3 points. Depiction of GCPs established surrounding SPR Mega Jaya was showed in Figure 4.

Figure 4. Ground control points established surrounding SPR Mega Jaya.

The information for each GCP and errors obtained from the field measurement were explained in Table 1. It needed to be put into a concern that the varied numbers of errors (0.7 m to 5.8 m) for each GCP were based on the coverage situation of the surrounding (i.e. vegetation, buildings, etc), as the GPS was mainly influenced by such conditions. However, despite the errors, it has been stated that consumer-grade receivers can provide acceptable accuracy and precision for environment as they are easy to use and inexpensive [7]. For this concern, the errors were not a significant considerations, as the main objective was to evaluate the efficiencies of commercial small drones for fodder cultivation. However, a further project-dependant study is required to overview the error results using consumer-grade GPS receiver.

Table 1. Ground control points established surrounding SPR Mega Jaya.

GCP Number	Latitude (°)	Longitude (°)	Altitude (m)	Error (m)
GCP 1	-7.0718	111.6661	95	2.2872
GCP 2	-7.0698	111.6618	103	0.7867
GCP 3	-7.0731	111.6608	87	5.8499

3.2. Data Processing

The processing of photogrammetric work was done with Agisoft Photoscan Professional 1.2.7 [8], as Agisoft is a common program to use still images of stationary objects to generate point clouds, 3D models and orthophotomaps [9]. Steps undertaken to finally yield orthophoto and digital elevation model was the establishment of dense cloud, mesh, texture, and tiled model, which was done in sequence. The resulting outputs for the point cloud, dense cloud classes, and tiled model are given below.

(point cloud) (dense cloud classes) (tiled model)

Figure 5. Photogrammetric process.

3.3. Orthophoto Results

From the above process, the orthophoto map acquired from drone survey could be attained. The orthophoto map was essential as the depiction resulted better understandings regarding the area. The representation of SPR Mega Jaya and its surrounding is given in the picture below.

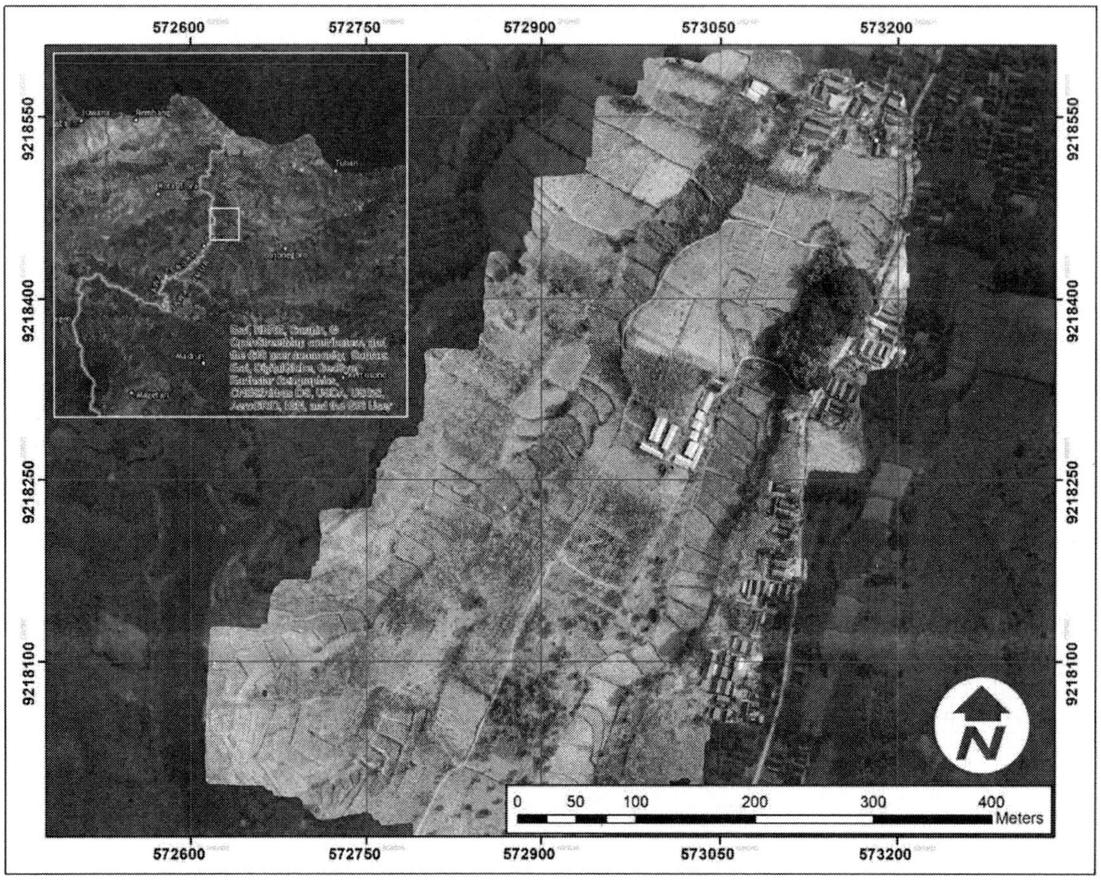

Figure 6. Orthophoto map of SPR Mega Jaya area.

3.4. Topographic Results

Finally, after all data acquisition and processing were done, the topographic results were yielded, i.e. contour and digital surface model. Both contour and digital surface model map of SPR Mega Jaya area are showed in Figure 6.

From maps in Figure 7, it could be informed that the overall survey of ±23 ha yielded elevations ranging from 59.965 m to 124.254 m (DSM-based topography). For the surrounding area of SPR alone (where existing buildings for cattles were positioned), the elevation approximately ranged from 96 m to 99 m.

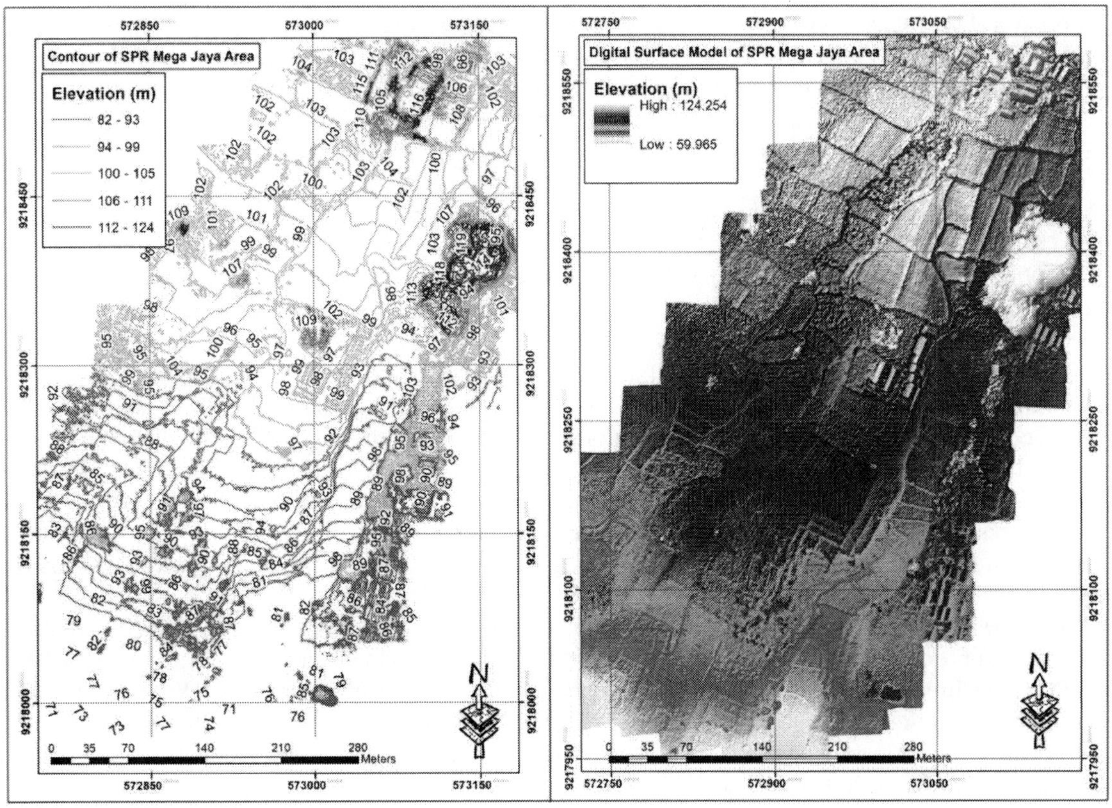

(contour map) (digital surface map)

Figure 7. Map of SPR Mega Jaya area.

3.5. Efficient Mapping

Compared to terrestrial mapping (using electronic total station or theodolite), drone mapping is one of such efficient approaches beneficial to be applied, especially for farmers which has low investment. This is mainly based on 3 aspects, i.e. cost, time, and man-power (related to the teamwork intricacy). Finally, the gross efficiency (percentage ratio of external work achieved compared to the total expenditure (project limit)) for both methods were calculated, based on the project limit for each aspect. The comparison between both methods, including the calculation results of the efficiency is given below.

Table 2. Efficiency results for terrestrial surveying and commercial small drone method.

No.	Description	Project Limit	Terrestrial Surveying	Gross Efficiency (%)	Commercial Small Drone	Gross Efficiency (%)
1	Cost for 23 Ha Survey	IDR 30,000,000	± IDR 25,000,000	16.67	± IDR 4,000,000	86.67
2	Time Consumption	3 days (24 working-hours)*	2 days (16 working-hours)*	33.33	14 minutes	98.54
3	Man-power Requirements	4 persons	3 persons	25.00	2 persons	50.00
	Gross Efficiency Average (%)			25.00		78.40

* Assumption of 1 day equals to 8 working-hours

IOP Conf. Series: Earth and Environmental Science **335** (2019) 012026 doi:10.1088/1755-1315/335/1/012026

From Table 2, there were some limitations regarding the aspects considered within this study. The cost for ±23 ha survey was limited to IDR 30,000,000 with 3 days (24 working-hours) and 4 persons of man-power. The efficiency calculated was mainly based on this limitations.

In cost perspective, terrestrial surveying commonly required IDR 1,000,000 – IDR 1,500,000 per hectare, depending on the location of the project. This cost was a common commercial value applied in surveying business environment in Indonesia. Within this study, the cost of terrestrial survey was estimated to be IDR 1,200,000. Therefore for a ±23 ha survey the total project was worth IDR ±25,000,000. However, in contrast drone mapping only cost IDR ±4,000,000, which included IDR 500,000 – IDR 1,000,000 for drone rent, IDR 200,000 for GPS rent, and for an estimated value for the qualification. From the project limit of IDR 30,000,000, the gross efficiency for both terrestrial surveying and drone survey amounted to 16.67% and 86.67%, respectively.

In terms of time consumption terrestrial mapping may require 5-10 ha per day, depending on the site condition. For this study, 10 ha per day was estimated for the terrestrial surveying, resulting the project to be accomplished in 2 consecutive days or around 16 working-hours. In contrast drone mapping only required 14 minutes of data acquisition for the same surface area. Therefore, the gross efficiency for field surveying and drone survey accounted to 33.33% and 98.54%, respectively.

With respect to man-power (which represented the teamwork intricacy or complexity), terrestrial surveying required 3 persons (1 surveyor and 2 helpers), where drone survey only needed 2 persons in charge for the data acquisition which consisted of 1 surveyor and 1 helper. Hence, the gross efficiency for terrestrial surveying and drone survey was 25% and 50%, respectively. Overall, the average gross efficiency for terrestrial surveying was 25%, whilst utilizing commercial small drone achieved 78.40%.

From the three aspects, it is apparent that drone mapping is one efficient approach to be utilized. For mapping agricultural areas/purposes, especially for activities with low investment, though had a fairly significant error for mapping (up to 5.8 m), considering the efficiencies, drone mapping could be a good choice compared to terrestrial mapping.

4. Recommendation

The derivation from digital surface model (DSM) to digital terrain model (DTM) is essential for future works. On the other hand, it is recommended that more information should be given to farmers regarding drone and its potential usage for field mapping. Remembering its efficient approach for future farming activities, it is expected that the usage of commercial small drone is possible to be undertaken by farmers as well.

5. Conclusion

In conclusion, small commercial drones were one way to acquire a high-resolution map of topography condition and undertake efficient mapping (in terms of cost and time) compared to field/terrestrial survey.

References

[1] Colomina L, Molina P 2014 Unmanned aerial systems for photogrammetry and remote sensing:
[2] A review. ISPRS J. Photogramm. Remote Sens. vol 92 p 79–97.
[3] Zheng S Y, Wang R R, Zhang Z X 2008 The three-dimensional scene modelling based on stereo camera. J Sur Map vol 11 p 20-29.
[4] Pix4D S A [internet] 2017 Pix4Dmapper 4.1 user manual; [updated available from:
[5] https://support.pix4d.com/hc/en-us/articles/204010419--iOS-Pix4Dcapture-Manual].
[6] Wing M G, Eklund A, Kellogg L D 2005 Consumer grade global positioning system (GPS)
[7] accuracy and reliability. J Forest. Vol 103 no 3 p 169-73.
[8] Rodriguez-Perez J R, Alvarez M F, Sanz-Ablanedo E, Gavela A 2006 Comparison of GPS receiver accuracy and precision in forest environments. Practical recommendations regarding methods and receiver selection. Shaping the Change XXIII FIG Congress Munich, Germany, October 8-13, 2006.

IOP Conf. Series: Earth and Environmental Science **335** (2019) 012026 doi:10.1088/1755-1315/335/1/012026

[9] Wing M G, Eklund A 2007 Performance comparison of a low-cost mapping grade global positioning systmes (GPS) receiver and cosumer grade GPS receiver under dense forest canopy. J Forest. Vol 105 no 1 p 9-14.
[10] Abdi E, Mariv H S, Deljouei A, Sohrabi H 2014 Accuracy and precision of consumer-grade GPS positioning in an urban green space environment. J Forest Science Tech. Vol 10 no 3 p 141-147.
[11] Agisoft LLC 2016 Agisoft PhotoScan user manual: professional edition, version 1.2.
[12] B, Tadeusz W, Karol D 2018 The analysis of the accuracy of spatial models using photogrammetric software: Agisoft Photoscan and Pix4D. E3S Web of Conferences 26: 00012.

IOP Conf. Series: Earth and Environmental Science **335** (2019) 012027 doi:10.1088/1755-1315/335/1/012027

Executive information system lecturer mobility and scientific reputation of higher education

Y D Indriani[1], K B Seminar[2] and H Sukoco[3]

[1]Management Business, Bogor Agricultural University, e-mail:julia@apps.ipb.ac.id
[2]Machine Technics and Biosystem Department, Bogor Agricultural University
[3]Computer Sciences Department, Bogor Agricultural University

Abstract. Lecturer mobility and scientific reputation is one of the key indicators for the establishment of world class university. The total number of papers published in the academic journals indexed by Elsevier's Scopus database per scholar, and scholarly publications are one of the parameters for universities in the world are ranked. Those are challenges for the Executives Boards in universities to develop strategies of the existing conditions in order to plan programs to support the vision of universities. Executive Information System is required to display information strategically based on the existing data conditions. The present research identified a web-based Executive Information Systems called SIMORI. It is the system development method using System Development Life Cycle (SDLC) approach through system investigation step with feasibility study, system analysis, design analysis, and implementation. Critical factors for the success of the system are good data governance, while the data currently displayed still need to be completed and improved. The strategy to be developed will be managing mobility data of scientific reputation by involving all Faculties and the Departments of university to up-date data through Academic Management Information System, to be verified by the executive and will be presented as strategic information for the university communities.

Keywords: Information system, System development life cycle

1. Introduction

The deepening of globalization and tough competition will increase performance of universities to improve their higher education systems and build world-class universities (WCU). According to [14] WCU can be achieved by utilizing three key indicators: a) high concentration of talented lecturers, students, researchers in internationalization activities, b) abundant resources in terms of funding so as to provide a comfortable learning environment and advanced research, c) favorable governance that fosters strategic vision, innovation, leadership and flexibility and enables institutions to make decisions and manage resources without being burdened by bureaucracy. The results of synergizing these three indicators are highly qualified graduates, widely published national and international research results, and technology exchange. According to [13], those indicators are used by international agencies in conducting surveys to rank the world's reputable university. [6] study stated that WCU is universally understood to mean achieving a globally ranked institution. Further, he compared indicators used by Times Higher Education Supplement (THES) and Shanghai Jiao Tong University (SJTU) for world class university rankings, among others are students, scholars, visiting lectures from abroad, as well as the use of information and communication technology. QS World University Ranks the world's top universities in individual subject areas, covering scientific reputation, alumni's working reputation, student and staff ratios, number of scopus-indexed citation works and student mobility. Another ranking that is also a concern of universities in the world is webometrics that assess the activities of universities through the use of the website as a storefront of information and number of papers published in the academic journals indexed by Elsevier's Scopus database per scholar.

An Executive Information System is really required to manage the data. According to [8] Executive Information System (EIS) incorporates various management information systems and decision support systems that facilitates and supports senior executive information and decision-making needs. It provides easy access to internal and external information relevant to organizational goals, and critical success factors (CSF). The System is equipped with a drill down menu to display detailed information in attractive interfaces, and is presented in the graphs and tables. With reference on the above statement, it is necessary to examine how to develop an executive lecturer's mobility and scientific reputation information system at universities. The System is expected to provide information for planning, controlling, reporting and evaluating mobility data (inbound and outbound) and university's scientific reputation and displaying information based on Systems Development Life Cycle approach. It is a term used in systems engineering, information systems and software engineering to describe a process for planning, creating, testing, and deploying an information system.

2. Research Method

The research was conducted at the Directorate of Data Integration and Information System of IPB, Office of Cooperation and International Program, Directorate of Student Affairs, and Directorate of Research and Innovation IPB, in April-November 2017. This research used descriptive qualitative research method with case study approach. The method used in the development of Executive Information System using System Development Life Cycle (SDLC) was [7] method. It is an Executive Information System (EIS) characterized by business oriented compared with transactional requirement, apply strategic decision, focus analysis to strategic business requirement, and the development process of SIE that focuses upon a cyclical process for assessment success rates.

Stages for the development of SIE with SDLC approach were focused on the description process and combined perfectly with object oriented approach. The stage of development of executive information system adopted system development cycle according to [9] consisting of following stages:

2.1. Stage of System Investigation

This stage includes the identification of business opportunities and priorities. They were developed in the system development of feasibility study and information needs to provide a framework for technical and non-technical activities to deliver a quality system which meets or exceeds a business or institution expectations or manage decision-making progression. It was executed through interviews, document review and direct observation of lecturer and student mobility as well as the scientific reputation of higher education through related units. The life cycle approach is used so users can see and understand what activities are involved within a given step. Feasibility studies in the investment stage include:

a. Organizational Feasibility: it will analysis how well the proposed system supports the company's business priorities;
b. Technical Feasibility: includes ability, reliability, and availability of hardware, software and network;
c. Operational Feasibility: it will analyze employee acceptance, management support, and government requirements and other requirements;
d. Economic Feasibility; it relates to cost savings, increased revenues, increased profits, reductions of necessary investments and other benefits expected to exceed the development costs and operating costs of the proposed system; and
e. Benefit Analysis; consists of tangible benefits (tangible benefits) and intangible benefits (intangible benefits).

2.2. Stage of System Analysis

In this stage, eliciting system functional requirements based on business requirements based on business object components as defined by UML (Unified Modeling Language). According to Wikipedia UML is an industry-standard language for visualizing, designing and documenting software systems and offering a standard for designing a system model. By using UML we can create models for all types of software applications, where they can run on any hardware, operating system and network, and written in any programming language [16]. But because UML also uses class and operation in concepts basically, it is more suitable for writing software in object-oriented languages such as C ++, Java, C # or VB.NET. The system analysis phase with UML approach generates the system model depicted in 13 different diagrams. The diagram of object approach (object oriented model) according to [1] and [11] which include:

a. Context Diagrams
   It is a diagram that defines the boundary between the system, or part of a system, and its environment, showing the entities that interact with it to;
b. Use case diagrams
   It describes an interaction between the system and its environment; the expected functions of a system. It also expresses the externally visible behaviour of the system and its interactions with actors, i.e. users or different systems.
c. Activity diagrams
   Activity diagrams, which are related to program flow plans or flowcharts, are used to illustrate activities. In some circumstances users the use activity diagrams for the description of business processes that describe the functionality of the business system;
d. Class diagrams (class diagrams; names, attributes, methods)
   Class diagrams is a type of static structure diagram that describes the structure of a system by showing the system's classes, their attributes, operations (or methods), and the relationships among objects. Class diagrams can be utilized by analysts, business modelers, developers, and testers throughout the software development lifecycle; and
e. Sequence diagrams, is the diagram that shows the interaction between the actors with the system and between system components.

2.3. Stage of System Design

System design determines how the system will meet the objectives of information system development. In the system design the three activities performed are data identification, user interface design, and process design. Outputs generated from this stage are screen design, report and dialog box, data structure, and programs / procedures design.

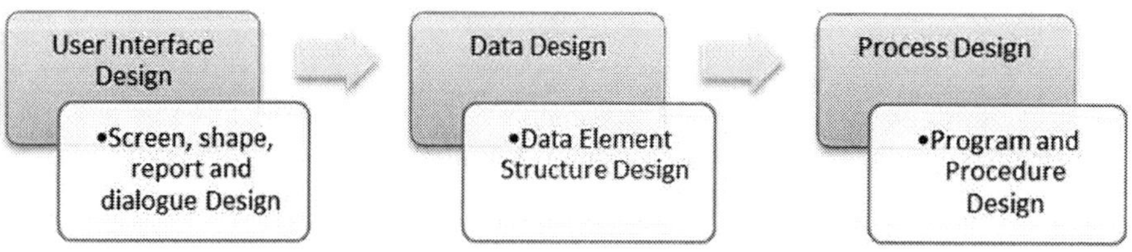

Figure 1. *Stage of System Design*

2.4. Stage for System Implementation

The system implementation stage is a process of transforming the system design into an application or software system. The present study showed that the application was developed for working prototype with development platform based on php programming language with sqlserver based database using query data from integrated database IPB. System testing is testing conducted on a complete, integrated system to evaluate the system's compliance with its specified requirements as desired. Testing the output of program packages and functions in the application using black box test. The test step includes the analysis of needs specifications, design specifications, model description, for each model with input parameters and outputs, for each parameter included range of data, and test cases for system functions developed (Bhasin et al 2014).

3. Results and Analysis

3.1. General Description

In the grand design of IPB ICT development 2013-2017 as reported by Seminar (2012), that to ensure secure operations of information systems and thus safeguard assets and the data stored in these systems, and to ensure that applications achieve their objectives in an efficient manner, an organization needs to institute a set of policies. Which will control directly the achievement of the objectives of Information Systems of the organization which include Value, Productivity, Quality, Simplicity, Efficiency & Effectivity, and Go Green. Integrated information system is supported by data, and any concerns on the integrating databases. They are arranged in an integrated database architecture which has the main function of integrating the data collection owned by universities and then the data is used as needed by the existing systems. So the systems do not need to reassemble the master data for the transaction process because the data has been integrated and consistent. The integrated database used by several systems are divided into 3 groups: Personnel System, Academic System, and Finance system. It is also used by single sign on (SSO) system and strategic information system such as SIMAKER, MSS, SIE and many others. Management of lecturer and student mobility data includes inbound and outbound activities currently inputted through the application of performance management information system (SIMAKER) and the data stored in the integration database for transactional and operational activities. However historical data are stored in the data warehouse to facilitate strategic information system displaying historical information as strategic decision making materials. Scopus-indexed scientific and google scholar reputation data is not yet integrated with existing information systems and even data is stored in spreadsheets in the work unit. The integrated university database is presented in figure 2.

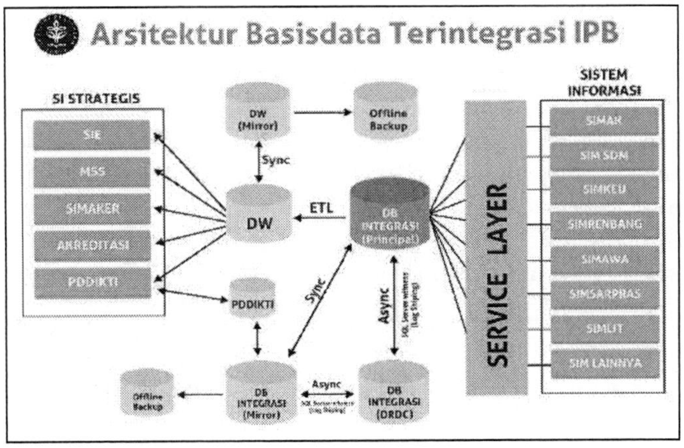

Figure 2. Integrated Database Architecture

Lecturer mobility both inbound and outbound programs for lecturers is managed by Directorate of Cooperation and International Program (Directorate of KSPI). It is stated in Article 68 that one of the functions of the international program sub-sector is to carry out the function of coordinating the implementation of international program activities and the provision of administrative services and travel information abroad for lecturers, education staff, students and alumni. The lecturer mobility activity is divided into 3, namely seminars (conferences), visiting professor, and follow-up actions of Memorandum of Understanding (MoU). While other academic activities include academic competitions / entrepreneurship / cultural / sports, internships / practical / teaching / coaching activities other than curricular activities. Moreover, Presentations in seminars / workshops / conferences, Exchange / credit earnings / summer course, Research, Others (Voluntary, service learning program, etc.) are also the responsibility this Directorate.

Scientific Reputation or also known as academic reputation which means that the university is widely known. A very good academic reputation was the most important factor influencing college decisions. Scientific reputation is important for all of us, but this come slowly, step by step and through the hard work, it is acquired over a lifetime and is akin to compound interest, the more you have the more you can acquire. The management of this international reputation relies heavily on the contributions of researchers / lecturers in publishing their scopus-indexed and google scholar-scanned works and dissemination efforts through various academic / academic community communities such as research gate, academia.edu, or linkedIn. The total number of papers published in the academic journals indexed by Elsevier's Scopus database per scholar, and scholarly publications are one of the parameters for universities in the world are ranked. The total number of scientific publications searched results of 1413 (one thousand four hundred and thirteen) faculty members of IPB have google scholar and scopus accounts. The scientific reputation search results are presented in Table 1.

Table 1. Search Results of Scientific Reputation of Lecturer with browser optimization application

Indexer	Number of Lecturers Profile	Ambiguous Account	Account not found
Scopus	714	89	665
Google Scholar	660	36	702

3.2. Stage of System Development

The above description describes how the condition of the system runs and its relevance to this present research. How the development of the Executive Information Systems through utilizing system development life cycle (SDLC) web-based methods had been carried. The steps of system development are presented as follows:

3.2.1. Stage of System Investigation. This stage includes organizational, technical and operational feasibility studies to assess the Executive Information Systems to be developed to support business priorities, reliability, availability of software and supporting hardware, and support of management / human resources that can meet the information needs of executives and deserve to be developed. Feasibility study analysis is presented in Table 2.

Table 2. Feasibility study phase of systems investigation and analysis system.

Feasibility Study	Mobility of lecturers and students as well as the scientific reputation of Higher Education
Organization	The feasibility of developing Executive Information Systems for lecturers and students mobility, as well as the scientific reputation of universities based on the criteria, as follows: 1. The Vision of Higher Learning Institution as stated in the Strategic Plan 2014-2018 is to Become the International Research-Based University, and Prime Mover of Agricultural Mainstreaming. 2. As stated in its Strategic Plan 2014-2018, one of the pillars in it its development strategy is Strengthening Reliability Management System. One of the points is strengthening the integrated Management Information System, through: a. Strengthening data management of the Information Systems covering the relevant institutions and working units. b. Updating the ICT' needs of working units.

	c. Improving the interoperability and role of working units in updating data and utilizing ICT services and information systems. 3. The main duties and functions of thc Directorate of Data Integration and Information Systems are to carry out strategic tasks in the development of Information and Communication Systems and data management databases on information and communication technologies. The task is emphasized in Article 74 which stipulate the development and upgrading of infrastructure resource of institute integrated communication network (hardware, software, netware, and electrical system). 4. IPB vision for 2019-2023 as planned by the newly elected Rector is "To develop IPB to be a research-based and leading-edge university in innovation for nation self-reliance towards techno-socio entrepreneurial university that excels at global level in agriculture, marine, tropical bioscience". One of the main programs of Information Technology and Communication is a Robust ICT System through Integration of information systems for big data management, governance improvements, and managerial capacity.
Technical phase	The availability and reliability of software, hardware and networks in the development of Executive Mobility of Information Systems and the scientific reputation of IPB determine whether or not it is feasible to develop SIMORI in order to enhance the reputation of IPB towards world-class universities. The technical support tools that IPB currently possesses are: 1. 300 *based controller access point* have been distributed for the Department, Faculties, Dormitories, and Auditorium in all the working units. 2. *Secure DNS appliance* and *Virtual Machine* has been provided to anticipate malware and improve access to information. . 3. Adequate Internet connection *bandwidth* 4. IP and licensed Topology of *Three Layer Hierarcycal Model* is applied to Public IP usage so that access and search accessibility is easier and faster. **5.** The strong specifications for database servers and are available in sufficient quantities
Economics	The analysis of financial data of budget allocation for Information and Communication Technology (ICT) development, which is executed by the Directorate of Data and Information System Integration, for 2015-2018 shows that institutional support for the management of information technology both hardware and software is significant enough. Data on budget allocation received from the Funding Assistance for State Universities Legal Entity (*Bantuan Pendanaan Perguruan Tinggi Negeri Badan Hukum* BPPTNBH), Community Fund (*Dana Masyarakat* DM) and Cooperation (Kerjasama KS), are presented as follows:

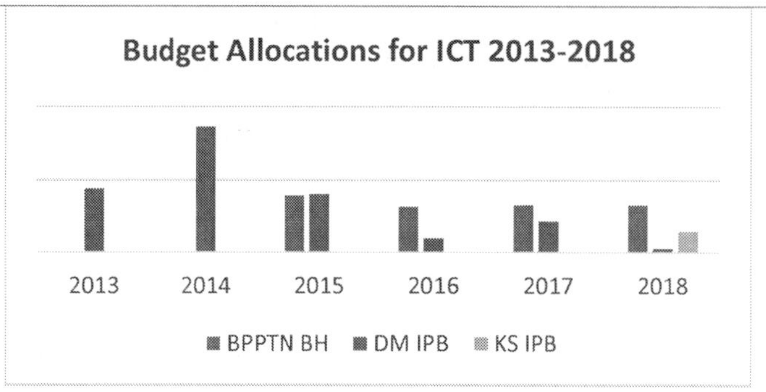

Figure 3. Budget Allocation for ICT 2013-2018.

Operational	The annual work program of HR strives to improve and strengthen the competence of human resources in Higher Learning Institutions. HRM organizes TEMATIC Training of Computer and Application Field annually. Besides that, there is also Socialization of Functional Appointment of Computer Functional Position for education personnel. Organizing competency-based online training course for learning process that utilizes electronic technology, and to track institutional learner progress.
The Analysis of tangible and intangible benefits	o Reducing the cost of information processing by reducing the printing cost and distribution, as they can be displayed online that choice will impact the website's success in nearly every area, included search engine optimization (SEO) and social media marketing (SMM). The total number of scientific publications displayed in the official website facilitates the tracking of information and improve the institution's position in the university rankings in the world. o Describing the information needs of researchers / lecturers in the academic community as well as information on the subject range of research categorized in the QS University Ranking into factors that can support the need for system development. By leveraging existing platforms and components in the development of other existing information systems, the SIE is feasible to be developed. o Improvement of the availability information and analysis required to support the lecturer and student mobility activities. Criteria for the needed information are, their country origin, home university, time of arrival and their return, visa identity, purposes their inbound / outbound visits, country distribution, their length of stay, purpose of activity (credit earning, internship , apprenticeship), purposes grouping mobility, accommodation while they do their international programs (in campus / off campus), email (for information dissemination), gender, and type of accommodation. o Human resources who manage such system does not require special qualifications because SIE is developed to provide easy access of information for leaders / executives. o The convenience for policy makers in their decision making process, as the information needed are displayed in graphics and tables. Those forms are very useful for displaying the relationship between quantities quickly and in an accessible form. Online display improves the institutional' image to the world.

3.2.2. Stage of System Analysis. This is the stage where the system analysis process is occurred. This stage describes the functional requirements of the system with object approach using UML (Unified Modeling Language) that include context diagrams, use case diagrams, activity diagrams, class diagrams and sequence diagrams [11].

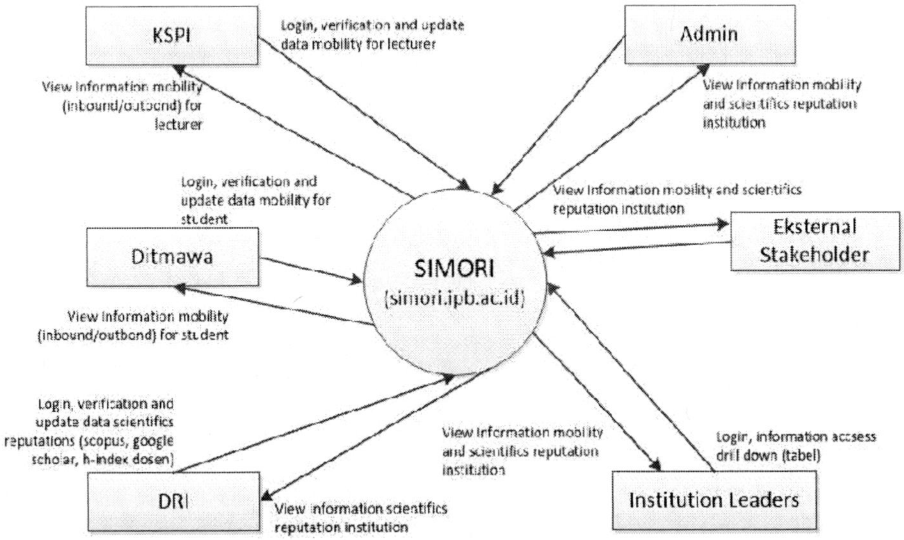

Figure 4. Context diagram

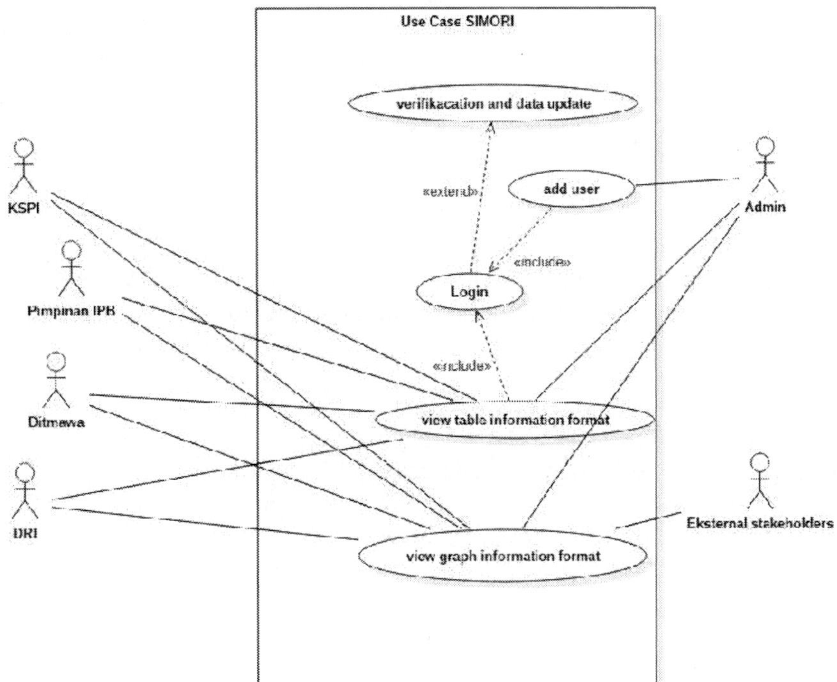

Figure 5. Use case diagram

IOP Conf. Series: Earth and Environmental Science **335** (2019) 012027 doi:10.1088/1755-1315/335/1/012027

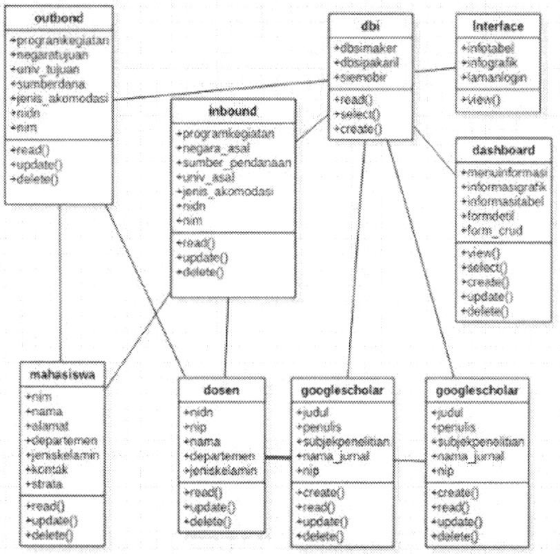

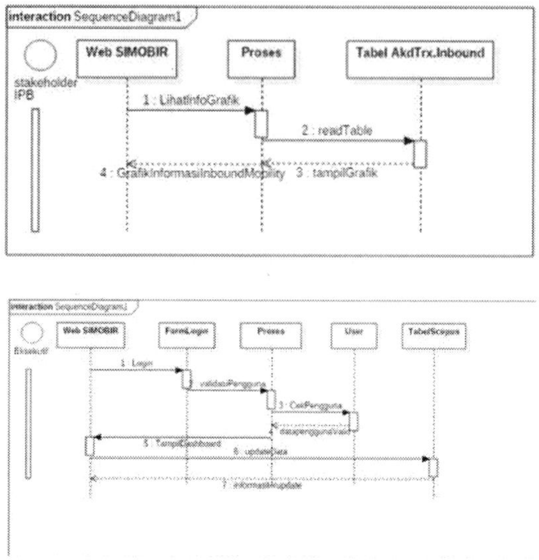

Figure 6. Use case diagram **Figure 7.** Use case diagram

3.2.3. Stage of System Design. System design determines how the system will meet the objectives of information system development. System Design focuses on how to accomplish the objective of the system. This stage mainly focuses on: 1). Database design in database integration for lecturer and student mobility data with sqlserver platform, while for scientific reputation data using mysql platform. 2) User interface design is divided into three parts. The first part contains information on graph analysis which is accessible by all stakeholders of higher learning institutions. The interface is displayed in tabular menu which consists of Lecturer Mobility, Student Mobility, Scopus Index and Google Scholar Index. The second part is the login interface by entering the username and password the user has added by the admin. The third part of the system is the dashboard, which displays all the information required based on feasibility analysis and information needs. 3). Process design, is the activity of determining the workflow, equipment needs, and implementation requirements for a particular process. In general analysis of system design was based on the institutional's current business processes. This web-based information systems webpages also have unique addresses URL to help people locate them. The web page displays the required information by executing the site address associated with the web server called query string [4].

3.2.4. Stage of System Implementation. The stage of system implementation is a process of transforming the system design into an application or software system. It is the implementation of the development of SIE web with PHP micro frameworks, and codeIgniter approach. CodeIgniter is a powerful PHP framework that can help users greatly speed up the development of web applications. The developed EIS with codeIgniter, SIMORI, can be accessed at address: the following address: ehttp://simori.ipb.ac.id, and is presented in figure 8.

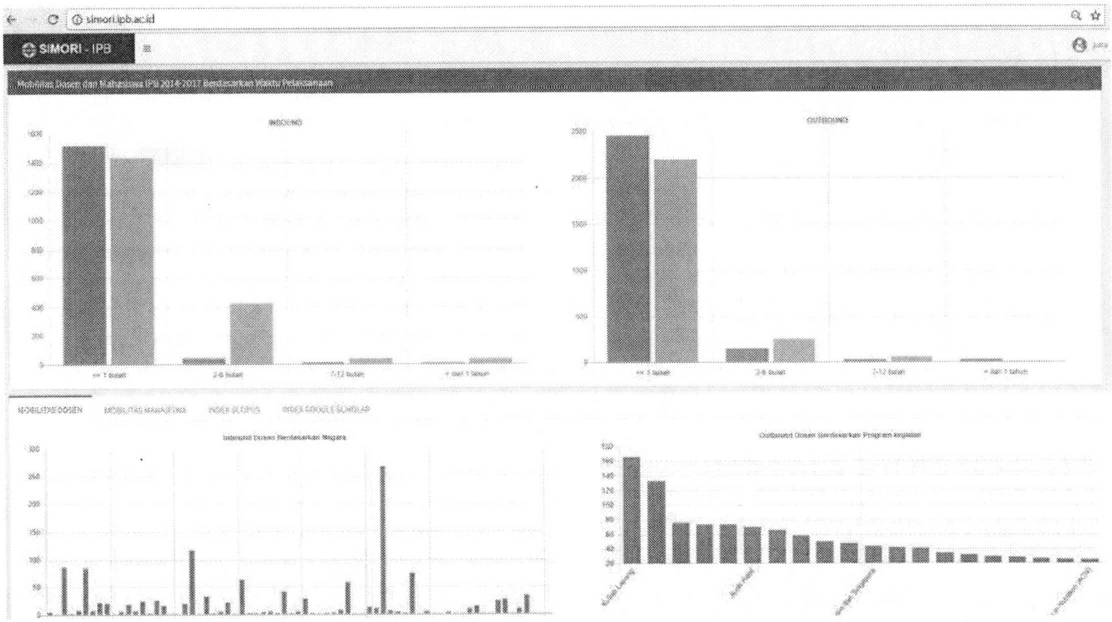

Figure 8. Executive Information Systems of Lecturers and Students Mobilities and Scientific Reputation

4. Conclusion

The formulation of managerial implications will be utilized as the basis of strategic recommendation for executives in Higher Education. With reference to the present research recommendations were formulated for higher education management:

1. The program of research and cooperation activities are still very few, so it is necessary to develop multi-institutional research consortium to establish networks with both national and international universities and research institutions to maintain and improve the scientific reputation of universities.
2. The most presented information on inbound lecturers are their main activities in delivering general lectures, therefore inviting foreign professors for research collaboration in the campus will generate new spirit to develop a global reputation to increase its World Reputation Rankings.
3. The total number of foreign students (inbound) from various countries for the most activities, the summer courses, show that the hospitality program for international students is an important program to increase the interest of international students. Such program is very beneficial in creating the international atmosphere as well as increasing the reputation of Higher Learning Institution toward World Class University.
4. Based on total number of scientific publications searched results, only 50% out 1413 (one thousand four hundred and thirteen) faculty members of IPB have google scholar and scopus accounts. Therefore, it is advisable for other faculty members to increase their profile accounts in various social media, and actively publish their scientific articles of their research programs through various scientific publishers.
5. Critical success factors of Executive Information Systems development for education management is data management [5]. The role of the Internal Verifier to manage data for success implementation of Executive Information Systems is really required. The Directorate of Cooperation and International Program will be responsible to verify the mobility of lecturers data. While for scientific reputation, as it slightly differs for its data management, therefore it is necessary to analyze the interoperability between executive information systems with the

scientific lecturers' management system to get the most advance and up-dated data on the scientific publications.

6. The challenge to manage critical organizational data is growing, and universities are embracing data governance (DG) strategies to protect the integrity of their valuable enterprise assets. According to [3] MongoDB expressions to control document structure can be taken as optional data governance. The flexibility of a dynamic schema in MongoDB can now easily be paired with data governance controls over an entire collection of the universities. Mobility data management and scientific reputation by accessing data through a data warehouse in accordance with the integrated database architecture according was developed according to to [15]. Data warehousing has been an important part of the data architecture and IT infrastructure of many organizations and institutions. While big data and analytics do have their place, data warehousing stands out as a practical, proven practice for analyzing structured business data in order to support data-driven decision making. The sql command executed to access the data warehouse real will be faster, the number of records processed less and the join process is reduced.

Suggestion

To make attractive presentation of the information accessible in the SIMORI, further research program is really required. The application software to synchronize data extracted from external servers such as scopus and google scholar need to be developed. The proposed developed system will display all titles published by lecturers and students of IPB. It is expected that the systems will cover the most titles cited, as well as the metadata of those publications indexed and displayed in the Executive Information System. Of course, it requires skillful EIS team who are skillful to communicate and improved user attitude towards EIS to influence EIS implementation. Good EIS team communication skills and a positive user attitude towards EIS are essential to promote scientific reputation of Bogor Agricultural University, to increase its ranks in World University Rankings.

References

[1] Bennet S, McRobb S, Farmer R. 2010. *Object-oriented system analysis and design using UML.* McGraw Hill Education.
[2] Bhasin H, Khana E, Sudha. Black box testing based on requirement analysis and design specifications. *International Journal of Computer Applications.* 2014. *(0975–8887) Volume 87 – No.18.*
[3] Heripracoyo S, Kurniawan R. Big Data Analysis with MongoDB for Decision Support System. *TELKOMNIKA.* 2016. *Vol.14, No.3, September 2016, pp. 1083~1089.*
[4] Karosekali AS, Pertiwi S. *Solusi sawit sebagai rintisan portal industri sawit. Prosiding Seminar Nasional Informatika Pertanian Indonesia. Bandung (ID): Kampus Unpad Dipatiukur, Jawa Barat.* 2011.
[5] Koh C, Watson H J. *Data management in executive information systems. Information & Management.* 1998. Volume 33, Issue 6, 22 June 1998, Pages 301-312.
[6] Levin HM. What is world class university?. *Conferences of Comparative and Education Society, Honolulu, Hawaii.* 2006
[7] Lungu I. Executive information systems development lifecycle**.** *Economy Informatics.* 2005. *1-4.*
[8] O'Brien, James A. *Pengantar sistem informasi perspektif bisnis dan manajerial.* Dewi Fitriasari dan Deny Arnos Kwary. 2005. Jakarta: Penerbit Salemba Empat. 2008:510-550.
[9] O'Brien J A, Marakas G M. *Management Information Systems (Tenth Edition).* 2011. McGraw-Hill Irwin.
[10] Seminar K B. *Grand design ICT di IPB.* Buku Laporan Grand Design ICT IPB. 2012.
[11] Sommerville I. *Software Engineering Tenth Edition.* England: Pearson Education Limited. 2016: 138.

[12] Sopingi, Utami E, Amborowati A. Prototype executive information system untuk mendukung evaluasi diri perguruan tinggi (studi kasus stmik duta bangsa surakarta). *Jurnal Ilmiah DASI.* 2015. *Vol. 16 No. 02 Juni 2015, hlm 66-74*
[13] Suyanti E. Strategi internasionalisasi dalam bidang penelitian pada Universitas Indonesia dan Institut Pertanian Bogor. *Jurnal manajemen pendidikan.* 2014.
[14] Salmi J. The challenge of establishing world-class universities. *The International Bank for Reconstruction and Development / The World Bank.* 2009.
[15] Warnars S. Tata Kelola Database Perguruan Tinggi Yang Optimal Dengan Data Warehouse. *TELKOMNIKA.* 2010. *Vol. 8, No. 1, April 2010 : 25 – 34.*
[16] Nugroho A. *Analisis dan Perancangan Sistem Informasi dengan Metodologi Berorientasi Objek.* Penerbit Informatika [ID]. 2005

IOP Conf. Series: Earth and Environmental Science **335** (2019) 012028 doi:10.1088/1755-1315/335/1/012028

ICTs for small scale farmers in Indonesia: how to make it possible?

A U[1] Seminar and S Sarwoprasodjo[1]

[1]Department of Communication and Community Development Sciences, Bogor Agricultural University, Dramaga, Indonesia, e-mail: seminar_annisa@apps.ipb.ac.id

Abstract. The effort to digitalized agricultural process has been started since the 1980s to maximize the food production, yet this effort majorly targeted agricultural industries or state-supported farmers group. In Indonesia, the number of small independent farmers is high. These small farmers have little access to ICT infrastructure to support their farming activities. This paper seeks to propose strategies for improving small independent farmers' access to and utilization of ICTs for agriculture. Qualitative method is used to identify the problems and needs of small farmers and their perception of ICTs for agriculture. Using multi-sites ethnography strategy in gathering data, we conducted in-depth interviews with 22 small farmers from 12 regions in Indonesia. We also observed the day-to-day activities of small independent farmers in two villages in Bogor and Yogyakarta, Indonesia. We found that several strategies have to be considered for developing ICTs for small scale farmers, which are (1) analysing small farmers perception about technology; (2) identifying local-specific issues; (3) involving small farmers in planning digital agriculture system; (4) facilitators of new technology must have communicative competences or communicative skills so the planning of digital agriculture system is done through participatory approaches.

Keywords: small scale farmers, local-specific issues, perception, ethnography

1. Introduction

ICT for agriculture has been developed to increase yields and outputs and improving the efficiency and effectiveness of agricultural inputs. However, the beneficiaries of this technology is majorly agriculture industries or government-supported farmers group. Digital divide or the globally skewed distribution of information and communication resources makes small farmers living in rural area have little to no access to technology infrastructure [1,2]. The world thought that the solution of this problem is by delivering infrastructure from rich to poor, instead of empowering the people by involving them in improving the quality of their own lives [3]. Simply distributing tools to people would not guarantee a prosper live. Small scale farmers seldom feel the impact of digital agriculture, either because they have no access to ICT or because they are poorly disseminated [2].

Research about ICT for small scale farmers has been majorly limited to improving the adoption rate of ICT for agriculture. Several studies found barriers in delivering ICT for agriculture, because small scale farmers are: illiterate and poor [2,4], have no trust to innovation [1], lack of awareness of ICT benefits and lack of confidence in using ICT [5] hence, they do not need complex digital agriculture infrastructure [2,6,7]. Several studies also stated that digital agriculture provider did not design the ICT to meet small scale farmers needs [6–10], it did not give impact to their live [11], it did not provide proper training for small scale farmers in using ICT [2,11,12], the ICT did not use local language so it was hard to use for small scale farmers [4,13], and agricultural ICT is not yet integrated with other development programs to address the numerous related problems faced by farmers [2].

The constraints showed that the problem of digital agriculture utilization by small scale farmers is the lack of small scale farmers input or participation in designing the ICT system itself. Many studies recommended that the design of ICT have to use participatory process where small scale farmers do not only act as end user but also part of planner of the ICT system planning [14–18], however there are several studies showed that participatory approaches is hard to do when actors involved have no commitment [16], farmers have no understanding in decision-making in practice as well as delivering their needs [17], the lack of competences, approaches and scientific disciplines needed in developing a sustainable ICT for agriculture [17], challenges on local adaptation and lack of farmers who could analyse data [19]. These findings shows that there are still challenges even though ICT for agriculture has been developed or assessed in participatory research, yet these challenges also explained that the purpose for doing participatory approach is to, first and foremost, to adapt local need and issues in the development of ICT for agriculture. This research project proposes an analysis of communicative competences needed to conduct participatory approach, especially with small scale farmers, so that there is a guide for facilitators in discussing the needs of small scale farmers in developing ICT for agriculture. This article is a part of bigger research that already published two articles [20,21].

Communicative competence is the ability to grasp language code and knowledge to deliver information in any kind of condition to someone based on their social and culture knowledge so that they could use and interpret the information in linguistic form effectively [22,23]. Communicative competences or communicative skills are needed by people who deliver the ICT to be able to do participatory approach in developing ICT. This article analysed a social movement organization of small scale farmers that put participatory approach as one of organization value. By using Ethnography Communication (EO), we observed organization meetings and day-to-day farming activities and also conducted in-depth interviews with the members to discover the communicative competences needed by the organization members to do participatory approach.

2. Methods

This study used qualitative case study method [24]. The case study of this article is Serikat Petani Indonesia (SPI), a social movement organization based in Jakarta, Indonesia. SPI consists of small scale farmers spread out through 19 provinces in Indonesia, from provinces in Sumatera Island to East Nusa Tenggara Island. SPI supports participatory approach in designing collective actions based on local needs to achieve organization's main goal: food sovereignty. SPI members are majorly traditional farmers who practice agroecology, which refuse to use agriculture industrial inputs. They promote the practice of local wisdom, local seed, and alternative markets (direct selling from producer to consumer).

Ethnography of communication (EO) is used to discover the meaning of ICT based on small scale farmers perspective, to discover the communication process inside farmers groups in discussing the use of ICT in their farming practices, and to discover communicative competences used by members of farmers group in the decision making process. Hymes stated that communication and interaction do not appear in a vacuum, but depend on a particular context or case [25]. In EO, observation and in-depth interview are techniques in collecting data. Observations were made during organizational meetings, such as public discussions, meetings, or conferences. Interviews, in EO, were conducted to deepen or clarify the observations results to participants.

The 'SPEAKING' grid is used as a systematic collecting data method [22]. The 'SPEAKING' is an abbreviation of: settings, participants, ends, act sequence, keys, instrumentalization, norms, and genre (table 1). The data collection based on the 'SPEAKING' grid could guide us to identify how to conduct a participatory approach: does it need to be in certain settings? Who are the most important participants to be involved? In what language or in what manner to conduct participatory approach? Are there any local norms that we have to understood to conduct participatory approach effectively? These type of questions could help us in breaking down things to be aware of when conducting participatory approach.

Table 1. The 'SPEAKING' Grid.

SPEAKING	Description
S: Setting and Scene	Situation and physical condition during a meeting
P: Participant	Describes the participant and its social status, also its relationship status
E: Ends	The purpose of the meetings and also the purpose of the participant
A: Act Sequence	Explains the sequence of the meetings and also the topic that played out through the meeting
K: Key	The tone and attitude of the meetings
I: Instrumentalization	The form and style of communication, such as written or oral, languages, dialects, and over-use phrase
N: Norm	Rules and values of the meetings
G: Genre	The type of communication, such as discussion or lecture.

Observation of day-to-day farming activities was conducted as a technique to gather data about the use of ICT in farmers live. The observation was conducted in Dramaga and Cijujung, Bogor and Sleman, Yogyakarta for 3 weeks. Other than day-to-day activities, we also observed SPI meetings (both in national and local level) in 2016-2017 to understand how participatory approaches is implemented as a decision making process. We also participated in Whatsapp group organized by the members of SPI. It consists of 256 members to discuss organic farming. This technique was used as a way to identify the information needed by farmers. Other techniques conducted was in-depth interviews with 22 farmers from 12 regions in Indonesia to understand their information needs and their perception about ICT for agriculture.

We used multi-sites ethnography approach to select the sites and informants. Multi-sited ethnography solves the need for a method to follow a story or a process and ideas that extend over multiple locations or through time and space [26]. Multi-sited ethnography allow us to understand a variety of perspectives from a single idea. In this paper, the multi-sites ethnography helped us to trace the various communication practices within SPI in doing participatory approach. This approach guided us in choosing study sites and informants. The study sites were determined based on meetings that are attended by small scale farmers in 2016 until 2017. We found five observable meetings (table 2) and through these meetings we met 22 local peasants from 12 regions from North Sumatra, South Sumatra, Bengkulu, Riau, Lampung, Jambi, West Java, D.I. Yogyakarta, Central Java, East Java, Southeast Sulawesi and Central Kalimantan, as our informants. The in-depth interviews were conducted in several ways: face-to-face, telephone, and instant messaging so we could consistently follow the stories of ICT for agriculture utilization in their daily life and also how did participatory approach conducted within SPI.

Table 2. List of observed meetings (2016-2017).

No.	Description of Meetings	Location
1	Natural Farming Training	Yogyakarta, Indonesia
2	Internal Meetings: Preparation for Plantation Conference	Jakarta, Indonesia
3	Internal Meetings: Preparation for Agroecology Conference held in Srilanka	Jakarta, Indonesia
4	Plantation Conference	Jakarta, Indonesia
5	Field Trip: Natural Farming Practices, Southeast Asia-China Sub-Regional	Yogyakarta, Indonesia

After collecting data, we did the coding process using QSR Nvivo Pro 12, a qualitative research software [27]. We did two steps of coding process: (1) the 'SPEAKING' grid and (2) Miles, Huberman, & Sadana coding process which includes descriptive, in vivo, process, evaluation, emotion, and values coding [28]. The coding process helped us in explaining activities that needed to be conducted to support participatory approach.

3. Results

3.1. SPI as an organization

The main goal of SPI is to make food sovereignty as the main paradigm of agricultural and food system in Indonesia. Food sovereignty is a concept where agricultural system are heavily based on farmers knowledge and experience in farming activities, instead of agriculture industrial approach [29–32]. Food sovereignty supports the use of natural agricultural inputs and local farming activities while refusing the use of agriculture industrial inputs [31]. SPI consists of groups of small scale farmers who differentiate themselves from state-supported farmers groups. This conditions makes them have little access to agriculture development projects from governments, since government prioritize their development projects to state-supported farmers groups. Members of SPI are small scale farmers that already have bad experience with government's projects and the use of agriculture industrial inputs. Some of them even mention they suffer from a "trauma" to participate in government projects or use agriculture industrial inputs as it gave them a big loss instead of benefits. The farm size of members are varies, majority members from Java Island have farm less than a hectare while majority members from Sumatra, Sulawesi, and Kalimantan Island are more than 5 hectare yet they are prone to agrarian conflict. Some of members of SPI do not actively work as farmer anymore since their land is grabbed by agriculture industry.

There are two type of members in SPI: small scale farmers and organization staffs that had experience in farming activities. The organization staffs act as the facilitator to farmers in SPI. Organization staffs arrange meetings, provide information about agricultural practices, provide training and education to its members, and give access to any kind of training for its members as a part to achieve the main goal of SPI: food sovereignty.

SPI organize themselves in several levels: national, regions, districts, and villages. The secretariat office is located at Jakarta where national meetings are usually held. At region levels, each region have to organize 3 districts at minimum and each district have to organize 3 villages at minimum. National meetings are arranged by organization staffs in national level, while meetings at local level (region, district, and village) are arranged by organization staffs in local level and the schedule of meetings is based on its members' needs. Other than meetings held by its own members, members of SPI also have access to public communicative spaces such as public discussion with government agencies, trainings or workshops held by other organization or NGOs, and also field trip to at local level, national, and global level. These communicative spaces, whether public or internal, act as place where farmers could gain knowledge and plan collective actions in order to achieve food sovereignty.

The main concept of food sovereignty is independency. Members of SPI strive to be independent from agriculture industrial inputs, so they could be more adaptive or resilient to local problems and opportunities to support their farming activities. We found several groups of farmers that applied "multi-role peasants" [21]. Multi role peasants, basically, is a role division system than makes farmers work collectively to fulfill the pre-production, production, and distribution process of their farming activities. There are farmers who responsible to produce natural fertilizer and microbes, other farmers responsible to produce seedlings and organize the post-harvest process, and other farmers who builds alternative markets by directly visiting possible customers, or other farmers who raise cattle could sell milk and used its manure for making natural fertilizer. Multi role peasants system was able to drive farmers to think critically about their farming activities so that they could overcome local problems by utilizing local opportunities and made them more adaptive and resilient.

SPI members had not yet use any ICTs for agriculture collectively, but they have practiced many innovations that could drive the members to be independent in conducting their farming activities. The only ICT they used to support their farming activities is through Whatsapp group. Members could ask anything in the group and if other members happened to know the answer or solution, then the members could explain it in the Whatsapp group. They also sell their products through this group. Even though our observation did not assess how to develop ICTs for agriculture in participatory approach, but we could identify how SPI conducted participatory approach in introducing innovations.

3.2. Information needed by SPI members: local-specific issues

Based on participation observation on Whatsapp group organized by SPI members, the Whatsapp group are majorly used to share knowledge about agroecology, to seek information and access to local seeds, and to promote their products. However, based on interview and observations, the most needed information by SPI members are very specific to local issues. Figure 1 shows the most dominant information needed by members from each region. For example, majority regions in Sumatra Island are really prone to agrarian conflict so the most needed information is about agrarian reform or legal steps to ensure that they have farm to cultivate. Members in Sumatra and Kalimantan needs informations about government policy and its law about agrarian reform. Several agrarian conflicts happened and made farmers lost their land and house because agriculture industry destroy their crops and houses without any prior effort to solve it in participatory manner. Agrarian conflict also limits farmers' access to food. Farmers rarely bought food as they can enjoy foods from their own farm, however agrarian conflict makes them have to buy food and the variation of food is also limited.

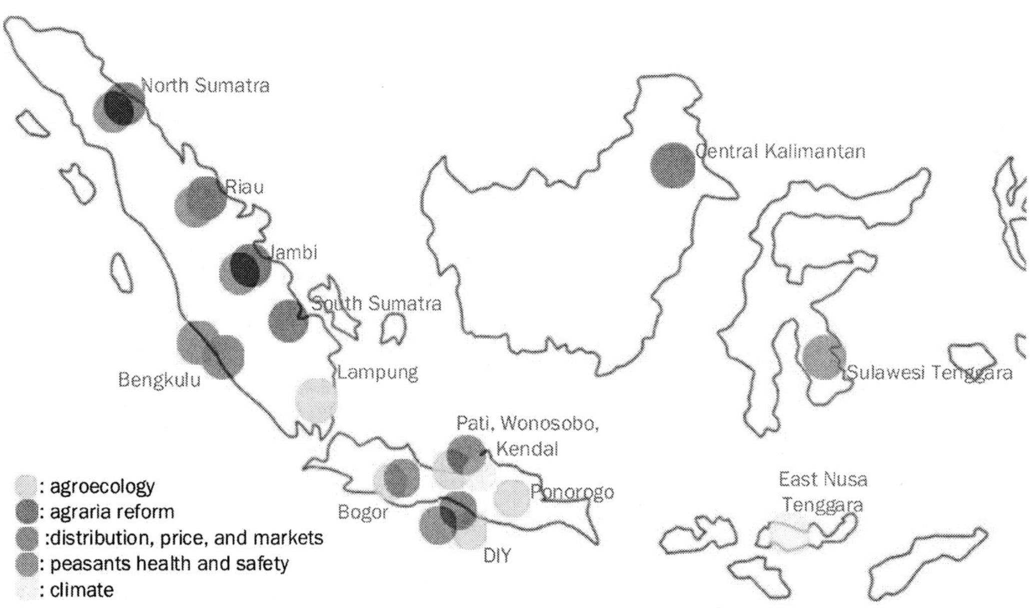

Figure 1. Information needed by members based on local specific issues.

In Java, the most needed information are agroecology practices and access to markets. Agroecology practices include polyculture practices, local seeds, organic farming, and also land preparation. Access to markets include distribution process, access to alternative markets, and price. While other regions, such as in East Nusa Tenggara the most needed information are climate as they usually suffered from climate change.

3.3. Small scale farmers' understanding on ICT for agricultures

Several members have access to ICT for agriculture, however they stated that the price is expensive even though they know the benefit of the technology. When asked whether they have used any ICT to support their farming activities, they only use Whatsapp and e-commerce apps to sell their products. They also answer that they have no knowledge to access other ICTs for agriculture and found that Whatsapp as the most convenient ICT as they could share knowledge with other farmers that they could not meet face-to-face. The use of e-commerce apps are also operated by organization staffs, since farmers stated that they do not really understand how to operate the apps. However, this answers could only be found in Java since they have more farming activities compared to other regions. They also have no access to government programs regarding the utilization of ICT for agriculture. Some farmers also stated that they have heard the use of ICT for agriculture, however they thought that the ICT are for agriculture industry as they have big land to cultivate not for small scale farmers who still farm traditionally.

3.4. Trusted sources of information

The most trusted sources of information are experienced farmers who have already practiced agroecology. SPI facilitate knowledge sharing between regions, where farmer-to-farmer learning method is used. Farmer-to-farmer learning method is a method where experienced farmers delivers their experience on innovation to other farmers. For example, experienced farmers from Bogor was invited to Sukabumi to discuss the preparation of agroecology practices. Other trusted sources of information are academic experts introduced by organization staffs or other experts who have the same concern with members. They stated that they do not think extension agents as the most trusted source of information since they do not understand anything about agriculture and they only sell agriculture industrial inputs, which do not fit with organization value. Even some of small-scale farmers have never met extension service agents. They won't seek more information if there were experts that promote the use of agriculture industrial inputs such as fertilizer, pesticides, and seeds. However, they do not limit their information seeking process as long as it serves the organization value, some members came personally to government agency to learn how to cultivate soil bacteria and microbes and as a result they could independently cultivate soil bacteria and microbes.

3.5. Communicative Competences for participatory approach in SPI

The success of SPI in driving its members to be independent are highly influenced by meetings and trainings organized by SPI. These meetings and trainings were conducted using participatory approaches where farmers could freely discuss their thoughts and doubts in doing collective acts. However, we found that communicative competences have to be possessed by organization staffs and also small scale farmers. The communicative competences possessed by these two types of member are categorized based on the purpose of meetings, which explained each purpose of meetings have its own communicative competences (table 3).

There are at least three purpose of meetings that we found, which are: (1) identifying needs of small scale farmers; (2) addressing misconceptions of the introduced innovations; and (3) decision making process. Building collective identity is an activities that needed to be embedded in each meetings as it is the activities that could drive the commitment and the sense of belonging for the members to do collective actions. In addressing misconception of the introduced innovations, the organization staffs could invite experienced small scale farmers who already used ICTs for agriculture so discussion about misconceptions with local small scale farmers could be conducted. The experienced small scale farmers also need to possess communicative competence, because this person have to explained his experience in using ICTs for agriculture clearly and objectively.

In practice, these meetings did not conducted sequentially nor could it happened in a short time. Sometimes, it is hard to gather people at designated time. Organization staffs have to personally visit farmers to explain the importance of the innovation and ask the farmers to come. Some of organization staffs might quote local wisdom saying or religion saying that could drive the members to participate in the projects. These efforts also show the communicative competences needed by organization staffs or facilitators in conducting participatory approach.

Table 3. Communicative competences needed by small scale farmers and organization staffs in conducting participatory approach.

Purpose of meetings	Communicative competences of organization staffs	Communicative competences of small scale farmers
Identifying needs of small scale farmers	- Able to explain to locals the purpose of the meeting - Able to grasp local languages or vocabularies - Able to use local languages or vocabularies when discussion occured, do not use terms that are hard to understand by locals - Able to act as a listener, asked only when locals have problems in articulating their thoughts. - Probe locals in explaining their needs by asking basic informations about farming activities in neutral manner, e.g.: Where did you buy seeds? Where did you seek informations about farming practices? - Able to emphatize with locals, by expressing concerns	- Able to explain local problems - Able to explain local opportunities - Able to explain local wisdom for farming activities - Able to explain solution taken to overcome local problems - Able to explain trusted sources of information
Addressing misconception of the introduced innovation *needs an experienced farmers that already implemented the innovation	- Able to explain example of other small scale farmers or communities that succeeded in using the innovation and give a description how this example could be applied to local conditions - Able to let farmers discuss the misconceptions of the introduced innovation, instead of driving or controlling farmers to specific issues - Able to explain local conditions, so that the experienced small scale farmers could empathize with conditions of local small scale farmers	Experienced small scale farmers: - Able to explain why he chose to use the innovation - Able to explain the benefits and challenges of the innovation - Able to compare the condition of farmers groups that used and did not use the innovation - Able to show proves such as videos or photos of the implementation of innovation - Quote statements from big figure in agriculture to ensure that the innovation is needed Small scale farmers: - Able to listen the experienced small scale farmers - Able to express doubts, so that farmers can discuss or get more proves on the implementation of innovation - Able to compare the implementation of innovation based on local conditions, e.g. explaining local conditions and ask whether this kind of conditions could also support the innovation

Decision making process in implementing the innovation	- Able to bring locals to have their conclusion about their needs in neutral manner, e.g.: so, based on previous discussions can we list problems and opportunities identified? Are there any misconceptions left? Are there still any doubts about the innovation? - Able to help the locals in planning collective action, e.g. Which problems that have to be overcome immediately? When will we start the innovation? For how long the trial of innovation will be conducted? Who will be responsible for this project? How do we divide our job?	- Able to choose which problems that have to be overcome first - Able to decide what kind of innovation will fit local conditions - Able to critically adjust the schedule of projects with their farming activities
Building collective identity	- Able to facilitate the meetings in semi formal manner, e.g. lunch session by asking small scale farmers to bring their own crops to be cooked and eaten together; refreshing session such as singing or dancing together to local songs - Able to drive small scale farmers to create easy to remember or localized slogans to emphasize the importance of farmers participation in the innovation, e.g. "women are the mother of food sovereignty!" - Able to drive small scale farmers to creat a slogan that could explain the importance of the innovation so farmers are attached to the idea of using the innovation e.g. "agroecology is the way to achieve food sovereignty"	- Able to show empathy to the organization staffs the facilitate the meetings - Able to use slogans when discussion gets boring or stuck

4. Discussion

The findings showed that, *first*, the development of ICTs for agriculture cannot be design uniformly in Indonesia. Each location have its own needs. The top-down approach of agricultural projects are still used by Indonesian governments, that makes farmers could not utilize these aids. Several aids from government such as distribution of 41,000 tractors to farmers in Indonesia [33], yet government do not do any initial mapping to understand the dynamic of local agricultural practices where tractors delivered could not be used by every farmers. This finding basically proved that top-down approach is not the effective way in delivering ICTs for agriculture. *Second*, there will be misconceptions from any kind of innovations, including ICTs for agriculture. To clarify the misconception, facilitator needs to invite trusted source of information. Several studies showed that trusted source of information for small scale farmers are: peer-group or fellow farmers groups, opinion leaders, extension service agents, popular radio that use local language, and NGO staffs [11,12]. Our findings showed that independent small scale famers do not consider extension service agents as trusted source of information. They only trust people who shared same notion with the struggle of small scale farmers to become independent.

Third, participants of the project of developing ICTs for agriculture need to possess basic communicative competence in doing participatory approach. Our findings showed that both organization staff and small scale farmers have its own communicative competences which explained that participatory approach could not be conducted from the perspective of facilitators only or small scale farmers only. It needs a collaboration from both type of participants to conduct an effective participatory approach. The designer of ICTs for agriculture could learn to possess the communicative competence of organization staff as they have to build ICTs for agriculture based on local small scale farmers' needs.

Fourth, there are lots of challenges that are already identified by previous studies, where farmers are illiterate, have no trust to innovations, and lack of local adaptation. These challenges shows that the participatory approach was not conducted properly or facilitator and small scale farmers have no communicative competence to do participatory approach. For example, if farmers are illiterate, teaching them to read and write is not the only solution. Facilitator might use symbols, picture, or videos to deliver information or in developing ICTs for agriculture. This is where the competence to empathize is important, what kind of solution is needed by farmers? If they are reluctant to learn how to read and write, then facilitators have to empathized and be more creative in addressing this challenge. Have no trust to innovations meaning that there are no communicative space to address any misconceptions about innovations, which shows that our conclusion in addressing misconceptions of innovation is highly needed in developing ICTs for agriculture. *Fifth*, several studies showed that farmers have lack of commitment in doing participatory approach, we showed that participatory approach is not only about identifying local needs and addressing misconception of innovation but also have to build collective identity to drive a sense of belonging in implementing the project. Efforts to emphasize the need to collaborate and to be committed in the project is also important.

Our findings might be limited since it does not clearly reflect the process of participatory approach in designing ICTs for agriculture, however SPI has introduced many innovations that support the idea of food sovereignty and could change the behaviour of its members that used to farm in conventional way to agroecology farming, or members used to have no knowledge how to get their land back but now they understand the legal process and could fight to have their land back, or at first they only think that their products have to be sell to market, but now they know how to approach other alternative markets. All of these behaviour changes could be achieved by using the participatory approach and we could learn from this case study to implement the participatory approach in developing ICTs for agriculture.

5. Conclusion

Our article showed at least two important things: participatory approach in designing innovation could lead small scale farmers to be independent and participatory approach gives more benefit and impact to small scale farmers than implementing top-down approach in designing innovation. The implication of this research that governments agency need to facilitate more communicative spaces that make small scale farmers could participate in designing agriculture projects, especially projects of developing ICTs for agriculture. Agriculture projects can not only be designed by policy makers or programmers, but also by involving small scale agriculture who knows the reality of their day-to-day farming activities.

Communicative competences that we found for conducting participatory approach also showed that competent facilitator is urgently needed in designing and delivering agriculture projects, especially ICTs for agriculture. Indonesia has been decreasing the number of extension services agents by implementing cyber extension, however the implementation of cyber extension in Indonesia has not yet showed satisfying results [34–36]. Government should also pay attention that Indonesia needs facilitator that not only possess agriculture knowledge, but also possess communicative competences so that they are able to empathize with local small scale farmers.

References

[1] Jamaluddin N 2013 Adoption of E-commerce Practices among the Indian Farmers, a Survey of Trichy District in the State of Tamilnadu, India *Procedia Econ. Financ.* **7** 140–9
[2] Ajani E N 2014 Promoting the use of information and communication technologies (ICTs) for agricultural transformation in sub-Saharan Africa: Implications for policy *J. Agric. Food*

Inf. **15** 42–53
[3] Hamelink C 2014 Equality and human rights *The Handbook of Development Communication and Social Change* ed K G Wilkins, T Tufte and R A Obregón (Oxford: Wiley-Blackwell) pp 72–91
[4] Nmadu J N, Aiyelitsoya F A and Sallawu H 2013 Use of ICT in Securing Marketing Information among Small Scale Farmers in Niger State, Nigeria *Int. J. Trade, Econ. Financ.* **4** 66–72
[5] Syiem R and Raj S 2015 Access and Usage of ICTs for Agriculture and Rural Development by the tribal farmers in Meghalaya State of North-East India *J. Agric. Informatics* **6** 24–41
[6] Meijer S S, Catacutan D, Ajayi O C, Sileshi G W and Nieuwenhuis M 2015 The role of knowledge, attitudes and perceptions in the uptake of agricultural and agroforestry innovations among smallholder farmers in sub-Saharan Africa *Int. J. Agric. Sustain.* **13** 40–54
[7] Krone M, Dannenberg P and Nduru G 2016 The use of modern information and communication technologies in smallholder agriculture: Examples from Kenya and Tanzania *Inf. Dev.* **32** 1503–12
[8] Ali J 2012 Factors Affecting the Adoption of Information and Communication Technologies (ICTs) for Farming Decisions *J. Agric. Food Inf.* **13** 78–96
[9] Mittal S and Mehar M 2015 Socio-economic Factors Affecting Adoption of Modern Information and Communication Technology by Farmers in India: Analysis Using Multivariate Probit Model *J. Agric. Educ. Ext.* **22** 199–212
[10] Wyche S and Steinfield C 2016 Why Don't Farmers Use Cell Phones to Access Market Prices? Technology Affordances and Barriers to Market Information Services Adoption in Rural Kenya *Inf. Technol. Dev.* **22** 320–33
[11] Aker J C, Ghosh I and Burrell J 2016 The promise (and pitfalls) of ICT for agriculture initiatives *Agric. Econ.* **47** 35–48
[12] Akullo W N and Mulumba O 2016 Making ICTs relevant to rural farmers in Uganda: A case of Kamuli district *Rev. Bras. Planej. e Desenvolv.* **5** 291–304
[13] Adejo P E, Idoka M H and Adejo E G 2013 Gender issues and access to ICTs for agricultural and rural development in Dekina local government area of Kogi State *J. Agric. Ext. Rural Dev.* **5** 77–82
[14] Van Meensel J, Lauwers L, Kempen I, Dessein J and Van Huylenbroeck G 2012 Effect of a participatory approach on the successful development of agricultural decision support systems: The case of Pigs2win *Decis. Support Syst.* **54** 164–72
[15] Matthews K B, Schwarz G, Buchan K, Rivington M and Miller D 2008 Wither agricultural DSS? *Comput. Electron. Agric.* **61** 149–59
[16] Barakabitze A A, Fue K G and Sanga C A 2017 The use of participatory approaches in developing ICT-based systems for disseminating agricultural knowledge and information for farmers in developing countries: The case of Tanzania *Electron. J. Inf. Syst. Dev. Ctries.* **78**
[17] Lindblom J, Lundström C, Ljung M and Jonsson A 2017 Promoting sustainable intensification in precision agriculture: review of decision support systems development and strategies *Precis. Agric.* **18** 309–31
[18] Joseph M K and Andrew T N 2008 Participatory Approaches for the Development and Use of Information and Communication Technologies (ICTs) for Rural Farmers *2008 IEEE Int. Symp. Technol. Soc.* 1–13
[19] Steinke J and van Etten J 2017 Gamification of farmer-participatory priority setting in plant breeding: Design and validation of "AgroDuos" *J. Crop Improv.* **31** 356–78
[20] Seminar A U, Sarwoprasodjo S, Santosa D A and Kinseng R A 2017 Agroecological Education Aimed at Achieving Food Sovereignty *J. Dev. Sustain. Agric.* **12** 34–44
[21] Seminar A U, Sarwoprasodjo S, Santosa D A and Kinseng R A 2018 Local Peasants' Understanding on Food Sovereignty: Case of Indonesian Peasants within Transnational Agrarian Movement *Makara Hubs-Asia* **22**

[22] Saville-Troike M 2008 *The Ethnography of Communication: An Introduction: Third Edition* (Oxford, UK: Blackwell Publishing Ltd)
[23] Kalou Z and Sadler-Smith E 2015 Using Ethnography of Communication in Organizational Research *Organ. Res. Methods* **18** 629–55
[24] Yin R K 2003 *Case Study Research Design and Methods* (Thousand Oaks, CA: SAGE Publications)
[25] Hymes D 1972 Toward Ethnographies of Communication *Language and Social Context* ed P P Giglioli (London: Penguin)
[26] Marcus G E 1995 Ethnography in/of the World System: The Emergence of Multi-Sited Ethnography *Annu. Rev. Anthropol.* **24** 95–117
[27] Bazeley P and Jackson K 2013 *Qualitative data analysis with Nvivo* (London: SAGE Publications Ltd)
[28] Miles M B, Huberman A M and Sadana J 2014 *Qualitative Data Analysis: A Methods Sourcebook* (Thousand Oaks, CA: SAGE Publications)
[29] Brent Z W, Schiavoni C M and Alonso-Fradejas A 2015 Contextualising food sovereignty: the politics of convergence among movements in the USA *Third World Q.* **36** 618–35
[30] Edelman M, Weis T, Baviskar A, Borras S M, Holt-Giménez E, Kandiyoti D and Wolford W 2014 Introduction: critical perspectives on food sovereignty *J. Peasant Stud.* **41** 911–31
[31] Altieri M and Nicholls C I 2012 Agroecology Scaling Up for Food Sovereignty and Resiliency *Sustainable Agriculture Reviews* vol 11, ed E Lichtfouse (Dordrecht: Springer) pp 1–29
[32] McMichael P 2014 Historicizing food sovereignty *J. Peasant Stud.* **41** 1–25
[33] The Jakarta Post 2015 Jokowi wants farmers to work harder for food sufficiency *Jakarta Post*
[34] Gultom D T, Sumardjo, Sarwoprasodjo S and Muljono P 2016 *Vegetable-Farmer Behavior Communication in Fulfilling the Need of Agricultural Information through Cyber Extension in Lampung Province* (Bogor Agricultural University)
[35] Wijaya A S, Sarwoprasodjo S and Seminar K B 2015 *The Utilization of Cyber Extension-Based Agricultural Extension as Information Media by Extension Officers in Bogor Regency* (Bogor Agricultural University)
[36] Mulyandari R S H, Sumardjo, Lubis D P and Panjaitan N K 2010 Implementation of Cyber Communication Innovation in Agricultural Extension *Inform. Pertan.* **19** 17–43

6. Acknowledgements

We would like to thank Serikat Petani Indonesia (SPI) for providing data and informants for this research.

Canopy cover of mangrove estimation based on airborne LIDAR & Landsat 8 OLI

L B Prasetyo[1,5], W I Nursal[2], Y Setiawan[1], Y Rudianto[2] , K Wikantika[3], B Irawan[4]

[1]Department of Forest Resources Conservation & Ecotourism, Forestry Faculty, Bogor Agricultural University, e-mail: lbprastdp@apps.ipb.ac.id
[2]Forests2020- Department of Forest Resources Conservation & Ecotourism, Forestry Faculty, Bogor Agricultural University
[3]Remote Sensing and Geographical Information Sciences, Faculty of Earth Science and Technology, Bandung Institute of Technology
[4]Department of Biology, Airlangga University, Surabaya, Indonesia

Abstract. Mangroves are very important ecosystems, because of their economic value and environmental services, including as a habitat for various wildlife species, storing carbon, and protecting land from sea abrasion. Indonesia is known to have large mangrove area and diversity. It is estimated that the area of mangroves in Indonesia in 2015 reached about 3 million hectares, with 15 families, 18 genera, 41 true mangrove species and 116 species of mangrove associations. Unfortunately, the area to continue to decline due to degradation and conversion to other land uses, especially ponds and built up areas. Usually, mangrove degradation assessment is carried out by field survey and relying on Normalized Difference Vegetation Index (NDVI) clustering derived from satellite image data. Field surveys require a large amount of time and cost, meanwhile NDVI clustering is either inaccurate or too rough. Therefore, exploration of another methods are needed. Our result showed that pixel value of Band 5, Band 6, NDVI and PC1 can be used to estimate canopy cover. Regression using quadratic equation is better than linear equations. However, we noticed limitations of optical Landsat 8 OLI data for canopy cover mapping, namely pixel saturation on high canopy cover and high pixel value of bush/shrubs/regrowth that was not always representing high canopy cover.

Keyword: airborne, canopy cover, mangrove

1. Introduction

It is estimated that the area of mangroves in Indonesia in 1990 reached about 3.6 million hectares, however, the area decreases to 2.94 million hectares in 2015, in which of about 1.4 million hectares is secondary mangrove forest [1]. The deforestation and degradation mangrove forest is due to anthropological factors such as logging, ponds development, and settlements [2][3]. This will lead to species extinction since mangrove consist of 116 species, 18 genera and 15 families of plant and host of aquatic and terrestrial wildlife [4][5].

The Ministry of Environment and Forestry measured the degradation of mangrove based on stand density in the field, in which stand density below 1000 individual/ha is classified as degraded mangrove

[6]. Meanwhile, [7] proposed another approach based on Normalized Difference Vegetation Index (NDVI) and Canopy cover. NDVI is an index derived from remote sensing data (band red and near infrared) and its value ranging from -1 to +1 [8]. Positive index represents vegetation, while below 0 indicates non vegetation. It is widely used for canopy cover estimation [9], forage [10] and vegetation changes [11]. Based on this method, the mangroves are heavily degraded if the NDVI value is in the range of -1 to 0.32 or canopy density is less than 50%. Meanwhile, mangroves are moderately degraded if NDVI ranges from 0.32 to 0.42 or canopy cover is between 50-70%. Field observation based data is precise for measuring degradation, but it is timely and costly, meanwhile, the NDVI clustering approach is either inaccurate or the classification is too rough. The same conclusion also mentioned [12], that there is limitation to use the NDVI to measure forest degradation. It is an urgent need to explore another approach to measure mangrove degradation.

Light detection and ranging (LIDAR) is widely used and accurate to estimate canopy cover [13][14][15], tree height and forest structure [16] and above ground biomass mapping[17][18]. However, LIDAR is very expensive and only cover limited area. The objective of the research is to look a relation between canopy cover derived from LIDAR and the pixel value of LANDSAT 8 OLI. The algorithm can be used for mapping mangrove degradation in area beyond the LIDAR coverage.

2. Material and Methods

2.1. Study area

Mangrove of the study area is located in Banyuasin District of South Sumatra Provinces (figure 1). Part of the mangrove forest is situated within Berbak-Sembilang National Park. The mangrove condition is very good with very limited disturbance. The common species in the ecosystem are *Rhizpora mucronata*, *R. apiculata*, *Bruguiera gymnorhiza*, *B. sexangula*, *B. paviflora*, *Ceriops tagal*, *Xylocarpus granatum* and nipah.

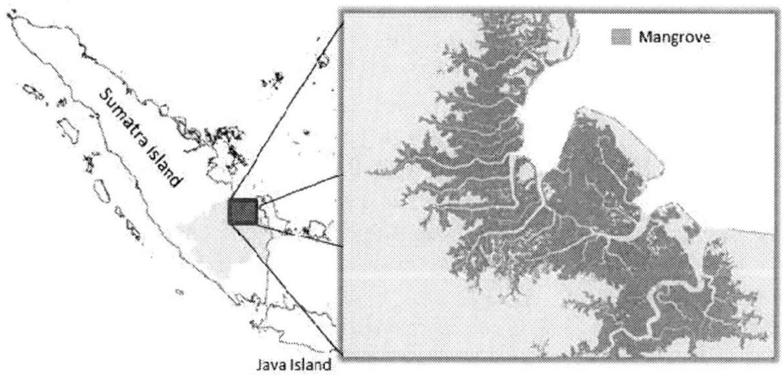

Figure 1. Study area

2.2. Landsat 8 OLI & LIDAR data acquisition

Landsat data over the area study with acquisition date on 25th July 2014 (Path/Row: 124/061) was downloaded from the United Stated Geological Survey (https://earthexplorer.usgs.gov/). Meanwhile, high resolution aerial photo and LIDAR data were taken by fixed wing from 800 m above ground level, with swath width of 851 m in October 2014 (figure 2). There were two Lidar acquisition mode namely full waveform (8-15 points/m^2) with 60% overlap and discrete return (6-8 points/m^2) with 80% overlap.

In this research discrete returns data were utilized. The raw LIDAR was taken by PT ASI Pudjiastuti Geosurvey and provided by GIZ-Bioclime project.

2.3. Landsat 8 OLI data processing
Prior further analysis, the Digital Number (DN) of Landsat 8 OLI were converted to spectral reflectance based on algorithm of [19], which was also adopted by [20]. An important part of this method is changing the DN value by taking into account the illumination of each pixel, which previously only used illumination values of the center scene of Landsat. NDVI, Principal Component 1 (PC1) and Principal Component 2 (PC2) were calculated by using ERDAS Imagine software.

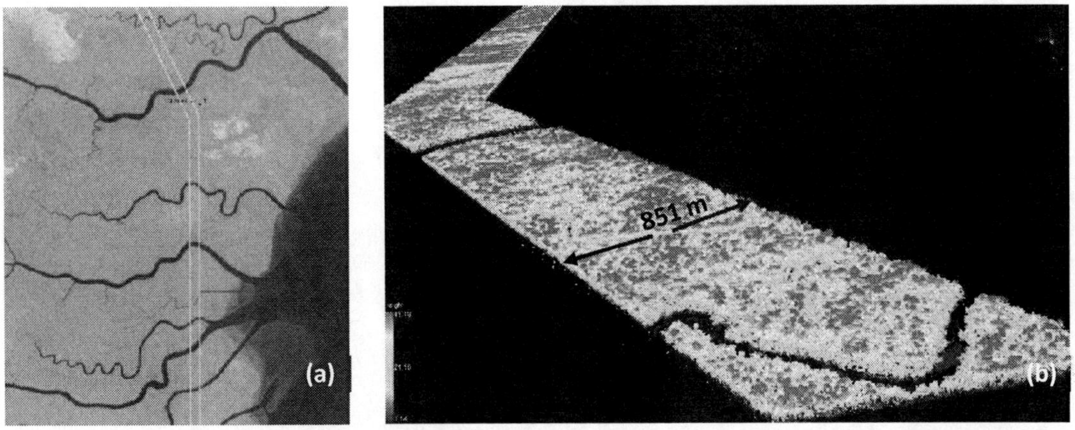

Figure 2. (a) Lidar Swath on overlaid with Landsat 8 OLI, (b) 3D of LIDAR Point clouds

2.4. Canopy cover estimation
Canopy cover were calculated by using open source software lidR package of R (https://github.com/Jean-Romain/lidR) based on First-Return Cover Index (FRCI) as shown in formula 1. The FRCI was adopted from [13] and it was selected since was more appropriate compare to other algorithm such as All-return cover index (ARCI) and CHM-based method. The FRCI was calculated for every 30 m x 30 m, in accordance to Landsat 8 OLI spatial resolution.

$$\mathrm{FRCI} = \frac{\Sigma First\ canopy + \Sigma Single\ canopy}{\Sigma First\ total + \Sigma Single\ total} \quad (1)$$

Where FRCI: First-Return Cover Index; ΣFirst canopy: Total point of canopy return; ΣSingle canopy: Total point of single canopy return; ΣFirst total: Total of point clouds; and ΣSingle total: Total point of single return.

Prior the FRCI calculation the point clouds of LIDAR data were normalized with Digital Elevation Model/ DEM (30 m x 30 m spatial resolution) derived from LIDAR. To ensure good fit of spatial position between Landsat 8 OLI & LIDAR, the boundary of FRCI calculation was made based on fishnet (.shp) of LANDSAT 8 OLI. The FRCI calculation result were extracted as point attribute data (.shp) of the center point of the pixel (figure 3).

2.5. Data Analysis
Total 2078 data were selected from 4562 original data points to ensure representativeness of data distribution. The selection is done due to the fact that more than 85% data are above 60% canopy cover. To determine the relationship between FRCI values as dependent variable with the NDVI, PC1, PC2

and reflectance values of each bands from Landsat 8 OLI as independent variables, correlation and regression analysis were performed using SPSS version 22. Prior to regression analysis data outliers' removal, and normality test based on standard residual histogram and P-P plot as well as heteroscedasticity test based on scatter plot of standardized predicted value and standardized residual analysis were conducted. Finally, after the above step 707 points were selected as data sample. The significant independent variables with high correlation (>0.5) were analyzed further.

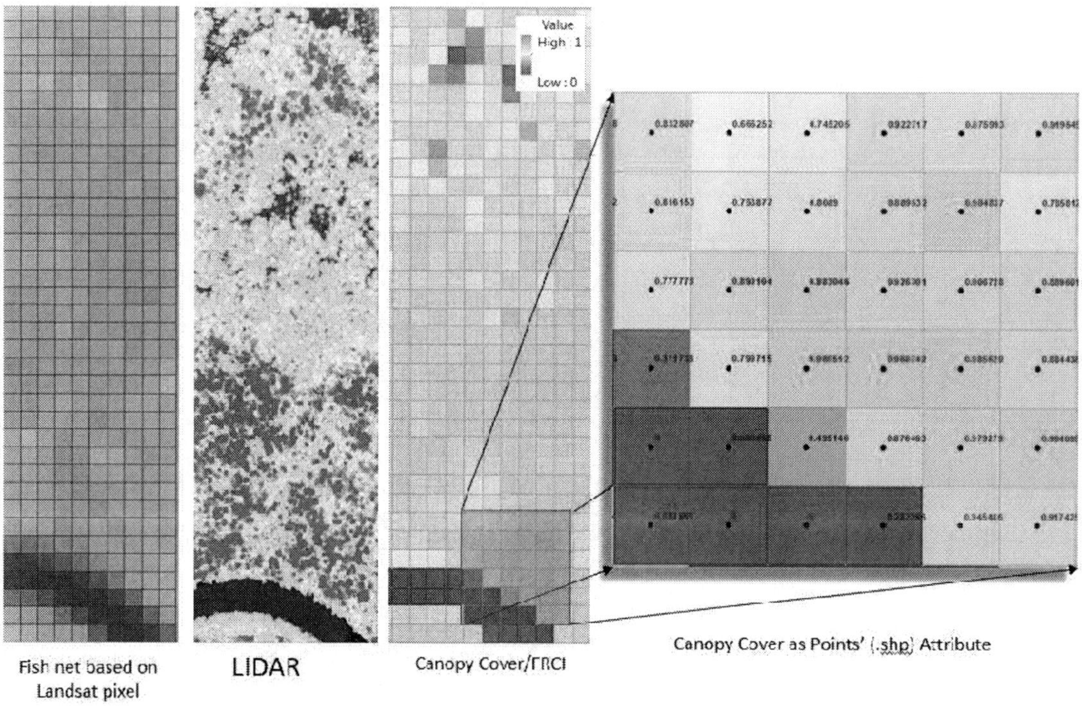

Figure 3. FRCI calculation based on polygon (fishnet) of Landsat

3. Results & discussion

Correlation analysis result showed that all independent variables (spectral value of Band 1, Band 2, Band 3, Band 4, Band 5, Band 6, Band 7, Band 8, NDVI, PC1 and PC2) have a significant correlation with the percentage of canopy cover (FRCI). However, correlation with NDVI (0.698) was the highest, followed by Band 5 (0.666), PC1(0.634), PC2 (-0.662) and Band 6 (0.545).

The scatter plot of the data showed that the FRCI increase with the increase of spectral values (B5, B6, NDVI, PC1), but when the FRCI close to 80% the spectral reach to saturation. Unlike the other responses, PC2 have a negative response to the increase of FRCI (figure 4). Therefore, the quadratic function model are more appropriate with the data distribution which is indicated by the highest coefficient of determination (R^2) (table 1).

Observing in more detail of figure 4, it appears that there are several cases where the FRCI value are low but the pixel value of band 5, band 6, PC1 and NDVI bands are high. This may be related to the condition of land cover in the mangrove gap. If the condition of land cover in the gap is water /river, then the pixel of NDVI/Individual bands value is low. Conversely, if the condition of land cover inside the gap is understory vegetation/bush/shrub, then the pixel value will remain high. This is due to a high rate of photosynthetic activity, in which vegetation absorb red as well as blue bands, but reflects green and near infrared bands. Variation of gap within mangrove is presented in figure 5.

4. Conclusion

Band 5, Band 6, NDVI and PC1 values can be used to estimate canopy cover. Regression using quadratic equations is better than linear equations. We noticed limitations of optical Landsat 8 OLI data for canopy cover mapping in forest mangrove ecosystem, namely pixel saturation on high canopy cover and high pixel value of bush/shrubs/regrowth that is not always representing high canopy cover.

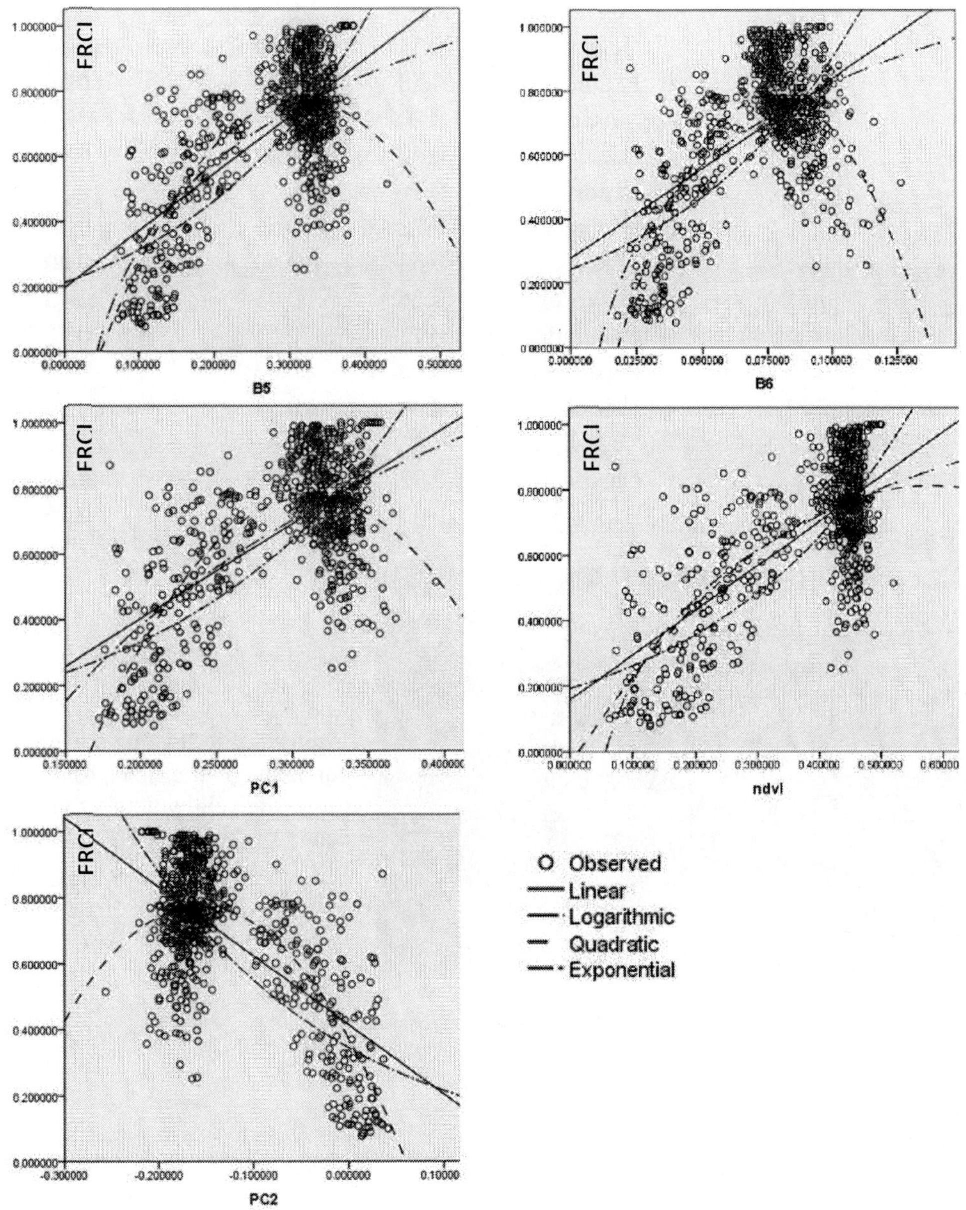

Figure 4. Fitting curve of regression

Table 1. Fitting of the equation

No	The independent variable	Model	R	R Square	Adjusted R Square
1	Band 5	Linear	0.67	0.44	0.44
		Logaritmic	0.69	0.48	0.48
		Quadratic	0.72	0.51	0.51
		Exponential	0.68	0.46	0.46
2	Band 6	Linear	0.55	0.30	0.30
		Logaritmic	0.62	0.38	0.38
		Quadratic	0.74	0.54	0.54
		Exponential	0.58	0.34	0.34
3	NDVI	Linear	0.70	0.49	0.49
		Logaritmic	0.70	0.49	0.49
		Quadratic	0.71	0.50	0.50
		Exponential	0.72	0.51	0.51
4	PC1	Linear	0.63	0.40	0.40
		Logaritmic	0.66	0.43	0.43
		Quadratic	0.71	0.51	0.51
		Exponential	0.66	0.43	0.43
5	PC2	Linear	0.66	0.44	0.44
		Logaritmic	-	-	-
		Quadratic	0.72	0.51	0.51
		Exponential	0.67	0.46	0.45

Figure 5. FRCI value regarding the gap land cover of mangrove (note: the yellow number in the center of the grids (30 x 30 m) is FRCI value, while the light blue color is NDVI, meanwhile the background image is a very high resolution air borne images)

References

[1] Ditjen Planologi Kehutanan dan Tata Lingkungan 2016 *Peta penutupan lahan tahun 2016*. Ditjen Planologi Kehutanan dan Tata Lingkungan. Jakarta

[2] Yusuf D N, Prasetyo L B, Kusmana C and Machfud 2017 Geospatial approach in determining anthropogenic factors contributed to deforestation of mangrove: A case study in Konawe Selatan, Southeast Sulawesi *IOP Conf. Ser.: Earth Environ. Sci.* 54 012049

[3] Mayuftia R, Hartoko A, and Hendrarto B 2013 Tingkat kerusakan dan karbon mangrove dengan pendekatan data satelit NDVI (Normalized Vegetation Index) di desa Sidodadi, Kabupaten Pesawara, Provinsi Lampung. *Diponegoro Journal of Maquares : Managemen of Aquatic Resources*, Vol. (4) : 146-154.
[4] Pramudji 2001 Dampak perilaku manusia pada ekosistem mangrove di Indonesia. *Oseana*, Volume XXV, Nomor 2, 2000:13-20
[5] Haris R 2014 Keanekaragaman vegetasi dan satwaliar hutan mangrove. *Jurnal Bionature*, Volume 15, Nomor 2, Oktober 2014, hlm. 117-122
[6] KLH 2004 Keputusan Menteri Negara Lingkungan Hidup. No. 201 tahun 2004. Kriteria baku dan pedoman penentuan kerusakan mangrove
[7] Kusmana C 2010 *Tingkat Kerusakan Mangrove berdasarkan Nilai NDVI dan Kerapatan kanopi. Jurnal Respon Mangrove Terhadap Pencemaran.* Dept. Silvikultur, Fakultas Kehutan IPB
[8] Xu D and Guo X 2104 Compare NDVI Extracted from Landsat 8 Imagery with that from Landsat 7 Imagery. *American Journal of Remote Sensing*. Vol. 2, No. 2, 2014, pp. 10-14. doi: 10.11648/j.ajr s.20140202.11
[9] Carreiras J M B, Pereira J M C and Pereira J S 2005. Estimation of tree canopy cover in evergreen oak woodlands using remote sensing. *Forest Ecology and Management* 223 (2006) 45–53
[10] Hogrefe K R, Patil V P, Ruthrauff D R, Meixell B W, Budde, M E, Hupp J W and Ward D H 2017 Normalized Difference Vegetation Index as an Estimator for Abundance and Quality of Avian Herbivore Forage in Arctic Alaska. *Remote Sens.* 2017, 9, 1234.
[11] Cui X, Gibbes C, Southworth J and Waylen P 2013. Using Remote Sensing to Quantify Vegetation Change and Ecological Resilience in a Semi-Arid System *Land* 2013, 2, 108-130; doi:10.3390/land2020108
[12] Tovar, M C L 2011 NDV I as indicator of degradation. *Unasylva* 238, Vol. 62, 2011/2
[13] Ma Q, Su Y, and Guo Q 2017. Comparison of canopy cover estimations from airborne Lidar, aerial imagery and sattellite imagery. *IEEE Journal of selected topics in applied earth observation and remote sensing*, Vol. 10 (9) : 4225-4236
[14] Rutledge A M and Popescu S C 2006. Using Lidar determining forest canopy parameters. *ASPRS 2006 Annual Meeting*, Reno, Nevada, May 1-5, 2006
[15] Korhonen L, Korpela I, Heiskanen J and Maltamoet M 2011 Airborne discrete-return LIDAR data in the estimation of vertical canopy cover angular canopy closure and leaf area index. *Remote Sensing of Environment* 115 (2011) : 1065-1080
[16] Lim K, Treitz P, Wulder B, St-Onge, Flood M 2003 LiDAR remote sensing of forest. *Progress in Physical Geography* 27 (1): 88–106
[17] Asner G P, Mascaro J, Muller-Landau H C,Vieilledent G, Vaudry R, Hall J S, and van Breugel M l. 2011. A universal airborne LiDAR approach for tropical. *Oecologia*. DOI 10.1007/s00442-011-2165-z
[18] Silva C A, Klauberg C, Carvalho P C, Hudak A T and Rodriguez L C E 2014 Mapping aboveground carbon stocks using LiDAR data in Eucalyptus spp. plantations in the state of São Paulo, Brazil. *Sci. For., Piracicaba*, Vol. 42 (104) : 591-604
[19] Tan B, Wolfe R J, Masek F, Gao F, and. Vermote E F 2010 An illumination correction algorithm on Landsat-TM data, *IEEE International Geoscience and Remote Sensing Symposium*, 2010, pp. 1964-1967
[20] Hudjimartsu S, Prasetyo L B, Setiawan Y, Suyamto D and Ikbal W. 2018. Automated Landsat 8 data preprocessing for national forest monitoring system monitoring system, *Proc. SPIE 10773, Sixth International Conference on Remote Sensing and Geoinformation of the Environment* (RSCy2018), 107730R (6 August 2018); doi: 10.1117/12.2326100

Acknowledgement
This research was supported by World Class University Research Program in 2018 (Program Riset Kolaborasi Indonesia-World Class University 2018). The data are supported by the UK Space Agency

(UKSA) International Partnership Program (IPP) under the Forests2020 program coordinated by Ecometrica, Ltd. The Forests 2020 is a collaborative program to advance Earth Observation applications to forests monitoring.

A preliminary investigation of various signal phenomena generated by termite infestation

M A Nanda[1], K B Seminar[1], D Nandika[2], and A Maddu[3]

[1]Department of Agricultural Engineering and Biosystem, Bogor Agricultural University, Bogor 16680, West Java, Indonesia, e-mail: seminarkudangboro@gmail.com
[2]Department of Forest Products, Bogor Agricultural University, Bogor 16680, West Java, Indonesia
[3]Department of Physics, Bogor Agricultural University, Bogor 16680, West Java, Indonesia

Abstract. Subterranean termites are major urban pests in Indonesia and lead damage to wooden structures and buildings. Termite pest management involves two sections: prevention and control. The most crucial part of termite control is to detect termites; however, due to their cryptic behavior, the termite infestation evidence on wooden structures is difficult to detect visually. When termite infestation is active and wood damage is occurring, the distinction of various signal phenomena are released including moisture content, temperature and acoustic emission. The identification of these phenomena becomes the basic foundation for designing a future termite detection system. The aim of this study is to identify various signal phenomena generated by termite infestation. In this investigation, the pine wood, as the medium for termite infestation, was divided into two groups, i.e., the wood infested by 220 termites ('infested') and the normal wood as a control ('uninfested'). Each signal phenomenon generated by termites was observed and analysed to produce six pieces of information, i.e., (a) moisture content; (b) temperature; (c) four information of acoustic emission such as energy, entropy, peak frequency and zero moment power. Based on statistical analysis of Duncan multiple range test (DMRT; two-way ANOVA, $P \leq 0.05$), it was revealed that the information that can distinguish significantly between infested and uninfested wood are moisture content, temperature, energy, entropy and zero moment power. Finally, various phenomena (i.e., moisture content, temperature and acoustic emission) can be embedded and integrated in the sensing technology of termite detection.

Keywords: acoustic emission, signal phenomena, termite infestation, termite detection

1. Introduction

Termites are the most destructive and economically important insect pest of wooden structures, buildings and other cellulose products [1, 2]. Living in large underground colonies, termites may infest any wood in contact with the soil and may even construct protective shelter tubes over non-wood materials to attack wood above ground [3]. These social insects having a soldier, worker, queen and king caste, are widespread to terrestrial environment [4]; the main executors of various building damage are the termite worker and soldier (figure 1). Of the >3000 termites species recognised globally, 80 are known to cause serious damage, and the genus *Coptotermes* contains the largest number among these serious pest species [5, 6]. In Indonesia, the economic cost of termite damage to wood used in and around buildings is great. It is difficult to establish the exact amount of damage caused by termites, but estimates of yearly damage of IDR 8.68 trillion are common [7]. These numbers may not include the cost of treatments (both preventive and curative), repairs to damage wooden structures and disadvantage of property value. To overcome termite attacks and avoid a higher wood damage, a non-destructive technique is required that can detect a termite infestation in wood as early as possible [8].

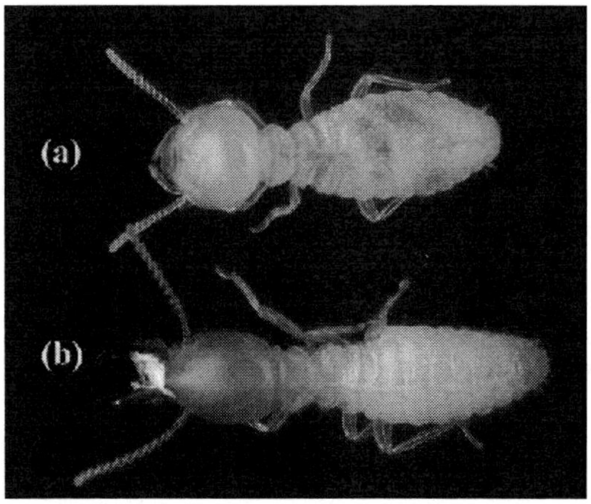

Figure 1. Colony structure of genus *Coptotermes* (a) worker, (b) soldier (this image provided by Lenhard, 2007 [9]). The quantity of individuals of each caste varies among species and also depends on the age and size of the colony.

Detecting termites are difficult. All termite species, be they drywood, dampwood or subterranean, are cryptic, because this is their primary defensive strategy against predation. Detecting infestations in structures and locating the entrance routes used by termites are often critical in effective management of pests. Various detection methods are used, with some varying for different types of termites. Basically, the termite detection system is now usually combined with the use of more sophisticated instruments, such as borescopes, moisture meters, electronic odor devices, acoustic emission and infrared heat detection; in addition, dogs have been trained to smell termites [10]. Recently, acoustic emission is a technique used to detect termites. When termite infestation is active and wood damage is occurring, microfractures are generated in the wood by the feeding of workers. As a result, elastic waves, or an acoustic emission phenomenon are generated, owing to the release of the strain energy stored in the wood [11]. In addition, other signal phenomena that occur due to termite infestation are moisture content and temperature. Clearly, there is a scope for improvement in termite detection, i.e., through a combination of various phenomena. In this study, considerable attention has been given to identify the characteristics of these phenomena including moisture content, temperature and acoustic emission. The results of this study become the basic foundation for designing a future termite detection system.

2. Materials and method

2.1. Timber Test

In this paper, for experimental design protocol, is provided by Nanda *et al.* [5, 8]. Firstly, Pine wood, used as a medium for the feeding activity test of termites *Coptotermes curvignathus*. This wood has a size of 20 (l) x 9.5 (w) x 2.5 (h) cm and the inside has a hole with size 12 (l) x 6 (w) x 0.5 (h). Furthermore, we divide the wood into two classes, i.e., the wood infested by 220 termites *C. curvignathus* ('infested') and the normal wood that is not infested by termites ('uninfested'). Each class consists of five wood samples. In addition, the termite infestation is carried out for two weeks.

2.2. Various Signal Phenomena

After the termite infestation process for two weeks, the characteristic phenomena in each class were acquired. The three signal phenomena revealed in this investigation are as follows: (1) Firstly, moisture content signal. It was measured by wood moisture meter GMK 1010 model (G-won Hitech CO., Ltd. of

Korea). Average moisture content can be expressed by Equation (1), where m_i is moisture content value on the wood surface at a point i (i=1, 2, 3, …, n).

$$\overline{mc} = \frac{1}{n}\sum_{i=1}^{n} m_i \tag{1}$$

(2) Secondly, the temperature signal. This signal was measured using thermometer infrared (Fluke 62 Max, Washington, USA). The mean temperature is calculated using Equation (2), where T_i is the temperature value on the wood surface at a point i (i=1, 2, 3, …, n).

$$\bar{T} = \frac{1}{n}\sum_{i=1}^{n} T_i \tag{2}$$

Furthermore, to provide visualisation of the contouring of moisture content and temperature on the wood surface, then both are plotted using the Surfer software® 13 (13.0.383 64-bit). This contouring is performed under the kriging method with easy built-in function [12], where every moisture content and temperature at each point is used as input. In a brief description [13], the kriging method is a linear interpolar that belongs to the best linear unbiased estimator (BLUE) family estimators. Thus, the primary purpose of the kriging method is to estimate a certain unknown variable as a linear combination of the known values. As in optimal interpolation, kriging requires that the observed process is second-order stationary, which means that the mean values and the standard deviation have to be independent of the location.

(3) The last signal, acoustic emission. The equipment used for the acoustic signal data acquisition is electret microphone sensor by ITead Studio, with a sensitivity of -56 dB (frequency of 0.1-10 kHz), which is connected to the microcontroller. This signal produces four informational pieces, i.e., energy, entropy, peak frequency and zero moment power. The energy (E) and entropy (H) can be calculated following Equations 3 and 4, respectively [3, 14, 15]. Where, E_i is energy, H is entropy, W_L is length of the frame size, e_j is the energy ratio of sub-frames with the total energy total and X_i is value of signal, n =1, . . ., W_L.

$$E = \frac{1}{W_L}\sum_{i=1}^{W_L} |X_i(n)|^2 \tag{3}$$

$$H = -\sum_{j=1}^{J} e_j \cdot \log_2 (e_j) \tag{4}$$

Unlike energy and entropy, which can be directly generated in the time domain, the peak frequency and zero moment power require the signal transformation first due to both parameters being in the frequency domain. One well-known method used for signal transformation is fast fourier transform. In this study, FFT was applied to produce peak frequency and zero moment power. The peak frequency (F_p) is determined based on the maximum peak of the overall signal length, whereas zero moment power (M_0) is defined as the area under the peak of magnitude. This can be computed using Equation 5 [8], where W_f= frequency length $(n = 1, \ldots, W_f)$, $P_{(n)}$ = magnitude and $P_{(n)\max}$ = maximum value of the magnitude.

$$M_0 = \sum_{n=1}^{W_f} \frac{P_{(n)}}{(P_{(n)})_{max}} \tag{5}$$

2.3. Statistical Analysis

Statistical significance of the difference between infested and uninfested class was determined using the analysis of variance (two-way ANOVA, $P \leq 0.05$) and comparison was done by Duncan multiple range test (DMRT 95%). All data were analysed using the XL-stat software ©Addinsoft 1995-2014.

IOP Conf. Series: Earth and Environmental Science **335** (2019) 012030 doi:10.1088/1755-1315/335/1/012030

3. Results and discussion

3.1. Moisture Content Signal

Figure 2, shows the contouring of moisture content on wood surface attacked by termites and normal wood. Visually, the infested wood has a higher moisture content than normal wood, i.e., maximum moisture contents for infested and uninfested wood is 12.9% and 11%, respectively. In infested wood, we considered that areas that have maximum moisture content indicate the center of termite attack inside the wood. An important consideration is the termites' dependency on moisture. Subterranean termites seek out moisture for survival [16]. Their high moisture requirements increase the likelihood that they will maintain contact with the soil or locate near areas where water collects, i.e., leaking pipes can keep wood and soil continually damp and create a perfect home for termites. Thus, it is not surprising that wood attacked by termites has a higher level of moisture than normal wood. They extract water from the wood on which they feed, and also produce water internally during the digestive process. They prefer wood with 12.01 ± 0.753% moisture content.

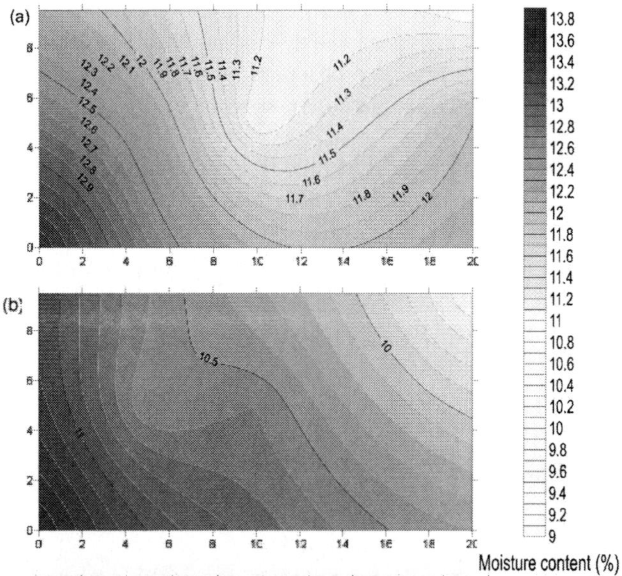

Figure 2. The contouring of moisture content on the wood surface by kriging method, (a) infested wood; (b) uninfested wood.

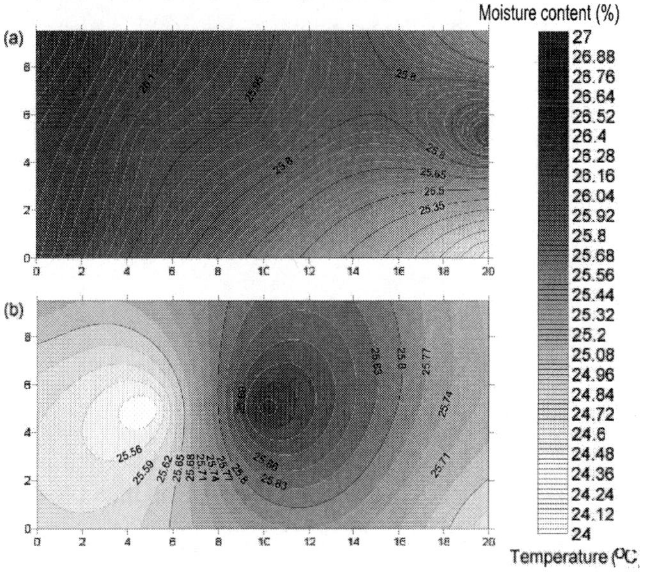

Figure 3. The contouring of temperature on the wood surface by kriging method, (a) infested board; (b) uninfested board.

IOP Conf. Series: Earth and Environmental Science **335** (2019) 012030 doi:10.1088/1755-1315/335/1/012030

3.2. Temperature signal

Figure 3, shows the contouring of temperature on wood surface attacked by termites and normal wood. Visually, the infested wood has a warmer temperature than the normal wood, i.e., maximum temperatures for infested and uninfested wood are 26.1 °C and 25.89 °C, respectively. In infested wood, we considered that areas that have maximum temperature indicate the center of termite attack inside the wood. According to Indrayani *et al.* [11], termites obtained their favorable temperature conditions indirectly via the available moisture from wood and metabolic process. As a result of termite attacks, the infested wood has a warmer temperature than the normal wood.

3.3. Acoustic emission signal

Figure 4, describes the acoustic emission signal in the infested and uninfested class on the frequency domain. Visually, the infested wood has a higher magnitude than uninfested wood. This is due to the activity of termites that produce acoustic signals; for instance, when termites are feeding and moving in the wood, they generate acoustic signals. In addition, acoustic signals can also be produced by termites as alarm signals through head banging to the wood [17]. Hence, in the last decade, the acoustic emission phenomenon becomes a foundation for designing a termite detection tools by various researchers worldwide [5, 18-21].

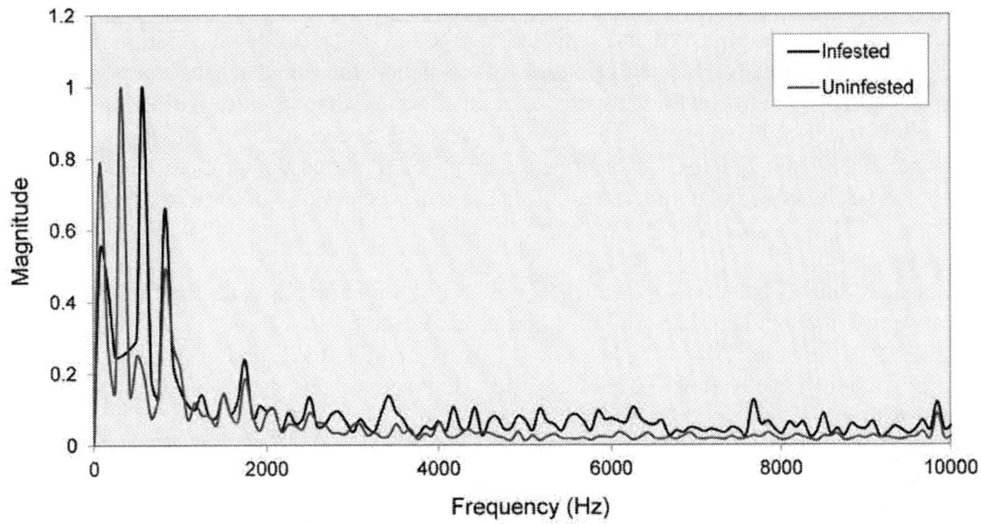

Figure 4. Acoustic emission signal on infested and uninfested wood.

3.4. Evaluation

Understanding the various signal phenomena and conditions that favour termite activities better prepares one to develop a termite detection system. Each signal phenomenon generated by termites was observed and analysed to produce six pieces of information, i.e., (a) moisture content; (b) temperature; (c) four pieces of information of acoustic emission such as energy, entropy, peak frequency and zero moment power. Characteristics of various signal phenomena generated by termite infestation are provided in table 1.

A statistical analysis of Duncan multiple range test (DMRT; two-way ANOVA, $P \leq 0.05$), revealed that the information that can distinguish significantly between infested and uninfested wood are $\overline{Mc}, \overline{T}$, E, H and M_0. While for information F_p, peak of frequency, is not able to distinguish significantly between infested and uninfested wood due to the standard deviation value being very wide. Finally, various phenomena (i.e., moisture content, temperature and acoustic emission) can be embedded and integrated in the sensing technology of termite detection. However, it needs to be reaffirmed that the availability of a transducer in the market, which is able to sense the moisture content, temperature and acoustic emission, needs to be investigated further. In addition, the transducer must be able to be integrated with each other to provide quick and accurate decisions, and it becomes a future research direction.

Table 1. Characteristics of various signal phenomena generated by termite infestation.

No.	Signal phenomena	Information	Wood condition	
			Infested	**Uninfested**
1	Moisture content (%)	[illegible]	12.01 ± 0.753^a	10.49 ± 0.589^b
2	Temperature (°C)	[illegible]	26.23 ± 0.205^a	25.68 ± 0.146^b
3	Acoustic emission	E	0.99926 ± 0.00034^a	0.99850 ± 0.00043^b
		H	-0.30062 ± 0.00034^a	-0.30002 ± 0.00024^b
		F_P	642.964 ± 477.372^{ns}	398.378 ± 451.021^{ns}
		M_0	6.65733 ± 3.112^a	5.34948 ± 2.258^b

Different notation on the same row indicated significant value, DMRT 95% ($P<0.05$).
ns = not significant.

4. Conclusion

This work has successfully investigated various signal phenomena generated by termite infestation including moisture content, temperature and acoustic emission signal. Each signal phenomenon generated by termites was observed and analysed to produce six pieces of information, i.e., (a) moisture content; (b) temperature; (c) four information of acoustic emission such as energy, entropy, peak frequency and zero moment power. Statistical analysis revealed that the information that can distinguish significantly between infested and uninfested wood are moisture content, temperature, energy, entropy and zero moment power. Finally, various phenomena (i.e., moisture content, temperature and acoustic emission) can be embedded and integrated into the sensing technology of termite detection.

References

[1] Arinana, Aldina R, Nandika D, Rauf A, Harahap I S, Sumertajaya I and Bahtiar E T 2016 Termite diversity in urban landscape, South Jakarta, Indonesia *Insects* **7** 20
[2] Nanda M A, Seminar K B and Nandika D 2017 Karakteristik Emisi Akustik untuk Mendeteksi Rayap Tanah pada Kayu (Characteristics of Acoustic Emission for Subterranean Termite Detection in Wood) *J. Tek. Pert.* **5** 227-34
[3] Nanda M A, Seminar K B, Nandika D and Maddu A 2018 Time domain features in combination with a support vector machine classifier for constructing the termite detection system *IOP Conf. Ser. Earth Env. Sci.* **157** 012037
[4] Nanda M A, Seminar K B, Nandika D and Maddu A 2018 Population Survey of Subterranean Termite *Coptotermes curvignathus* (Isoptera: Rhinotermitidae) on Infested Pine Boards *J. Entomol.* **15** 93-100
[5] Nanda M A, Seminar K B, Nandika D and Maddu A 2018 A Comparison Study of Kernel Functions in the Support Vector Machine and Its Application for Termite Detection *Information* **9** 5
[6] Kuswanto E, Ahmad I and Dungani R 2015 Threat of subterranean termites attack on the wooden buildings in asian countries and their control: A review *In. J. of Entomol.* **77** 356-62
[7] Nandika D, Rismayadi Y and Diba F 2015 *Rayap: Biologi dan Pengendaliannya, Edisi ke-2 (Termite: Biology and its Control), 2nd ed* (Surakarta: Universitas Muhammadiyah Surakarta)
[8] Nanda M A, Seminar K B, Nandika D and Maddu A 2018 Discriminant Analysis as a Tool For Detecting the Acoustic Signals of Termites *Coptotermes curvignathus* (Isoptera: Rhinotermitidae) *Int. J. Tech.* **9** 840-51
[9] Lenhard G 2007 Formosan Subterranean Termite (Coptotermes formosanus) Available online: http://animal.memozee.com/view.php?tid=3&did=23430. (accessed on 11 August 2018)
[10] Lewis V R 1997 Alternative control strategies for termites *J. Agric. Entomol.* **14** 291-307

IOP Conf. Series: Earth and Environmental Science **335** (2019) 012030 doi:10.1088/1755-1315/335/1/012030

[11] Indrayani Y, Yoshimura T, Yanase Y, Fujii Y and Imamura Y 2007 Evaluation of the temperature and relative humidity preferences of the western dry-wood termite *Incisitermes minor* (Hagen) using acoustic emission (AE) monitoring *J. Wood Sci.* **53** 76-9

[12] Cattle J A, McBratney A and Minasny B 2002 Kriging method evaluation for assessing the spatial distribution of urban soil lead contamination *J. Env. Qual.* **31** 1576-88

[13] Orús R, Hernández-Pajares M, Juan J and Sanz J 2005 Improvement of global ionospheric VTEC maps by using kriging interpolation technique *J. Atmos. Sol-Terr. Phy.* **67** 1598-609

[14] Giannakopoulos T and Pikrakis A 2014 *Introduction to Audio Analysis: A MATLAB® Approach* (Oxford: Academic Press)

[15] Nanda M A, Seminar K B, Solahudin M, Maddu A and Nandika D 2018 Implementation of Genetic Algorithm (GA) for Hyperparameter Optimization in a Termite Detection System. In: *Proceedings of the 2nd International Conference on Graphics and Signal Processing*: ACM) pp 100-4

[16] Peterson C, Wagner T L, Mulrooney J E and Shelton T G 2006 *Subterranean termites-their prevention and control in buildings* (United States: Home and Garden Bulletin 64)

[17] Hager F A and Kirchner W H 2013 Vibrational long-distance communication in the termites Macrotermes natalensis and Odontotermes sp *J. Exp. Biol* **216** 3249-56

[18] Dunegan H 2001 Elimination of extraneous noise sources from acoustic emission based termite detection instrument by use of modal ratios *Dunegan Engineering Consultants Inc., PO Box* **1749**

[19] Mankin R, Osbrink W, Oi F and Anderson J 2002 Acoustic detection of termite infestations in urban trees *J. Econ. Entomol.* **95** 981-8

[20] Fujii Y, Fujiwara Y, Yanase Y, Okumura S, Narahara K, Ngatsuma T, Yoshimura T and Imamura Y 2007 Nondestructive detection of termites using a millimeter-wave imaging technique *Forest Prod. J.* **57** 75

[21] Mankin R and Benshemesh J 2006 Geophone detection of subterranean termite and ant activity *J. Econ. Entomol.* **99** 244-50

Acknowledgment

This study was performed with the financial support from the Ministry of Research, Technology and Higher Education of the Republic of Indonesia (project no. 129/SP2H/PTNBH/DRPM/2018) through the accelerated master program leading to doctorate research grants (PMDSU).

Utilisation of fish skin waste as a collagen wound dressing on burn injuries: a mini review

A Afifah [1], O Suparno[1], L Haditjaroko[1], K Tarman[2]

[1]Department of Agroindustrial Technology, Bogor Agricultural University, Bogor, Indonesia,
e-mail: amaliaafifah111@gmail.com
[2]Department of Aquatic Products Technology, Bogor Agricultural University, Bogor, Indonesia

Abstract. The fish processing industries produce wastes such as skin, fish scales, stomach contents, and bones. Those organic wastes has not been widely used. The biggest content of fish processing industry is collagen. Fish skin has the highest collagen content (5%-30%). Collagen has been widely used in pharmaceuticals, cosmetics, and food. One of the benefits of collagen is as a wound dressing on burn injuries. Collagen as the most abundant fibrous protein in fish is the main constituent of the extracellular matrix (ECM). ECM is a dynamic substance between cells in a tissue and has the ability to improve the tissue functions. Burn injuries are one of the types of wounds that has a long healing phase, which requires intensive treatment. Severe burn injuries could make people who experience it have traumatic and paralysis that affects the entire organ system. Collagen wound dressing was expected to be able to engineer tissue and accelerate cell regeneration to replace or regenerate damaged tissues. This paper will describe the potency of fish skin waste, problems in healing burn injuries, collagen extraction, and collagen application as wound dressing.

Key Words: burn injuries, collagen, fish skin waste, wound dressing.

1. Introduction

The fish processing industries produce wastes such as skin, fish scales, stomach contents, and bones. Those organic wastes have not been widely used. All of these wastes have high collagen content, and the highest is found in the skins. Collagen has been widely used in pharmaceuticals, cosmetics, and food. Out of all the collagen applications, the highest economic value lies in the field of cosmetics and pharmaceuticals. However, in terms of benefits, collagen in the pharmaceutical field is very useful in its role as healing assistance. In pharmaceutical, collagen acts as a homeostatic agent, regeneration of bone tissue, membrane oxygenator, contraception, implants, and drug delivery systems [1]. In addition, the problem in the pharmaceutical world are about *halal*. The collagen from fish processing industry is expected to solve this problem due to its *halal* source.

Collagen is a protein found in vertebrate and invertebrate animals. Collagen is found in the skin, tendons, and bones. The collagen has about 25%-35% of the total protein in the body. In the body, collagen is in the form of fibers and it plays a role in the formation of cell structures in the extracellular matrix to maintain tissue shape. Collagen is also a major component in building bones, teeth, joints, muscles, and skin [2]. Collagen is not toxic and has no side effects, thus it is compatible to be utilized in the pharmaceutical field. According to Kartika [3], the field of regenerative medicine has not focused on biomaterials as a stimulator of cell growth. They are still focusing on substances that trigger tissue regeneration. Collagen-based biomaterials can be used as materials to grow cells, combining this function with the biodegradable and biocompatible properties. Burn injury is a wound that has a long healing process, so it requires wound dressing that can help speed up the healing of the wound. With the above-mentioned characteristics, collagen is expected to be applied as a wound dressing to accelerate healing of burns.

Currently, collagen is mostly extracted from the skin of mammals, such as cattle and pigs. Collagen extraction from the skin of these mammals still causes problems related to religious beliefs, such as Islam. Thus, the collagen extraction research from fish has been developed to solve this problem. Kittiphattanabawon *et al.* [4] conducted collagen extraction on bamboo shark and blacktip shark with acid soluble collagen (ASC) and pepsin soluble collagen (PSC) methods. Singh *et al.* [5] extracted collagen in the skin of *Pangasinodon hypophthalamus* using ASC and PSC methods. Skierka and Sadowska [6] extracted *Godus morhua* fish skin using ASC and PSC methods. Huang *et al.* [7] found a new method of extraction using high-temperature water in a short time, on *Tilapia* fish. The research was conducted to produce type I collagen with good physicochemical properties and high yield [7].

Red snapper, tuna, mackerel, cork, and belida are fish that are widely used as fillets. However, the skin from the waste of these fishes has not been utilized optimally. The collagen extraction process from these fishes needs to be optimized, especially during pretreatment by using NaOH, and during extraction. Pretreatment using NaOH is a process for removing non-collagen proteins contained in the ingredients and is influenced by different types of fish. There are several collagen extraction methods, i.e. acid soluble collagen (ASC), pepsin soluble collagen (PSC), papain soluble collagen (PaSC), and extrusion-hydro-extraction (EHE). However, because collagen utilization as a biomaterial, wound dressing requires high level of collagen purity, the use of enzymes as a method of extracting collagen can produce purer collagen.

2. Potency of fish skin waste

Fish waste has the potency to be utilized as a product of high economic value considering the national fisheries production in Indonesia in 2016 that reached 23.52 million tons [8]. The level of fish consumption (kg/cap/year) is always increasing every year. The Central Bureau of Statistics [8] predicted the level of fisheries production to reach 33.53 tons, and the level of fish consumption to reach 50.65 kg/cap/year. Of all fish waste, skin waste contains collagen with a yield value of 11%-63%, which is highly dependent on the type of fish, extracting material, and collagen extraction techniques [9]. Collagen can be used as a wound dressing with high economic value. Fish skin consists of two tissues, i.e. epidermal tissue and dermis tissue. Dermis is a layer of skin, which consists mostly of collagen fiber tissue that is constructed by binding fabric [10]. The fish skin composition is presented in table 1 [11].

Table 1. Chemical composition of fish skin [11].

Composition	Total (%)
Water	69.6
Protein	26.9
Fat	0.7
Ash	2.5

3. Collagen

Collagen is a protein that makes up 30% of all protein in the body and is found in vertebrate and invertebrate animals [12]. To date, there are 28 types of collagen identified, i.e. type I to XXVIII and more than 90% as collagen I. Collagen is a fibrin protein (a protein in the form of fibres) that plays a role in the formation of the largest cell structure in the extracellular matrix that maintains tissue shape. Collagen is found in fibrous tissues such as tendons, ligaments, skin, cornea, cartilage, bones, blood vessels, and intestines. Collagen is also a major component in building bones, teeth, joints, muscles, and skin. Collagen can expand due to its weakening binding capacity of the molecular structure when treated at pH below 4 or pH above 10 [2].

Collagen can be utilized in various fields, such as leather manufacturing [13,14,15,16], gelatine, adhesive, food, pharmaceutical and cosmetic industries. The use of collagen in the pharmaceutical and cosmetic fields have high economic values. However, for the field of cosmetics, collagen utilization is still limited to few groups. In the pharmaceutical field, collagen utilization is needed for people who experience various kinds of large open wounds.

4. Collagen extraction

Commercial collagen is obtained from cow hide, pork skin, or chicken skin, but its use still causes problems considering religious considerations and biological contamination such as MCD (Mad Cow Disease), TSE (Transmissible Spongiform Encephalopathy), FMD (Foot and Mouth Disease), etc. [17]. Therefore, the research is focusing on extracting collagen from fish skin. Collagen can be obtained by chemical hydrolysis or enzymatic hydrolysis [18], Collagen can be extracted by using three methods, i.e. Acid Soluble Collagen (ASC), using enzyme or Pepsin Soluble Collagen (PSC), and hydro-extraction [7]. Extraction by using enzymes can be conducted not only by using pepsin enzyme but also by enzyme papain. The most used is the chemical hydrolysis method, but enzymatic hydrolysis produces better collagen characteristics with high nutritional value for ingredients that want to maintain the functionality [19]. The enzymatic process also produces less waste and can reduce processing time. However, enzymes are still relatively expensive. Before collagen is extracted, pretreatment must be conducted. The material commonly used for pretreatment is NaOH. Pretreatment is done to eliminate non-collagen substances in order to obtain high collagen yield.

The pretreatment process creates crosslinking bond between collagen and other open substances, thus separating the collagen. In the acid extraction process, there are several types of acids to be utilized, such as acetic acid, citric acid, lactic acid, and inorganic acids. However, organic acids are better in dissolving collagen [6]. The acid that is often used for collagen extraction is acetic acid [20]. ASC extraction is more efficient but some enzymatic extractions can control the hydrolysis of enzymes that results in collagen with a higher level of purity. The extraction method must consider the final characteristics of the collagen produced, such as thermal stability, molar mass, and water holding capacity [21]. The hydro-extraction process begins with pretreatment using a base. Afterward, the process continued with treatment using acid, after which the material containing collagen is extracted by using water at high temperatures and regulated pressure. The results from this extraction show a higher yield compared to acid and enzyme extraction [7]. The current development of collagen extraction from fish skin is not only with one method but is combined between these methods. Kittiphattanabawon *et al.* [4] extracted collagen in bamboo shark and blacktip shark using acid soluble collagen (ASC) and pepsin soluble collagen (PSC) methods. Singh *et al.* [5] extracted collagen in the skin of *Pangasinodon hypophthalamus* using ASC and PSC methods. Skierka and Sadowska [6] extracted *Godus morhua* fish skin using ASC and PSC methods.

5. Problems in healing burn injuries

The general burn injuries problems are morbidity and long-term disability. Burn wound healing involves several phases, i.e. coagulation, inflammation, granulation, proliferation, matrix synthesis and deposition, fibrogenesis, angiogenesis, wound contractions, and re-epithelization, which are the problems of healing those wound. The use of synthetic drugs in some burn cases causes reactions such as allergies and drug resistance. Thus, the drugs need to be derived from natural ingredients [22]. Healing burns is quite slow and sometimes causes infections, pain, and continuous hypertrophic scarring [23]. Burns are divided into three levels based on the depth of injury, i.e. first-degree, second-degree, and third-degree burns. First-degree burn affects the outer epidermis. Second-degree burn injures deeper layers of the epidermis and part of the dermis, accompanied by blisters. Third-degree burn affects all layers of the epidermis and dermis, and looks like dry wounds [24].

6. Wound healing phase

Wound healing is divided into three phases, i.e. inflammation, proliferation, and maturation. The inflammatory phase occurs from day 0 to day 5. In this phase, there is a direct response in the form of blood clots to prevent blood loss. Hemostasis occurs in this phase and will be prolonged if an infection occurs. This phase is characterized by tumors, rubor, dolor, and calor. The proliferation phase starts from day 3 to day 14. The proliferation or epithelialization phase is also called the granulation phase because of the formation of granulation tissue. This phase is marked by red and shiny wound. Granulation tissue is a combination of fibroblasts, inflammatory cells, new blood vessels, fibronectin, and hyaluronic acid. Epithelialization occurs during the first 24 hours marked by thickening of the epidermal layer at the edges of the wound. The maturation phase (remodeling) lasts from several weeks to two years. In this phase, new collagen formation changes the shape of the wound and increases tissue strength (tensile strength). Scar tissue is formed around 50%-80%, as strong as the previous tissue [3].

According to Nystrom [3], the wound healing process consists of 4 phases unlike Kartika [11]. Before the inflammatory phase, the wound healing phase is preceded by the hemostatic phase. This phase occurs at the beginning of injury. In this phase, the platelets cells play a role, where clots occur to prevent the bleeding from continuing. Cytokines are released by thrombocytes as growth factors that collect cells to help the next healing phase. After this phase is complete, the next phases are inflammation, proliferation, and maturation [3, 11].

7. Wound dressing

Wound dressings have long been used to bandage temporary wounds and burns. Wound dressings are used to replicate skin function to the maximum extent possible during healing. There are various types of wound dressings based on their functions, i.e. as antibacterial, water vapour transmission, and initial healing of wounds. The wound dressing has many advantages, such as its easiness, inexpensive use, plenty of availability, and easy technology. However, wound dressings are inadequate in mimicking skin function, making it difficult to solve the problem of morbidity and mortality that occurs in serious injury to the skin [25]. Biosynthetic dressings are being developed to mimic skin function and replace the epidermis and dermis. Burns are easily infected. Therefore, antimicrobial dressings are widely used to prevent infection and minimise bacteria. Antimicrobial dressing products contain silver, iodine kadexomer, and honey, in the treatment of burns [23].

Injuries can occur due to damage to normal tissue structures, both inside and outside the body. The wound is divided into closed wounds and open wounds. Closed wound is a wound that does not damage the surface tissue, such as sprains and broken bones. Open wounds usually occur accidentally due to accidents which are often called traumatic injuries, e.g. penetrating wounds, stab wounds, torn wounds, cut wounds, and burns. Burns have similar conditions as amputation wounds, i.e. its slow healing property [26]. The wound dressing is used to protect the wound from infection and trauma. Wound healing is 50% faster when the wound is moist instead of dry. Humid atmosphere is an optimal condition for accelerating wound healing and it can trigger new tissue growth [27].

8. Collagen application as wound dressing

Shettigar *et al.* [28] investigated collagen sanitary napkins by extracting collagen from sheep intestines. Collagen is made with a cross-linked technique using ultraviolet light and tested in rats with second-degree and third-degree burns. The collagen applied to the wound makes the epithelial rate faster. There is also a slight inflammatory change in fibre growth in the subdermal layer. However, this study only aimed to test the effectiveness of collagen in healing wounds. Therefore, it is unknown whether the use of collagen can cover the wound faster because the examination of the wound was not conducted every day [28]. Histopathology of wound dressing with collagen after 10 days of dressing is presented in figure 1.

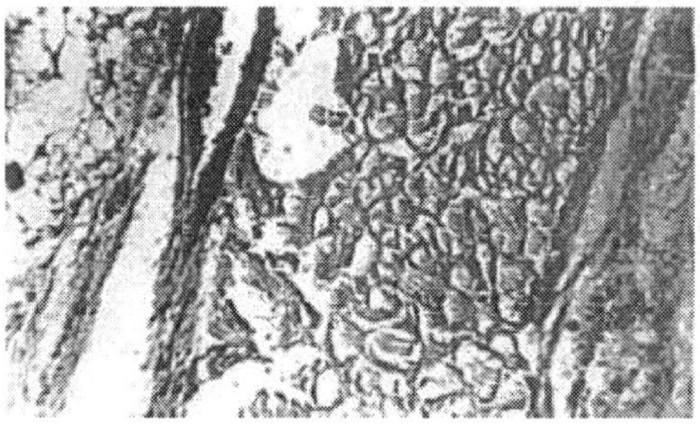

Figure 1. Histopathology of wound dressing after being wrapped with collagen for 10 days [28].

Yang *et al.* [29] developed the research on collagen dressings. Collagen is extracted from pigskin and made into collagen sheets, which are then applied to burns. The design of dressing in this study is called young collagenous wettable membrane (YCMW). YCMW is in the form of sheet and contains type I and type II collagen, wet but will dry when applied to wounds. The wound dressing does not form antigenicity, more sterile, can help in healing wounds, and cheap. The YCWM's weakness is that there are maceration and delays in separating eschar, so YCMW is only suitable for skin with shallow scarring and uninfected. YCMW is susceptible to infection, thereby needs additional material to overcome this deficiency [29]. YCMW is only suitable for dry and clean wounds. YCMW illustration can be seen in figure 2.

Figure 2. YCMW products and packaging [29].

Duraipandy *et al.* [30] studied the easy contamination in the dressing. In this study, collagen was combined with plumbagin, which is a naphthoquinone compound with pharmacological properties. Plumbagin is cross-linked with nanosilver, which also has antimicrobial properties called PCSN (caging plumbagin on silver nanoparticle). Plumbagin that has been cross-linked with nanosilver, is then cross-linked with collagen. Cross-linking between collagen and PCSN can improve physical, thermal, and mechanical properties between collagen molecules, and improve their antimicrobial properties. SEM test result (figure 4) shows a uniform pore structure. This pore structure determines the ability to transport substances on the scaffold. The result of antibacterial activity testing (figure 3) shows cross-link of PCSN and collagen that were able to inhibit *Escherichia coli* and *Bacillus subtilis* bacteria, with a minimum concentration of 1.25 M. PCSN-collagen has more effective wound healing ability than pure collagen [30].

IOP Conf. Series: Earth and Environmental Science **335** (2019) 012031 doi:10.1088/1755-1315/335/1/012031

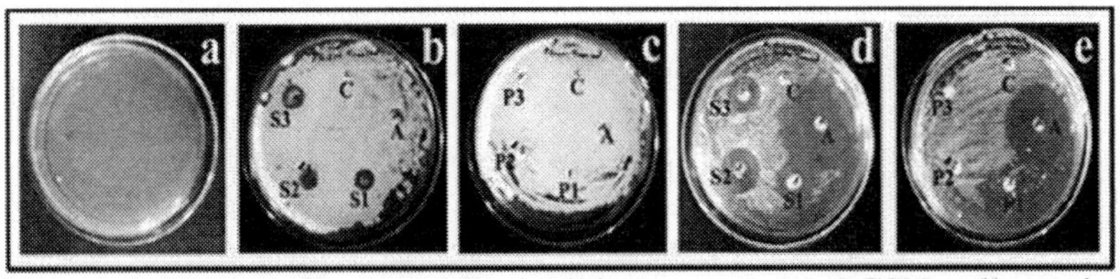

Figure 3. Test result of negative inhibitory zones (a), PCSN-collagen in *E. coli* (b), plumbagin-collagen in *E. coli* (c), PCSN-collagen in *B. subtilis* (d), plumbagin-collagen in *B. subtilis* (e) [30].

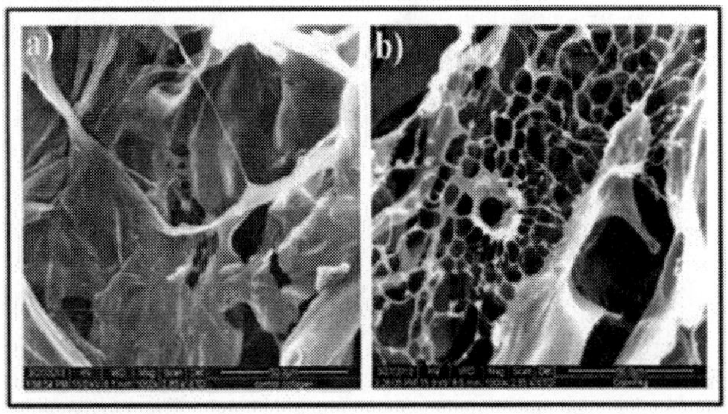

Figure 4. Pure collagen (a), Cross-linking collagen and PCSN (b) [30].

Collagen dressing continues to be developed to improve its properties and improve its capabilities. Xie *et al.* [31] conducted research by developing composites from chitosan, collagen, and alginate (CCA). CCA creation aims to act as wound healing and to inhibit the sea water. The application of dressing from collagen is developing towards the specific use. Collagen has low antigenicity, low inflammation, good biocompatibility, and good cell growth stimulation and cell proliferation. Chitosan has good properties, such as biocompatible, non-toxic, antibacterial, and hemostatic. Alginate is used in dressings because of its ability to absorb water and trigger wound healing. CCA composites are able to absorb exudate wounds well and create a relatively moist wound environment. CCA composites can also prevent injury from immersion in seawater for at least 4 hours [31]. The wound healing process and the rate of wound healing by CCA in mice are presented in figure 5 and figure 6.

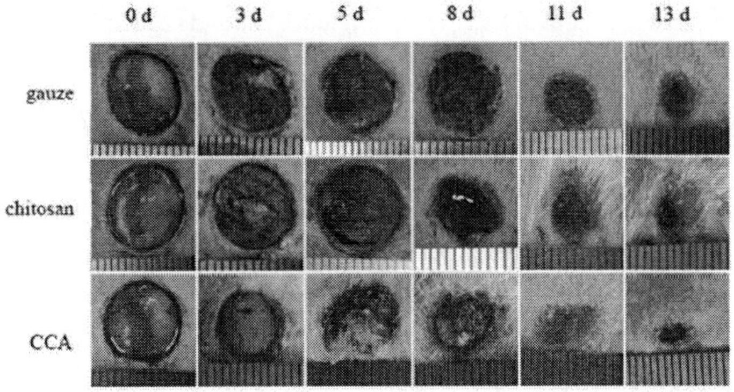

Figure 5. The wound healing process by CCA, pure chitosan, and gauze [31].

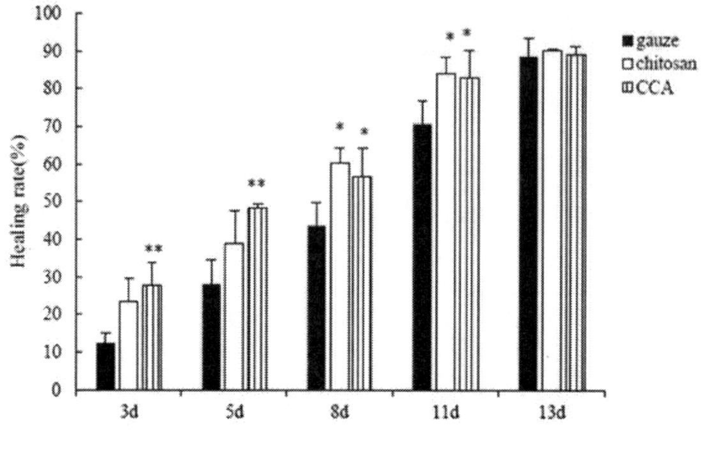

Figure 6. Wound healing rates by CCA, pure chitosan, and gauze [31].

9. Conclusion

Utilization of fish skin solid waste can produce products with high economic values. The wound dressing as an application of collagen has advantages compared to conventional wound dressing, i.e. its practicality. The wound healing process is relatively faster, and it does not cause pain. The wound dressing made from collagen using cross-linking and other ingredients have better mechanical and antibacterial properties. Wound treatment using collagen wound dressing will make the wound heal faster than the use of conventional wound dressing. Burns is one of the wounds that takes a long time to heal. Thus, collagen wound dressing combined with other material may be used to heal burns in a better way.

References

[1] Schrieber R dan Gareis H 2007 *Gelatin handbook: theory and industrial Germany (DE).* Wiley-VHC

[2] Peranginangin R, Murniyanti, Nurhayati W, Rahmad 2014 *Processing collagen from tilapia skin.* (Jakarta: Penebar Swadaya) (In Indonesian)

[3] Kartika RW 2015 *Chronic wound care with modern dressings. CDK.* **42(7)** 546-550. (In Indonesian)

[4] Kittiphattanabawon P, Benjakul S, Visessanguan W and Shahidi F 2009 *Isolation and characterization of collagen from the cartilages of brownbanded bamboo shark (Chiloscyllium punctatum) and blacktip shark (Carcharhinus limbatus). LWT-Food Science and Technology* **43(5)** 792-800

[5] Singh O, Benjakul S, Maqsood S, Kishimura H 2011 *Isolation and characterization of collagen extracted from the skin of striped catfish (Pangasianodon hypophthalamus). J. of Food Chemistry* **124(1)** 97-105

[6] Skierka E, Sadowska M 2007 *The influence of different acids and pepsin on the extractability of collagen from the skin of Baltic cod (Gadus morhua). Food Chemistry* **105(3)** 1302-1306

[7] Huang CY, Kuo JM, Wu SJ, Tsai HT 2016 *Isolation and characterization of fish scale collagen from tilapia (Oreochromis sp.) by novel extrusion-hydro-extraction process.* **190** 997-1006

[8] Central Bureau of Statistics [Indonesia]. 2017. *Fisheries production.* Jakarta: CBS

[9] Nagai T and Suzuki N 2000 *Isolation of collagen from fish waste material-skin, bone, and fins. Food Chemistry* **(68)** 277–281

[10] Padmasari R 2002. *Utilization of red tilapia skin waste (Oreochronis sp.) As raw material for making siomay* (Bogor, Indonesia: Bogor Agricultural University) (In Indonesian).
[11] Oosten JV 1969 *Skin And Scale*. (New York: Academic Press Inc.).
[12] Nystrom A 2016 *Collagens in wound healing, In: Wound healing biomaterial-volume 2.* (Cambridge (USA): Woodhead Publishing Series i Biomaterials).
[13] Covington AD, Evans CS, Lilley TH, Suparno O 2005 *Collagen and polyphenols: new relationships and new outcomes. Part 2. Phenolic reactions for simultaneous tanning and coloring. Journal of the American Leather Chemists Association* **100(9)** 336-343.
[14] Suparno O, Covington AD, Evans CS 2007 *Novel combination tanning using diphenols and oxazolidine for the production of high stability leather. Journal of the Society of Leather Technologists and Chemists* **91(5)** 188-192.
[15] Suparno O, Kartika IA, Muslich 2009 *Chamois Leather Tanning Using Rubber Seed Oil. Journal of the Society of Leather Technologists and Chemists* **93(4)** 158-161.
[16] Suparno O 2010. *Optimization of Chamois Leather Tanning Using Rubber Seed Oil. Journal of the American Leather Chemists Association* **105(6)** 189-194.
[17] Setyowati H dan Setyani W 2015 *The potential of nanocolagen for fish scales waste as cosmeceutical. J. of Science and community pharmaceutical* **12(1)** 30-40. (In Indonesian).
[18] Zavareze ER, Silva CM, Mellado MS and PrenticeHernández C 2009 *Funcionalidade de hidrolisados proteicos de cabrinha (Prionotus punctatus) obtidos a partir de diferentes proteases microbianas. Química. Nova* **32(7)** 1739-1743.
[19] Martin L, Wilson CG, Koosha F, Tetley L, Gray AI, Senel S 2002 *The release of model macromolecules may be controlled by the hydrophobicity of palmitoyl glycol chitosan hydrogels. J Control Release* **(80)** 87-100.
[20] Liu D, Wei G, Li T, Hu J, Lu N, Regenstein M. and Zhou P 2015 *Effects of alkaline pretreatments and acid extraction conditions on the acid-soluble collagen from grass carp (Ctenopharyngodon idella) skin. Food Chemistry* **(172)** 836-843.
[21] Schmidt MM, Dornelles RCP, Mello RO, Kubota EH, Mazutti MA, Kempa AP, Demiate IM 2016 *Collagen extraction process. Int. J of Food Research* **23(3)** 913-922.
[22] Zhang H, Chen J, Cen Y 2018 *Burn wound healing potential of a polysaccharide from sanguisorba officianalis L. in mice.*
[23] Wang Y, Beekman J, Hew J, Jackson S, Fisher ACI, Parungao R, Lajevardi SS, Li Z, Maitz PKM 2017 *Burn injury: Challenges and advances in burn wound healing, infection, pain and scarring.*
[24] Sabiston 1992 *Surgical Textbook* (Jakarta: Buku Kedokteran EGC) (In Indonesian).
[25] Garfein ES, Orgill DP, Pribaz JJ 2003 *Clinical application of tissue engineered constructs. J. of Clin. Plast Surg* **30(4)** 485-498.
[26] Stevens PJM, Bordui F, Weyde JAG 1992 *Nursing Science* (Jakarta: Penerbit EGC) (In Indonesian).
[27] Noer MS, Perdanakusuma DS, Saputro ID, Rizaliyana S, Basoeki AP, Airlangga PS, Semedi BP 2014 *Emergency management of major burn* (Surabaya: PT Revka Petra Media).
[28] Shettigar UR, Jagannathan R, Natarajan R 1981 *Collagen film for burn wound dressings reconstituted from animal intestines. J. of Artificial Organs* **6(3)** 256-260.
[29] Yang JY 1990 *Clinical application of collagen sheet, YCMW, as a burn wound dressing. J. of Burns* **16(6)** 457-461.
[30] Duraipandy N, Lakra R, Vinjimur SK, Korrapati PS, Ramamoorthy U, Kiran MS 2012 *Plumbagin caged silver nanoparticle stabilized collagen scaffold for wound dressing. J. of Materials Chemistry B* **00(1-3)** 1-11.
[31] Xie H, Chen X, Shen X, He Y, Chen W, Luo Q, Yuan W, Tang X, Hou D, Jiang D, Wang Q, Liu Y, Liu Q, Li K 2017 *Preparation of chitosan-collagen-alginate composite dressing and its promoting effects on wound healing. Journal of Biological Macromolecules,* p 1-37.

Acknowledgments
The authors highly appreciate The Indonesian Ministry of Research, Technology and Higher Education (*Kementerian Riset, Teknologi dan Pendidikan Tinggi R.I.*) for its research funding and support.

Utilization of rice husk as a tanning agent in the tanning process of leather (A mini review)

A W Nugraha[1], O Suparno[1], N S Indrasti[1] and Hoerudin[2]

[1] Department of Agroindustrial Technology, Bogor Agricultural University, Indonesia.
[2] Indonesia Center for Agricultural Post Harvest Research and Development, Bogor, Indonesia

Abstract. Rice is a food crop which is grown in Indonesia widely and used as a staple food crop for filling domestic food needs. Rice husk is a solid waste that formed from rice grain milling. Along with the increase in national rice production per year, the rice husk also increase. There is silica content in rice husk have a potency that used as a tanning agent. It can be used as a material substitution of chrome sulfate as conventional tanning agent that have carcinogenic properties. This article will explain about tanning process, the potency of rice husk in Indonesia, rice husk utilization, and silica extraction process also nano-silica processing from rice husk. Besides that, it will describe silica and nano-silica usage in the tanning process, mechanism tanning nano-silica and leather properties that produced from the tanning process with nano-silica.

Keywords: rice husk, tanning agent, silica content

1. Introduction

In the process of making leather, the tanning process is one of the key steps to produce good leather. Currently, the commonly used material for tanning agent is chromium sulfate. This is caused the leather produced can be utilized as various skin-based products. However, the use of chromate sulfate also has an adverse effect on the environment because of its carcinogenic property [1]. In the tanning process, 60% of chrome bonds with the skin, and the rest 40% is discarded along with wastewater [2]. This is very dangerous for the living beings and the people who live in the surrounding of the wastewater disposal. There are still many leather tanning industries that do not have a wastewater treatment plant (WWTP), and some industries with WWTP are still ineffective in preventing chromium from being discarded along with wastewater. According to [3], chromium in wastewater produced during the tanning process was 3-350 mg/L. According to [4], the content of Cr^{6+} in the wastewater of tanning industry was 0.03-2.09 mg/L.

Chromium is discharged into wastewater with a valency of Cr^{3+} and Cr^{6+}. However, Cr^{6+} has toxicity level of 100 times Cr^{3+}. Accumulation of Cr^{6+}in the organs may damage metabolic functions and also causes carcinogenic, mutagenic, and teratogenic effects [5]. According to [6], the chromium compounds may enter the body in several ways, i.e. through chromium contaminated food, and exposure or direct inhalation. Also, the chromium that enters the body through food has higher concentration compared to that which enters through inhalation or skin contact [7, 8]. This is because the food itself (plants) has been contaminated with chrome. [6] stated that chrome accumulates in each part of the plant. Therefore, certain handling is needed to ensure that chrome does not pollute the environment.

According to [9], there are several types of leather tanning processes, and the tanning agent commonly used is sulfate chromium. But there are other materials that can be used in the tanning process, such as is silica. Silica is a mineral material that can be renewed and is abundantly available in

Indonesia. It is an environmentally friendly material. The ability of silica to stabilize raw skin can reduce the use of chrome in the skin tanning process [10]. Silica tanning by using sodium silicate can increase skin shrinkage temperature up to 65 °C. The increase in shrinkage temperature indicates that silica can bond with collagen in the skin/hide. A smaller particle of silica is needed in order to penetrate the skin better [10], one of the means of which is by making nanoparticle of silica.

2. Tanning Process

Tanning process is the activity of using animal skin/hide waste (raw skin) into leather with higher selling value. This process causes properties changes in skin/hide. Leather is resistant to biological, chemical, physical, and mechanical damage. There are several stages to process rawhide into leather, i.e. beam house, tanning, and post tanning [9].

The objectives of the tanning process are: 1) To stabilize the skin against enzymatic degradation and increase resistance to chemicals. 2) To increase shrinkage temperature and skin resistance to hot water. 3) To increase skin/hide strength characteristics. 4) To increase porosity in skin/hide fiber texture. 5) To reduce the ability of the skin to expand [11]. The purposes mentioned are the impact of the cross-linking between collagen and tanning agent.

The tanning agent commonly used in this process is chromium salt. Chromium salt is easy to obtain, cheap, and it yields good results on the skin. It also can be used in various types of skin articles that will be produced. Chrome penetrate to skin/hide tissue by diffusion and reacts with carboxyl groups in collagen. Inter cross-linking and intramolecular cross-linking produce physical, chemical, and biological stability. This process includes basification by using weak alkaline chemicals to improve the anion properties of the carboxyl collagen groups, and increase the traction to Cr^{3+}, which causes covalent bonds [12]. Figure 1 is a reaction that occurs between chromium and carboxyl groups [9].

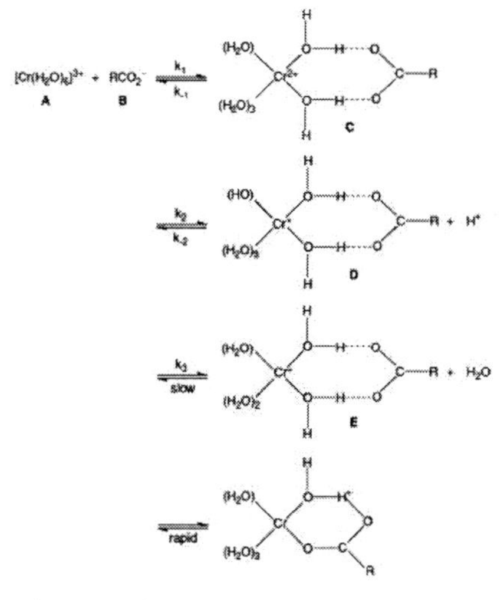

Figure 1. Reaction that occurs between chromium and carboxyl groups.

On the other hand, the use of chromium sulfate as tanning agent causes an increase in BOD and COD in wastewater [13]. In addition, it also causes the residual appearance of chromium in wastewater. It is caused only some part of chrome that bound to the collagen in skin/hide. This case will be very dangerous for the environment, chromium is carcinogenic material.

Chromium is a complex compound that can easily change shape simply by oxidation [14]. Chromium has an oxidation number (-2 to +6), but chromium 6+ and chromium 3+ are the most stable forms in nature [5] and have different biochemical properties [6]. However, both types have different level of toxicity. Cr^{6+} has toxicity level of 100 times of Cr^{3+}. The presence of chromium in waste of leather tanning process will endanger living beings, especially humans, as chromium compounds can cause cancer [1].

Some of the measures to prevent chromium pollution were by handling wastewater that has been formed, with flocculation, filtration, coagulation [2, 15, 16], and by recovering chrome from wastewater and solid waste [17, 18, 19, 20, 21, 22, 23]. In addition to these methods, other measures that can be done to deal with the problem of chrome in the process of tanning leather is by combining chrome tanning materials with a more environmentally friendly tanning material or by replacing chrome tanning materials with other tanning material. One of the ingredients that can be used is silica [10, 24, 25].

3. The Potency of Silica in Indonesia

Silica is a renewable and abundant material in Indonesia. In Indonesia, silica can be obtained from rice husk and sugarcane bagasse. Paddy is a food plant which seeds are the staple food source in Indonesia. The seeds of paddy plants are called grains, consisting of husks and rice grains (rice). [26] states that 20% of grain is husk. According to [27], dried paddy rice produced by Indonesia has increased from 2011-2015, from 65 to 75 million tons (Figure 2), making the potential of husks produced from dried milled grain approximately reaches 15 million tons. Thus, the increase in dried milled grain can also increase national silica production. The content of silica in rice husks is 87-97% after the rice husk is processed through perfect combustion [28].

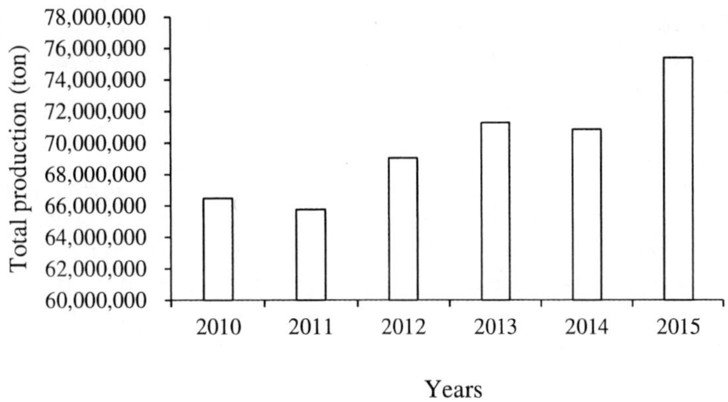

Figure 2. The total dry grain rice prodution.

4. Silica Utilization

The use of silica has now developed in various sectors of life. The common use of silica is as an air absorber to prevent high humidity in the environment surrounding the silica. Generally, Silica is used in the form of silica gel. It is used to maintain humidity in shoes, textiles, food, bags, etc. According to [29], silica can be used in various fields such as fertilizers, insulators, insect controllers, adsorbents, as well as catalysts and cement or concrete.

In addition to its utilization as an air moisture treatment, silica is also used in concrete-making mixtures. The use of silica in concrete mixes aims to deal with the crack problem while the concrete is still fresh. Silica addition in a concrete mixture can reduce cracks in concrete [30]. Silica that bonds with calcium hydroxide (its by product) may reduce the gas and water permeability and reduce cracks in concrete [30, 31] and increase the durability of concrete [31, 32]. In addition, the use of silica also aims to reduce the effects of greenhouse gases caused by the cement industry [33]. This industry contributes to CO_2 emissions that amount approximately 6-7% per year [34].

In other industries, silica is used as an absorbent of liquid waste [35, 36]. In [35] research, silica gel waste is used as an absorbent of phenol contained in a liquid. Meanwhile, silica gel waste is utilized to absorb cationic surfactants in textile wastewater [36].

Silica is also used as an ingredient in fertilization for plants [37, 38]. This is because silica is a component found in plant cells [39]. The use of silica as plant fertilizer can increase plant resistance to biotic and abiotic pressures [40, 41, 42], stimulate growth [43], increase resistance to toxic mineral pressures [44, 45] and improve photosynthesis [46, 47].

In the rubber industry, silica is used as a filler in the manufacture of rubber products. This is due to the addition of silica that provides unique properties, i.e. resistant to heat and cold, long lasting when exposed to the atmosphere, and physically inert to the rubber produced. It is also resistant to heat in a wide range of temperatures [48]. The unique properties of silica rubber provide convenience in utilizing the rubber products, one of which is insulator [49, 50, 51], and tires [52].

5. Silica and Nano-silica Extraction Process

Rice husk is a material that contains silica [53]. The content of silica in rice husks increases along with the process of rice husks ashing. The content of silica in rice husk ash is 60-97% [28, 54]. The extraction process of silica from rice husk can be done by the sol-gel method [55]. This extraction method is carried out based on the solubility of amorphous silica in alkaline solutions, such as KOH, NaOH, and Na2CO3. Furthermore, the dissolved silica is precipitated using acids, such as HCl, H2SO4, citric acid, and oxalic acid [28]. This is due to the nature of silica which will form gel under neutral conditions [55]. According to [56], this is also due to the deprotonation of siloxy and silanol groups in a large amount of silica filtrate.

In general, silica from rice husk extraction is nano-silica. The presence of agglomeration causes the size of the silica particles to become large. The process of making nano-silica is carried out by continuing the sol-gel extraction process of silica. After the precipitation process using acid, silica is aged in a few days. The gel obtained was centrifuged and then dried afterward. Furthermore, silica size was reduced to obtain nano size [57, 58].

6. Silica in Tanning Process

In the tanning process, silica becomes an option [12]. Silica in the form of sodium silicate can increase skin shrinkage temperatures up to 65 °C and can reduce the use of chrome in the tanning process. The increase in shrinkage temperature indicates that silica can bind to skin collagen. The bond that occurs is depicted in Figure 3 [10]. Hydroxyde ions in silica will form hydrogen bonds to collagen. It is very different with covalent bonding in chromium tanning. Hydrogen bonding is a weak bond that result low shrinkage temperature to leather.

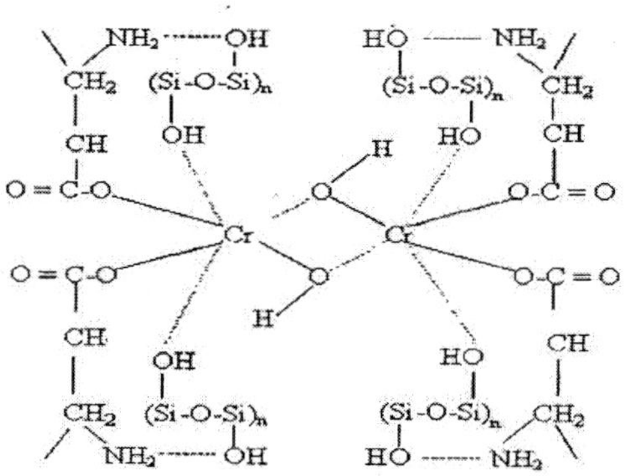

Figure 3. The interaction among silica, chromium, and collagen.

The use of sodium silicate in the skin tanning process has not completely replaced chrome in the tanning process. In addition, it has the disadvantages of low shrinkage temperature and weak mechanical properties of the skin [59]. To addicting of the bonding that occur between silica and collagen, [24, 25] did nanosilica tanning. It was done to determine the effect nanosilica on leather. According to [10], that smaller size of silica may produce more reaction with collagen. [24] said that addicting nanosilica 1% could keep stability of leather to temperature (Figure 4).

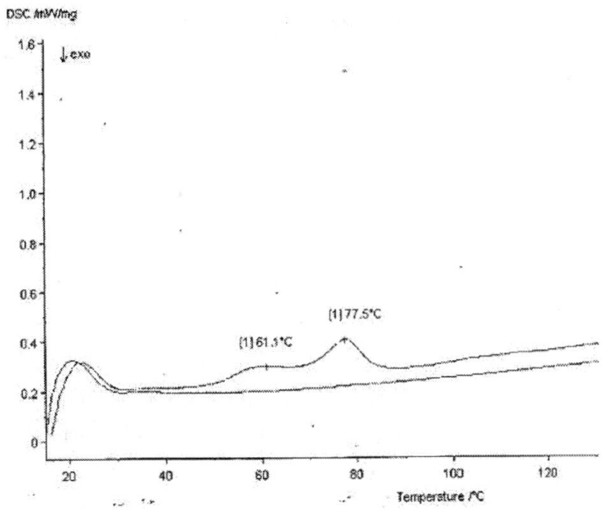

Figure 4. DSC patterns (heating curve) of unmodified collagen (top) and modified collagen (bottom).

The research showed that the use of nano-silica can increase shrinkage temperature up to 95 °C and have superior physical and mechanical properties compared to chrome tanning (Table 1) [24, 25]. However, before nanosilica tanning process, the pelt was treated by oxazolidine 2%. This temperature increase is due to the nano-silica bond with hydroxyl groups and with -C=N- in arginine, histidine, and tryptophan of collagens. Standart of shrinkage temperature in chromium tanning is 100 °C, [25] reported that a combination of tanning using nano-silica 7% and oxazolidine 2% produced leather with shrinkage temperatures reaching 107 °C (Figure 5).

Table 1. Comparation physical strength of crust leather with chromium and oxazolidine – nanosilica.

Tanning system	Tensile strength (Mpa)	Extension at break (%)	Tear strength (N/mm)	Burst strength (kg/cm^2)
Combination tanning	14.7	83.2	49.6	12.2
Chrome tanning	15.9	78.8	52..5	13.4
Standart in china	>6.5	25 – 60	>18	-

pH and nanosilica concentration affects the size of nano-silica size particles during the tanning process. The optimum pH in the nano-silica tanning process is 3-3.5, pH under these conditions will cause agglomeration of nano-silica particles. Meanwhile, the nano-silica concentration during the tanning process is 2% of the weight of pickle pelt. Increased concentration causes an increase in the size of nano-silica particles [24].

Up to date, nanosilica research as a tanning material uses synthetic tetraethoxysilane. In addition, the characteristics of nanosilica used as tanning material are still unknown. Rice husk is a byproduct of rice mills, but it contains silica, which can be used as a tanning agent. According to [28], rice husk is the largest source of silica product (amounted to 87-97% after experiencing perfect combustion). In addition, this method is relatively easy and cheap in cost [60].

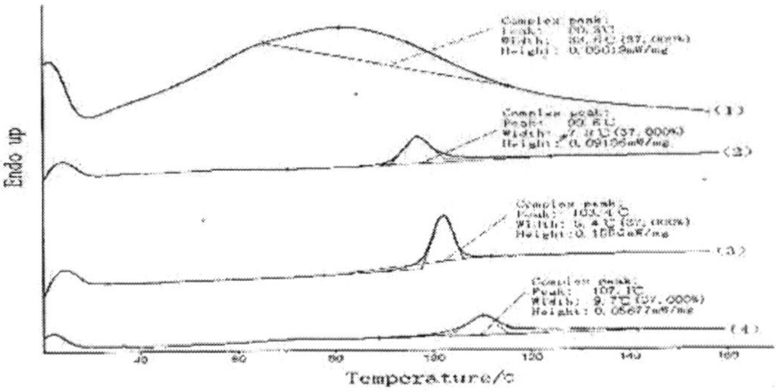

Figure 5. DSC patterns of hide powder treated with: 1) oxazolidine 2%, 2) oxazolidine 2% and nanosilica 3%, 3) oxazolidine 2% and nanosilica 5% and 4) oxazolidine 2% and nanosilica 7%.

Combination tanning can be done to achieve the same shrinkage temperature of leather as chrome tanning. Nanosilica tanning with oxazolidine has been shown to increase shrinkage temperatures reaching 95 °C and 107 °C [24, 25]. However, the result and characteristics of nano-silica single tanning are still unknown in order to produce leather that conforms to the standard, i.e. shrinkage temperatures of 100 °C.

7. Conclusions

Rice husk is a potential waste that can be used as a renewable source of silica. Silica can be utilized as fertilizer, cement/concrete, insulators, and adsorbents. However, silica from rice husk has the potential to be used as an environmentally friendly tanning material too. This is due to the ability of silica in increasing the shrinkage temperature of leather.

References

[1] Costa M and Klein CB 2006 *CRC Critical Revies in Toxicology* 36(2) 155 – 63.
[2] Chowdhury M, Mostafa MG, Biswas TK and Saha AK 2013 *Water Resources and Industri* 3 11 -22.
[3] Midha V and Dey A 2008 *Int. J. Chem. Sci* 6(2) 472 – 86.
[4] Nugraha AW 2018 *Strategy Development for Implementation of Cleaner Production in Leather Industries* (Bogor, Indonesia: Bogor Agricultural University) (In Indonesian).
[5] Ashraf A, Bibi I, Niazi NK, Ok YS, Murtaza G, Shahdid M, Kunhikrishnan A, Li D and Mahmood T 2017 *Int. J. Phytoremediat.* http://dx.doi.org/10.1080/15226514.2016.1256372
[6] Shahid M, Shamshad S, Rafiq M, Khalid S, Bibi I, Niazi NK, Dumat C and Rashid MI 2017 *Chemosphere* 178 513 - 33.
[7] Wang ZZ, Chen JQ, Chai LY, Yang ZH, Huang SH and Zheng Y 2011 *J. Hazard. Meter* 190 980 – 5.
[8] Xiong T, Leveque T, Shahid M, Foucault Y, Mombo S and Dumat C 2014 *J. Environ. Qual.* 43 1593 - 600.
[9] Covington AD 2009 *Tanning Chemistry: The Science of Leather* (Cambridge: RSCPublishing).
[10] D'aquino A, D'elia G, Seggiani M, Vitolo S 2004 *Journal of the American Leather Chemists Association* 99 26 – 35.
[11] [BASF] *Badische* Anilin und *Soda Fabrik 2007 Pocket Book for the Leather Technologist Fourth Edition* (German: BASF) p 454.
[12] Andualem G 2014 *Combinating based on tara dan sodium metasilicate a new chrome free tanning* (Ethiopia: Addis Ababa University).
[13] Zhang Y, Ingham B, Levencur J, Cheong S, Yayo Y, Clarke DJ, Holmes G, Kennedy J and Prabakar S 2017 *Royal Society of Chemustry Advances* 7 11665 – 71.
[14] Prado C, Ponce SC, Pagano E, Prado FE and Rosa M 2016 *Aquat. Toxicol* 175 213 – 21.

[15] Lofrano G, Belgiorno V, Gallo M, Raimo A and Meric S 2006 *Global NEST Journal* 8(2) 151 – 8.
[16] Ulfin I, Harmami and Rahmawati E 2014 *Prosiding Seminar Nasional Kimia UNS.* ISBN : 978-602-0951-00-3
[17] Voice TC, Davis ML, Johnson GB and Sturgess LL 1988 *Hazardous Waste & Hazardous Materials* 5(4) 343 – 52.
[18] Esmaeili A, Nia AM and Vazirinejad R 2005 *Americal Journal of Applied Sciences* 2(10) 1471 – 3.
[19] Vieira MS and Marcilio NR 2005 *2nd Mercosur Congress on Chemical Engineering and 4th Mercosur Congress on Process Systems Engineering* (Rio de Janeiro: ENPROMER).
[20] Guo ZR, Zhang G, Fang J and Dou X 2006 Journal of Cleaner Production 14 75 – 9.
[21] Wiryodiningrat S, Kismolo E and Prayitno 2007 *Majalah Kulit, Karet, dan Plastik* 23(1) 13 – 7 (In Indonesian).
[22] Kanagaraj J, Babu NKC and Mandal 2008 *Journal of Cleaner Production.* 16 1807 – 13.
[23] Sutyasmi S 2015 *Majalah Kulit, Karet dan Plastik* 31(2) 107 – 14 (In Indonesian).
[24] Fan H, Li L, Shi B, He Q and Peng B 2005 *Journal of the American Leather Chemists Association* 100 22 – 8.
[25] Yan L, Zhaoyang L, Haojun F, Yuansen L, Hui L, Biyu P and Bi S 2008 *Journal of the Society of Leather Technologists and Chemists* 92 252.
[26] Yoshida S. 1981. *Fundamentals of Rice Crop Science.* Los Baños, Philippines: IRRI.
[27] [BPS] Badan Pusat Statistik 2016 Tabel Dinamis Accessed at 2018 July 26th in https://www.bps.go.id/site/resultTab
[28] Handayani PA, Nurjanah E and Rengga WDP 2014 *Jurnal Bahan Alam Terbarukan* 3(2) 19 – 24 (In Indonesian).
[29] Adam F, Appaturi JN and Iqbal 2012 *Catalysis Today* 190 2 – 14.
[30] Maheswaran S, Bhuvaneshweri B, Palani GS, Nagesh RI and Kalaiselvam S 2013 *Research Journal of rRecent Sciences* 2 17 - 24.
[31] Atmaca N, Abbas ML and Atmaca A 2017 *Construction and Building Materials* 147 17 – 26.
[32] Khan MI and Siddique R 2011 *Resources, Conservation and Reycling* 57 30 – 35.
[33] Quercia G and Brouwers HJH 2010 *8th fib PhD Symposium in Civil Engineering: Proc (Kgs. Lyngby)* (Denmark: Technical University of Denmark).
[34] Biricik H and Sarier N 2014 *Material Research* 17(3)570 - 82.
Article I. [35] Idris A and Saed K 2003 *Environmentalist* 23(4): 329 – 34.
[36] Koner S, Pal A and Adak A 2011 *Desalination* 276 142 – 7.
[37] Marafon AC and Endres L 2013 *Amazonian Journal of Agricultural and Environmental Sciences* 56 (4) 380 – 8.
[38] Sukkaew E, Amkha S, Inboonchoy T and Mala T 2016 *Modern Applied Science* 10(11) 264 – 72.
[39] Ma JF, Miyake Y, and Takahashi E 2001 *Silicon as a beneficial element for crop plants. In*: Silicon in Agriculture, Eds. L. E. Datnoff, G. H. Snyder and G.H. Korndorfer, 17–39.
[40] Ma J and Yamaji N 2006 *TRENDS in Plant Science* 11 392 - 7. http://dx.doi.org/10.1016/j.tplants.2006.06.2007
[41] Hunt JW, Dean AP and Webster RE 2008 *Annals of Botany* 102 653 –6.
[42] Rodrigues FÁ and Datnoff LE 2015 *Silicon and Plant Diseases, 1st ed.* (Berlin/Heidelberg, Germany: Springer) p 67–100.
[43] Ma J and Takahashi E 2002 *Soil Fertility and Plant Silicon Research in Japan* (Amsterdam: Elsevier Science).
[44] Horiguchi T 1988 *Jpn. J. Soil Sci. Plant Nutr* 34 65–73.
[45] Savant NK, Snyder GH and Datnoff LE 1997 *Adv. Agron* 58 151 – 99.
[46] Epstein, E., & Bloom, A. J. (2005). *Minerals Nutrition of Plants: Principles and Perspective.* (2nd Ed.) Sinauer Associated, Inc. p. 400. http://dx.doi.org/10.1002/jpin.19721320211
[47] Gong HJ, Zhu XY and Chen KM 2005 *Plant Science* 169 313 – 21.

[48] Kopylov VM, Kostyleva EI, Kostylev IM and Koviazin AV 2011 *International Polymer Science and Technology* 38 (4) 35 – 47.
[49] Jalali B, Yegnanarayanan S, Yoon T, Yoshimoto T, Rendina I and Coppinger F 1998 *IEEE Journal of Selected Topics in Quantum Electronics* 4(6) 938 – 47.
[50] Celler GK and Cristoloveanu S 2003 *Journal of Applied Physics* 93 (9) 4954 – 78.
[51] Khan H, Amin M, Ahmad A and Yasin M 2017 *Research & Development in Material Science* 2(3) 1 – 8.
[52] Zafarmehrabian R, Gangali ST, Ghoreishy MHR and Davallu M. 2012 *E-journal of Chemistry* 9(3) 1102 – 12.
[53] Liou HT and Yang CC 2011 *Material Science and Engineering B* 176 521 – 9.
[54] Selvakumar KV, Umesh A, Ezhilkumar P, Gayatri S, Vinith P, and Vignesh V *International Journal of ChemTech Research* 6 (9) 4455 – 9.
[55] Suka IG, Simanjunta W, Sembiring S and Trisnawati E 2008 *MIPA* 37(1) 47 – 52 (In Indonesian).
[56] Sofyan IGG, Alauhdin M and Susatyo EB 2013 *Indonesian Journal of Chemical Science* 2 (In Indonesian).
[57] Andreas A, Kristianto H and Kurniawan DF 2016 *Prosiding seminar nasional teknik kimia "kejuangan", pengembangan teknologi kimia untuk pengolahan sumber daya alam indonesia, , pengembangan teknologi kimia untuk pengolahan sumber daya alam Indonesia, Yogyakarta.*
[58] Hayati D, Pardoyo and Azmiyawati C 2017 *Jurnal Kimia Sains dan Aplikasi* 20(1) 1 – 4 (In Indonesian).
[59] Fathima NN, Madhan B, Rao JR and Nair BU 2003 *Journal of the American Leather Chemists Association* 98 139 – 46.
[60] Kalapathy U, Proctor A and Shultz J 2000 *Bioreource Technology* 73 257 – 62.

Acknowledgements
We highly appreciate The Indonesia's Ministry of Research, Technology and High Education (*Kementerian Riset, Teknologi dan Pendidikan Tinggi Republik Indonesia*) for its research funding and support.

IOP Conf. Series: Earth and Environmental Science **335** (2019) 012033 doi:10.1088/1755-1315/335/1/012033

Design for improvement of sugar factory performance based on statistical thinking

A Kustiyo[1], Y Arkeman[2]

[1]Department of Computer Science, Bogor Agricultural University, e-mail: kustiyoaziz@gmail.com
[2]Department of Agroindustrial Technology, Bogor Agricultural University

Abstract. The amount of sugar production is determined, among others, by the quality of sugar cane and the performance of sugar factory equipment. The quality of sugar cane as an agricultural product is influenced by the quality of sugar cane seeds, soil conditions and climate. The differences in the factors affect the quality of sugarcane production. Sugarcane milling is not based on the quality of sugarcane but on the arrival time to the factory. Under this condition, different sugarcane qualities will be milled together. As a consequence, it may not yield the optimal sugar production. In the sugar production process, equipment performance must always be monitored. Equipment failure will reduce the amount of sugar produced. Therefore, efforts to increase sugar production must consider both factors, namely the variation of input quality and monitoring of equipment performance. One approach in solving this problem is statistical thinking approach. The purpose of the paper is to design the performance improvement of sugar factory based on statistical thinking. The improving activity based on statistical thinking will concentrate on manufacturing process enhancement and problem breaking that rely on realizing and degrading variation. The study shows that the statistical thinking approach potentially able to increase sugar production using process enhancement framework and overcome factory equipment failure using problem breaking framework.

keywords: sugar cane, statistical thinking, improvement, system

1 Introduction

The amount of sugar production is determined, among others, by the quality of sugarcane and the performance of sugar factory equipment. The quality of sugarcane as an agricultural product is influenced by the quality of sugarcane seeds [1], soil fertility, fertilizer management, sugarcane variety, harvesting process, time lag between harvest and milling [2] and climate [3]. The quality of sugarcane seeds is determined, among others, by the age of sugarcane seeds. The age of sugarcane seeds affects the production of sugarcane and the yield produced [1]. Soil conditions and fertilization management can have a direct effect on the quality of sugarcane syrup which can have an impact on the acquisition of sugar in the process of sugarcane processing into sugar [2]. Weather parameters such as rainfall, temperature, sunlight, humidity and so on affect the growth of sugarcane, the sugar produced and the quality of sugarcane syrup [3].

Sugar production process consists of these processes, namely: milling to extract juice from sugarcane, juice clarification, clear juice evaporation, crystallization, centrifugation, drying and packaging. The process of sugarcane milling is not based on the quality of sugarcane but according to the time of arrival of sugarcane to the factory. In this condition, the quality of different sugarcane will be milled together so as to produce juices with poor quality. As a result, this condition will affect the results of the next processes so that the sugar produced is not optimal.

During the sugar production process, factory equipment performance must always be monitored. The equipment performance monitoring will ensure that the equipment is always in good performance and prevent equipment failure. Equipment failure will reduce the amount of sugar produced [4].

Therefore, efforts to increase sugar production must consider both factors, namely variations in input quality and equipment performance monitoring. One approach to solving this problem is the statistical thinking approach [5]. Statistical thinking views all works as a system built by processes and realizing and degrading variation inside the processes are essentials for success. [5].

In this context, this paper first studied the sugar manufacturing process as a system of interconnected processes, followed by the variation sources that exist in the system, and then the frameworks for understanding and reducing the variation, and finally the conclusion.

2 Material and method

The research method is based on statistical thinking [5] and literature study. The steps in implementing statistical thinking are shown in figure 1 [5].

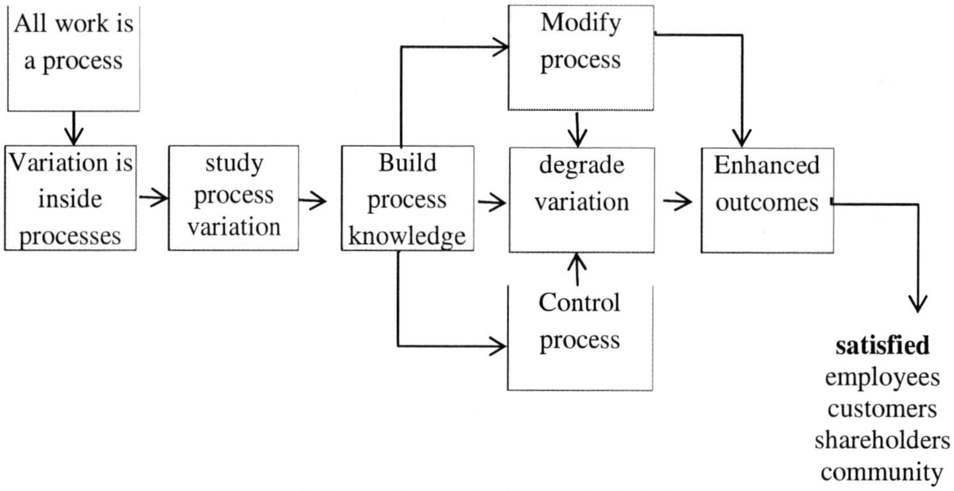

Figure 1. Stages in employing statistical thinking [5].

3 Results and Discussion

3.1 The first idea of statistical thinking is that all work happens inside a system built of processes that connected each other

The manufacturing process of raw sugar is a system that consist of the following processes: sugar cane milling, juice clarification, clear juice evaporation, crystallization, sentrifugation, and sugar packaging [6, 7]. In the milling process, the cane will be shredded and crushed to extract juice from the fibrous material. The juice clarification produce clear juice with the lowest concentration of insoluble and soluble impurities. Water is evaporated out of the clear juice by steam to form syrup in the evaporation stage. During the crystallisation stage, the syrup is further concentrated by boiling and seeded with fine sugar crystals. The raw sugar crystals are separated from the mother liquor and then conditioned using air in a rotating drum to yield the raw sugar in sentrifugation stage. The by-product generate from sugar manufacturing process are bagasse (from the milling stage), filter muds (from the clarification stage) and molasse (from the sentrifugation stage). The manufacturing process of raw sugar is shown in figure 2. The inputs, outputs and by-products of the sugar manufacturing process are shown in table 1.

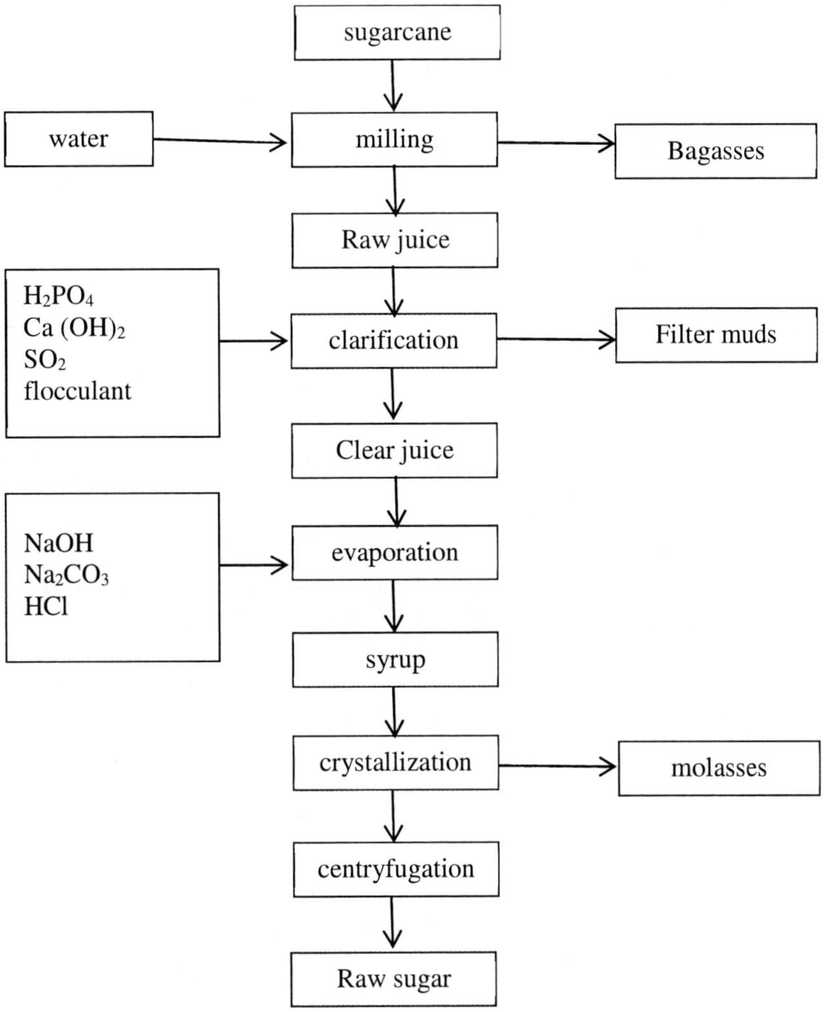

Figure 2. Manufacturing process of raw sugar {[6], [7]}.

Table 1. The inputs, outputs and by-products of sugar manufacturing process{[6],[7]}.

Stage	Input	Output	By-product
Milling	Sugarcane	raw juice	Bagasse
Clarification	raw juice	Clear juice	Filter mud
Evaporation	Clear juice	Syrup/liquor	
Crystallization	Syrup/liquor	Raw sugar crystal	
Sentrifugation	Raw sugar crystal	Raw Sugar	Molasse

3.2 The second idea of statistical thinking is that there are variation inside all process

From literature study, there are several sources of variation in the sugar manufacturing process. The variation comes from the quality of sugar cane and from the processes during the manufacturing stage.

The quality of sugarcane depends on several factors like seed cane, varieties, age, soil characteristics, and climatic conditions [1,2,3,4, 8]. Sugarcane varieties, weather, harvest handling and post-harvest storage procedures are factors that affect the content of sugarcane impurities such as phenolic acids, pigments and flavonoids. [9].

The quality of cane juice depends on sugarcane quality and the method of extraction [6]. Among the factors that may influence the composition of cane juice are variety, age, climatic condition, cultural practices, soil condition and harvesting time [10]. Harvesting technique have a direct impact on cane juice quality [11]. Further, cane juice quality is affected by soil salinity [12] and sulphur content in the soil [13].

The sugar juice ingredient, the content of inorganic phosphate, the type and state of liming, and the juice components interaction are factors that influence the efficiency of sugar cane purification [14]. These factors are potentially the sources of variation in clarification process.

Variation in evaporation, crystallization and sentrifugation stage come from dextran and starch content in cane juice. The existence of dextran inside sugar cane juice will degrade evaporation rate and increase duration of separation in centrifuges [15]. Existence 50-100 ppm of starch in cane juice cause reduction of crystallization and centrifugation rates [15].

3.3 The third idea of statistical thinking is that realizing and degrading variation inside the processes are essentials for success.
The variation that are considered are variation that occur accidentally. Furthermore it is analyzed and utilized to enhance performance. In order to understand the variation and processes that generate it, the variation must be recognized, categorized dan measured. With this knowledge the process will be changed to reduce its variation [5].

Two types of variation that is considered are common cause and spesial cause variation. Common-cause variation is natural variation that exist inside every process while special-cause variation is beyond natural variation that leads to extreme values. The distinction between these types of variation is shown in table 2. Common-cause approach is used to enhance performance by susbtantially modifying the process while special-cause approach is used to recovery the unstable process [5].

Table 2. The distinction between common-cause and special-cause variation [5].

Common-cause variation	Special-cause variation
Always exist	Nonpermanent and unforeseeable
Many sources with minor effect	Few sources with major effect
Portion of the natural behavior of the process	frequently associated with a certain case
Process is stable	Process is unstable

3.4 Statistical engineering
Statistical engineering is a technique to apply ideas, methods, and instruments and integrate them with information technology and other relevant sciences to upgrade outcomes. Statistical engineering is the tactical component that gives integral approaches to break unresolved problems that are appropriate with the ideas of statistical thinking. There are two kinds of statistical engineering frameworks to utilize the strategy of statistical thinking. The framework of process improvement is put forward to solve the common-cause variation in processes that are stable but on unsatisfying states. The framework is shown in figure 3. The problem solving framework is proposed to resolve the special-cause variation. The framework is shown in figure 4 [5].

3.4.1 Design of process improvement framework in sugar manufacturing process. The design of process improvement is based on the framework in figure 3. The stages of improvement are as follows:

3.4.1.1. Understanding the sugar manufacturing process. The process was described in section 3.1 and the flowchart of the process is shown in figure 2. Table 1 desrcribes the inputs, outputs and by-products of the process.

IOP Conf. Series: Earth and Environmental Science **335** (2019) 012033 doi:10.1088/1755-1315/335/1/012033

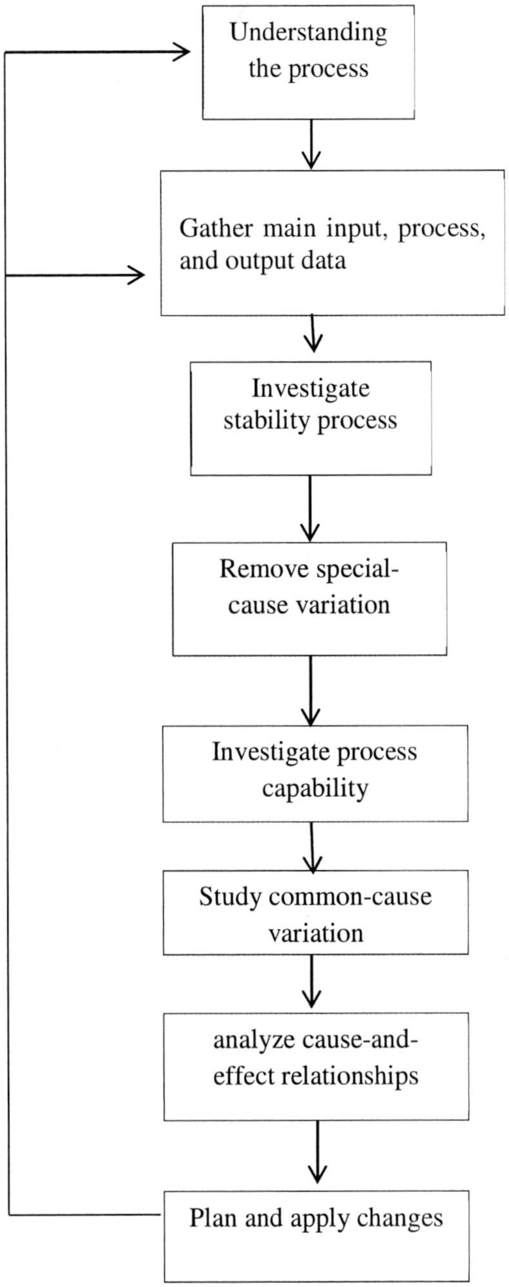

Figure 3. Basic process improvement framework [5].

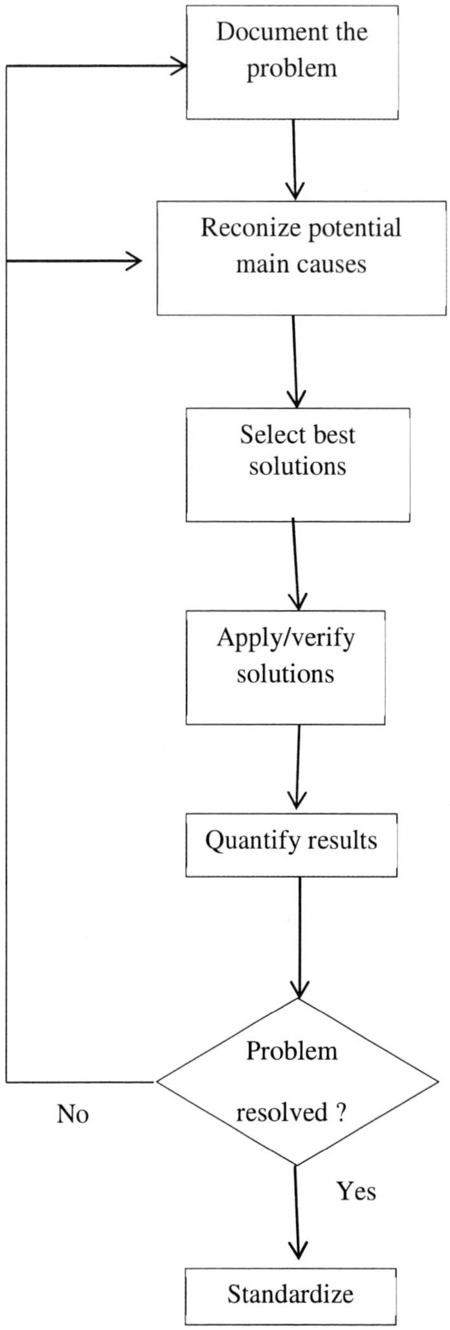

Figure 4. Framework of basic problem solving [5].

3.4.1.2. Gather data on input, process, and output of the sugar manufacturing process. Data is recorded from the sugar plant based on table 1. The data is recorded by sugar plant periodically. For example, data collected by sugar plant every day are the quantity of sugarcane, bagasses, filter muds, molasses and raw sugar.

3.4.1.3. Investigate the stability of the sugar manufacturing process. Process stability can be asses using run chart [5]. Run chart for raw sugar production and by-product generation every day will show the stability of the production process. The by-products include bagasses, filter muds and molasses. Stable process means consistent process. Unstable process usually indicates the existence of special causes that their root causes need to be recognized and removed. [5].

3.4.1.4. Remove special-cause variation. It will be described in section 3.4.2.

3.4.1.5. Investigate the capability of the sugar manufacturing process. The tools for process capability evaluation are histograms and standards [5]. The standards of the sugar manufacturing process can be measured using mass balance model [16].

3.4.1.6. Study common-cause variation. Common-cause variation can be analyzed using stratification and statistical inference [5]. The by-product generated in sugar manufacturing process are bagases, filter muds and molasses (figure 2). The generation of this by-product shows sugar losses in related stages. So, the identification of common-causes variation can be focused on the related stages. For example, if sugar content in bagasses tends to increase then the problem may be in the milling stage.

3.4.1.7. Analize cause-and-effect relationships. The relationships can be studied using experimental design and modeling techniques[5]. The application examples of modeling techniques are artificial neural network model for forecasting sugarcane juice purification by hydrogen peroxide [9] and simulation model for optimization of evaporation process in sugar industry [17]. The variation sources described in section 3.2 can be considered in experimental design and model buiding.

3.4.1.8. Plan and apply changes. After implementation of the improvement framework, data was collected to verify the impact of the framework. It may need to apply the framework several times to reach the desired results.

3.4.2. Design of problem solving framework in sugar manufacturing process. The primary isue of the problem solving is on identifying basic causes of unusual behavior, so the focus is on the unusual behavior and not on the entire process. The design of problem solving is based on the framework in figure 4. The stages of problem solving are as follows [5]:

3.4.2.1. Document the problem. The run and control chart may assist in detecting timely when the special caused happen. In addition, the charts able to show step by step or abrupt changes. [5]. In sugar manufacturing process, the sudden change in sugar production may be due to the unplanned breakdowns of the crushing mills [4].

3.4.2.2. Recognize potential root causes. Cause-and-effect diagram and "5 Whys" can be used to identify the root causes [5]. First, critical failures of the cane milling are separated from trivial failures. Then the root cause analysis is based on these critical failures. The factors that caused these failures were investigated. The important components of these factors are evaluated. The results obtained from the analysis of root causes are validated using the effectiveness of the overall equipment [4].

3.4.2.3. Select best solutions. The best solution can be determined by an interrelationship digraph and multivoting [5Evaluation was carried out to select an effective maintenance strategy by comparing the strengths and weaknesses of the three practical strategies applied to sugarcane milling. The strategy is based on three different aspects, namely condition, total productivity and reliability. The comparison is based on the methods used, objectives and employee support [4].

IOP Conf. Series: Earth and Environmental Science **335** (2019) 012033 doi:10.1088/1755-1315/335/1/012033

3.4.2.4. Apply/verify solutions. The chosen strategy in the previous stage will be tested using experimental design. The influence of several factors can be studied using experimental design [5]. If it is not possible to conduct experiments then simulation approach or model building may be applied.

3.4.2.5. Quantify results. After applying the changes, run and control charts are used to show whether the implemented solution is able to resolve the problem. [5].

3.4.2.6. Standardize. To avoid the same problem will occur in the future, the solution must be standardized. Flow charts is used in documenting the standard operating procedures and the employees training is based on this procedure. [5].

4 Conclusion

In this paper a design of performance improvement for sugar factory based on statistical thinking is presented. The study shows that the statistical thinking approach potentially able to increase sugar production using process improvement framework and overcome factory equipment failure using problem solving framework.

References

[1] Ayele N, Getaneh A, Hagos H and Birruma M 2014 Effect of age seed cane on yield and yield components of sugarcane at Tendaho sugar factory *The J. Of Agric. and Nat. Res. Sci.* **1** 165

[2] Meyer J H and Wood R A 2001 The effects of soil fertility and nutrition on sugarcane quality: a review *Proc. S. Afr. Sug. Technol. Ass.* **75** 242

[3] Srivastava A K and Rai M K 2012 Review: Sugarcane production: impact of climate change and its mitigation *Biodiversitas* **13** 214

[4] Nganga C J and Byiringiro J B 2018 Effective maintenance strategy of cane crushing mills for improvement of sugar production in Kenya *IOSR JMCE* **15** 1

[5] Hoerl R and Snee R 2012 *Statistical Thinking Improving Business Performance 2nd Edition* (New Jersey: John Wiley & Sons, Inc.) p 511

[6] Indrasti N S and Fauzi A M 2009 *Produksi Bersih* (Bogor: IPB Press) p 340

[7] Purwaningsih I W 2016 *Penilaian Daur Hidup (Life Cycle Assessment) Gula Tebu di PG Subang, Jawa Barat* (Bogor: Institut Pertanian Bogor) p 69

[8] Umesh Kumar P K and Chand K 2015 Application of response surface method as an experimental design to optimize clarification process parameters for sugarcane juice *J Food Proc. Tech.* **6** 422

[9] Sartori J A S, Ribeiro K, Teixeira A C S C, Magri N T C, Mandro J L and Claudio L A 2017 Sugarcane juice clarification by hydrogen peroxide: predictions with artificial neural networks *Int. J. of Food Eng* **13** 199

[10] Rodrigues R, Sperandio L C C and Andrade C M G 2017 Investigation of color and turbidity in the clarification of sugarcane juice by ozone *J Food Process Eng* **12** 661

[11] East C 2013 *The Effect Of Juice Composition On The Scaling Rate And Scale Composition In Sugar Mill Evaporators* (Quensland: Queensland University of Technology) p 224

[12] Lingle S E, Wiegand C L 1997 Soil salinity and sugarcane juice quality *Field Crops Research* **54** 259

[13] Naga Madhuri K V, Jayaprakash M and Sarala N V 2011 Effect of sulphur on quality of sugarcane juice and jaggery *IJABT* **2** 218

[14] Doherty W O S 2011 Improved sugar cane juice clarification by understanding calcium oxide-phosphate-sucrose systems *J.. Agric. Food Chem.* **59** 1829

[15] Fard Z M, Alamo A and Askary H 2014 Investigation of the effects of polyelectrolyte, phosphate and pH values on the removal of impurities in the cane sugar juice clarification *J. of Appl. Chem.* **8** 41

[16] Bantacut T and Novitasari D 2016 Energy and water self-sufficiency assessment of the white sugar production process in Indonesia using a complex mass balance model *J. of Cleaner Prod.* 1**26** 478
[17] George S and Kyatanavar D N 2015 Optimization of evaporation process in sugar industry for developing intelligent control strategies *Int. J. of Modern Trends in Eng. and Res.* **2** 998

Acknowledgment
The authors wish to acknowledge the financial support provided by LPDP through BUDI DN 2016 scholarship.

IOP Conf. Series: Earth and Environmental Science **335** (2019) 012034 doi:10.1088/1755-1315/335/1/012034

Formulation, configuration, and utilization of Laser Induced Breakdown Spectroscopy (LIBS) to analyze nitrogen particles in soil

A R Angkat[1], K B Seminar[1], M Rahmat[2], A Sutandi[3], M M Suliyanti[4]

[1]Department of Mechanical and Biosystem Engineering, Bogor Agricultural University, Indonesia, e-mail: rarangkat80@gmail.com
[2] Department of Mechanical Engineering, Ibnu Khaldun University, Indonesia
[3] Department of Soil and Land Resources, Bogor Agricultural University, Indonesia
[4] Research Center for Physics, Indonesian Institute of Sciences, Indonesia

Abstract. The use of Laser Induced Breakdown Spectroscopy (LIBS) method for real-time analsysis of Nitrogen particles in soil is proposed. This study aims to find the optimal LIBS formulations and configurations for analyzing soil particles, mainly Nitrogen particels. The LIBS hardware used in this reearch Nd-YAG Laser model Q-Smart 850. The optimization was done through laser pulse shooting to a pellet-shaped soil sample within a 5.75 torr pressurized vacuum chamber with 532 nm wavelength, 5.5 ns pulse duration 5.5 ns, 10 Hz repitition rate, Q-switch delay 150 μs and energy of 15 mJ/pulse.

Keywords : laser pulse, soil plasma, soil organic particle

1. Introduction

Laser Induced Breakdown Spectroscopy (LIBS) is an atomic emission spectroscopy equipment that uses lasers as a source of energy ablation [1]. The laser is fired towards the surface of the samples, some of which were distilled to form plasma. A plasma contains electrons, neutral atoms, ions and excited atoms. Excitation process is one of the methods in which material absorbs electromagnetic energy pulse (photon). In atoms, the excitation energy is absorbed by orbiting electrons which are lifted to different higher energy levels. These excited atoms in a very short time cause the electrons to return to the ground state while emitting photon emissions. When the electrons in the excited atom go down to the ground state it releases energy in the form of photons and is captured by the spectrometer displayed in the intensity of the wavelength function [2].

The LIBS method can be used to determine the composition of an element in solids, liquids and gases by focusing high-power laser pulses on a sample to make a plasma or spark a laser to be captured by a detector. Therefore, the LIBS method provides an alternative opportunity to analyze organic particles in soil samples in the field without having to go through the drying and sifting process which is usually done in laboratry based soil ananlysis.

Organic particle analysis in soil samples using LIBS will be effective if an appropriate LIBS configuration is obtained which can give rise to a soil plasma containing electrons, neutral atoms, ions and optimally excited atoms [3, 4,5]. Therefore, this study aims to find the right LIBS formulations and configurations for LIBS utilization in analyzing soil particles.

2. Material and Methods

2.1. Soil Sample Preparation

The soil samples were shaped into pellets with diameter and thickness of 3.2 cm and 0.5 cm respectively, using a manual hydraulic press with a pressure of 29.272 Mpa. These pellets are required in laser firing in order to get the LIBS spectrum of soil.

2.2. Formulation and LIBS Configuration

Formulation and configuration optimization of LIBS is done by firing laser pulses on soil samples using the Nd-YAG Q-Smart 850 Laser model with output wavelengths of 1064 nm, 532 nm and 266 nm. Sample soil pellets in pressurized space, were shot by laser pulses using wavelength of 532 nm, a pulse duration of 5.5 ns, a repetition rate of 10 Hz with a focal length of 10 cm. The soil pellets shot with laser pulses produce plasma due to damage to chemical bonds so that they undergo an ionization process. Plasma containing information about soil elements was concentrated in the lens collector by measuring light intensity in the wavelength range of 190 nm - 1000 nm. The best LIBS configuration is obtained by identifying the maximum spectra produced in different Q-switch conditions and chamber pressures.

3. Result and Discussion

Plasma formation occurred when a laser fired to a soil pellet reaches a certain threshold for optical damage. The threshold is mostly influenced the environment and the target material [6]. If the constituents of the material analyzed are known, LIBS can be used to evaluate the relative abundance of each organic nutrient element, or to detect the presence of nutrients in soil. In practice, the detection limit is a function of plasma excitation temperature, light collecting window, and the strength of the transition line visible.

3.1. Second Harmonic

The acquisition of ground LIBS spectra uses second harmonics at 532 nm waves to minimize the energy required when firing lasers in order to obtain a certain energy output. The second harmonic of two photons (ωp) with the same frequency interacts with nonlinear material and produces new photons (2ωp), whereas in fundamentals one photon (ωp) will produce one new photon (ωp) (figure 1). So that the second energy input is harmonic (532 nm) half of the energy required when using the fundamental (1064 nm) to produce the same energy output.

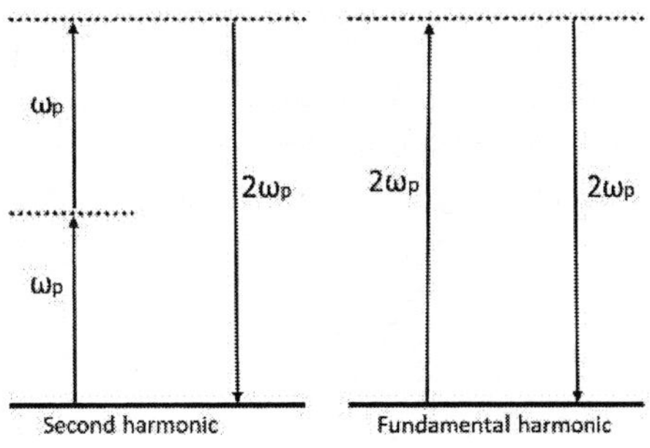

Figure1. Second harmonic and fundamental harmonic [7]

Besides that, second harmonic at 532 nm wave is a green light in a visible light spectrum that is easily found as a laser source when making tools (figure 2). The use of second harmonics can prevent the use of infrared as a laser source which is more difficult to obtain and more expensive.

IOP Conf. Series: Earth and Environmental Science **335** (2019) 012034 doi:10.1088/1755-1315/335/1/012034

Figure 2. Acquisition of ground LIBS spectra using second harmonic (532 nm)

3.2. Pulse Duration

Pulse duration 5.5 ns is the default setting when taking soil samples using Nd-Yag. Pulse duration is the time interval required by the pulse in the process of increasing the amplitude from a certain level to the level of a particular fraction and the time to decrease the amplitude of the pulse to the same level. Usually 50% of the maximum amplitude is used as the starting point and the endpoint determines the duration of the pulse (figure 3).

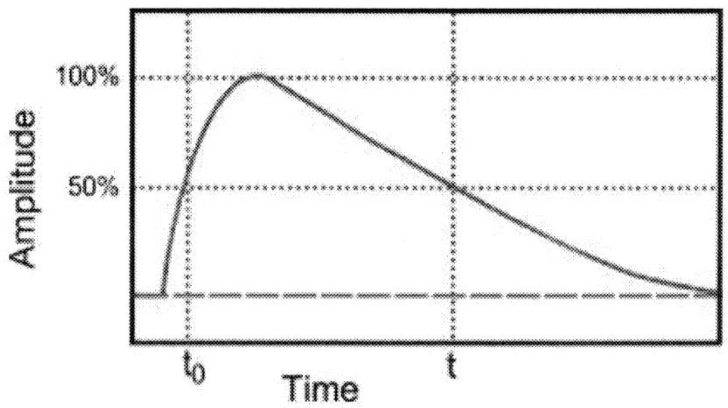

Figure 3. Pulse duration (t-t_0) [8].

Pulse duration is an important parameter in micromaching applications. Shorter pulse duration generally results in better spatial resolution, depth control, edge quality and minimizes perifier thermal damage. One reason is that each material has an ablation threshold (minimum peak intensity needed to remove material). For pulse energies which are given shorter pulses, they are desirable because they are more efficient at material removal in particular, shorter pulses give more total energy above the threshold strength, where absorption becomes increasingly nonlinear because of the higher electric field generated from higher peak forces. Conversely, most of the energy from longer pulses is wasted and is only used to heat the material (figure 4) [9].

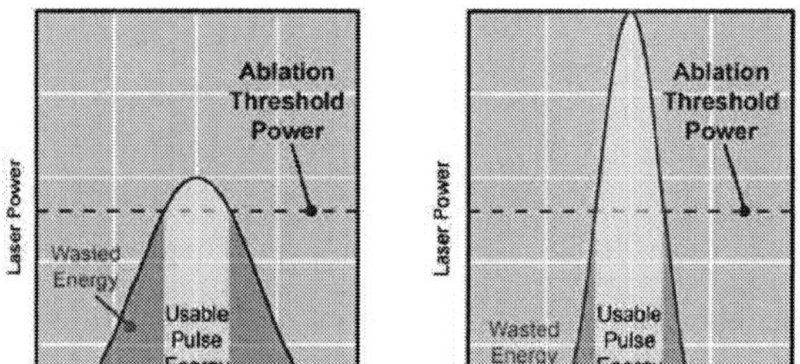

Figure 4. The impact of pulse width on LIBS spectra retrieval process [10].

3.3. Q-Switch Delay

The optimal Q switch delay to be used during the LIBS ground spectra retrieval process of 150 μs with an energy of 15 mJ (figure 5). Q-switch delay of 150 μs has the most intensity based on experiments conducted using 3 variations of Q-Switch delay, namely 150 μs, 160 μs and 180 μs. Q-switch delay that is smaller than 150 μs has energy greater than 15 mJ which causes soil samples to break. The smaller the Q-switch value , the greater the energy and vice versa.

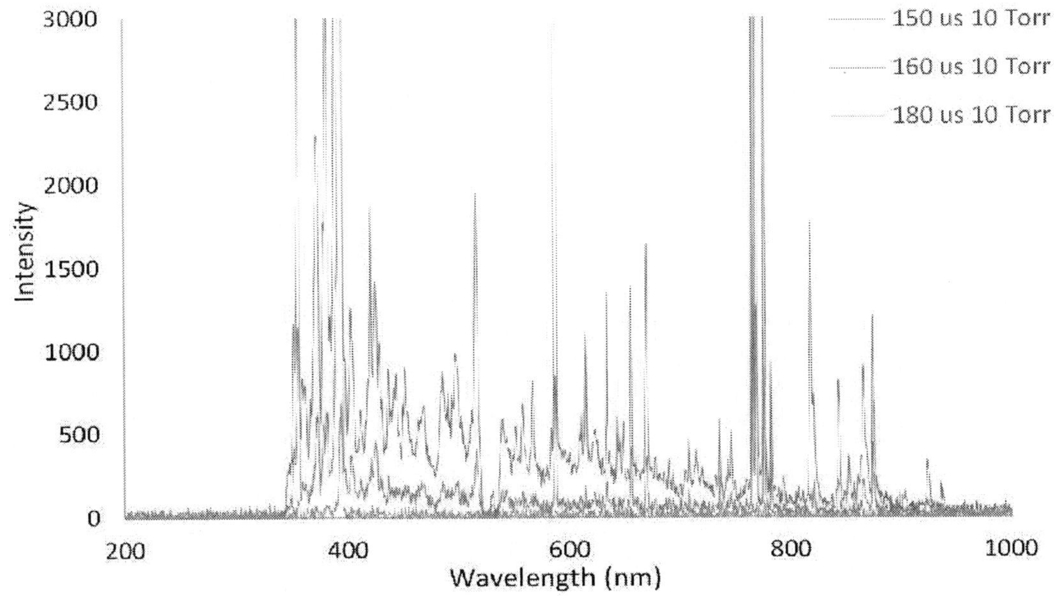

Figure 5. Soil LIBS spectrum on different Q-Switch delay but same pressure.

Laser pulses are usually made with the aim of producing very large laser power with a short radiation time. The Q-switch process occurs in the laser resonator from the modulation of the electrooptic effect which aims to produce laser light in the form of short pulses with high power.

Initially the laser media was pumped while the Q-switch was set to prevent an increase in light feedback to the media (resulting in a low Q-laser resonator). This results in population inversion, but the laser cannot be operated due to no feedback from the resonator. Due to losses from spontaneous emissions and other processes, the stored energy will reach the maximum level at a certain period. In the period that the medium reaches saturated conditions, the Q-switch device quickly changes from low to high allowing feedback and the amplification process by stimulating emissions to start. Due to a large amount of energy is stored in media amplification, the light intensity in the laser resonator accumulates very quickly, causing the energy stored in the medium to run out as soon as possible. The end result is a short pulse of light output from a laser, known as a giant pulse, which may have a very high peak intensity (figure 6). Factor value of Q can be explained through the following formula [11]:

$$Q(\omega) = \omega \times \frac{\max energystored}{powerloss}$$

Where ω is the angular frequency in rad / s when measuring energy stored in the system and the energy lost in joules. The Q value is affected by the Q-switch delay time, the longer the time delay the Q value gets smaller and vice versa. A higher Q value produces a shorter duration pulse and vice versa.

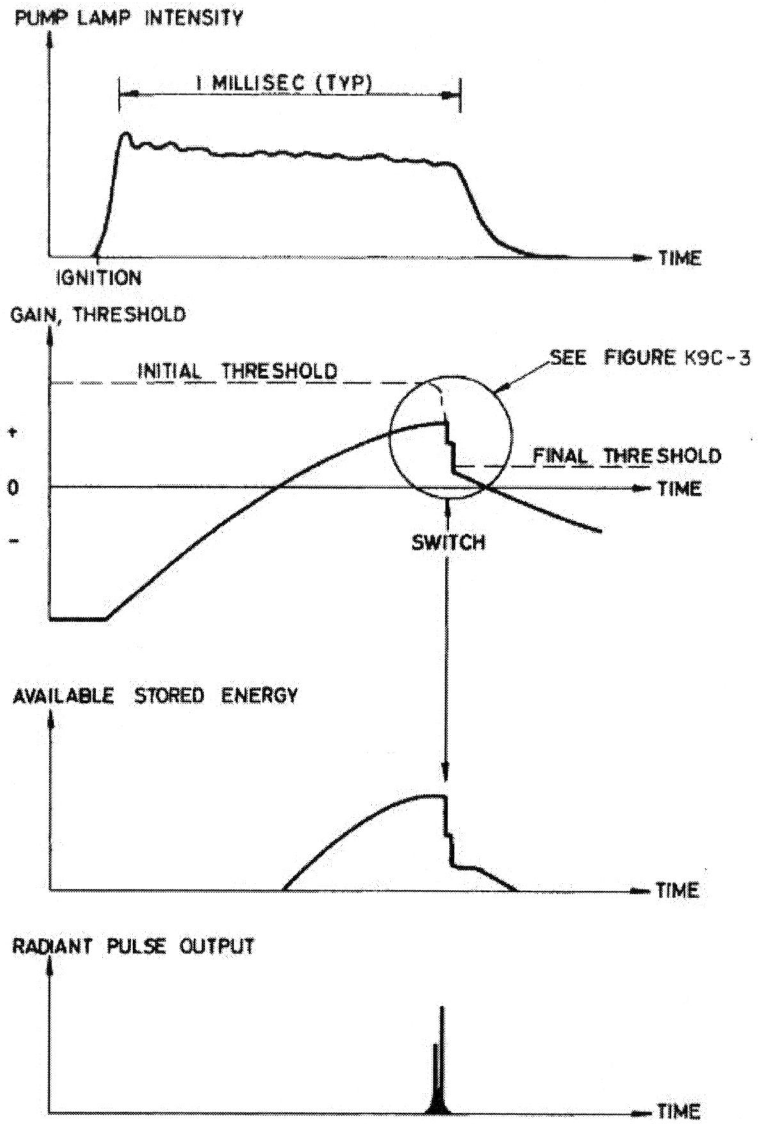

Figure 6. Process of laser pulse generation using Q-switch delay [12].

3.4. Pressure inside Vacuum Chamber

Pressure in the vacuum chamber plays an important role in producing a plasma with maximum wave intensity to be captured by the detector. Based on the experiments carried out, it can be seen that the more vacuum the sample space is the greater the intensity of the wave and vice versa (figure 7). So that the pressure in the sample room should approach the vacuum. The difficulty lies in the time needed to make a vacuum sample room of 3 Torr takes quite a long time. Through repeated experiments it is known that the sample room pressure is smaller than 5.75 Torr has almost the same wave intensity. In taking the next soil sample spectra, the maximum sample chamber pressure is 5.75 Torr to produce a plasma with maximum wave intensity.

IOP Conf. Series: Earth and Environmental Science **335** (2019) 012034 doi:10.1088/1755-1315/335/1/012034

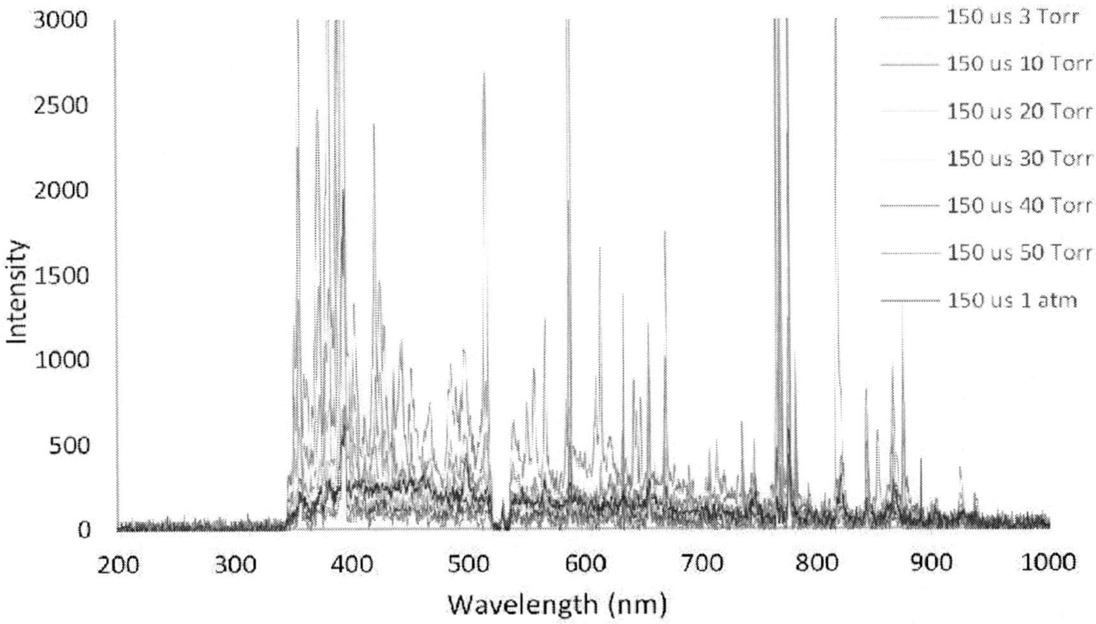

Figure 7. Ground spectra on same Q-switch delay but different pressure.

4. Conclusions

With the Nd-YAG Q-Smart 850 Laser model, the formulation, configuration, and optimal use of LIBS to obtain maximum soil plasma is carried out by photographing laser pulses to soil pellets in a 5.75 torr pressurized vacuum using a wavelength of 532 nm, pulse duration 5 , 5 ns, 10 Hz repitition rate, 150 µs Q-switch delay and 15 mJ/pulse energy.

References

[1] Radziemski, Leon J, Cremers, David A. 2006. Handbook of laser-induced breakdown spectroscopy. New York: John Wiley. ISBN 0-470-09299-8

[2] Kurniawan KH, Tjia MO, Kagawa K. 2014. Review of laser-induced plasma, its mechanism, and application to quantitative analysis of hydrogen and deuterium. *Applied Spectroscopy Review*. 49:5

[3] Yang N. 2009. Elemental analysis of soils using laser-induced breakdown spectroscopy (LIBS) [thesis]. Knoxville (USA): University of Tennessee

[4] Knadel M, Gislum R, Hermansen C, Peng Y, Moldrup P, Jonge LW, Greve MH. 2017. Comparing predictive ability of laser-induced breakdown spectroscopy to visible near-infrared spectroscopy for soil property determination. *Biosystems Engineering* 156:157-172

[5] Haddad JE, Kadri MV, Ismael A, Gallou G, Michel K, Bruyere D, Laperche V, Canioni I, Bousquet B. 2013. Artificial neural network for on-site quantitative analysis of soils using laser induced breakdown spectroscopy. *Spectrochimica Acta Part B* 79-80:51-57.

[6] Singh JP, Thakur SN. 2007. *Laser-Induced Breakdown Spectroscopy* 1st ed. Elsevier

[7] Pantazis P, Maloney J, Wu D, Fraser SE. 2010. Second harmonic generating (SHG) nanoprobes for in vivo imaging. *Proceedings of the National Academy of Siciences of the United States of America*. Pasadena-United States of America

[8] Wikipedia. 2015. Pulse duration [Internet]. [diunduh 19 Oktober 2018]. Tersedia pada: https://en.wikipedia.org/wiki/Pulse_duration

[9] Goh, KJ. 2011.Fertilizer Recommendation Systems for Oil Palm Estimating the Fertilizer Rates. *Agronomic principles and practices of oil palm cultivation*. 338-388

[10] Haupt O, Muller D, Gabler F. 2013. Shorter Pulse Widths Improve Micromachining [Internet]. [diunduh 19 Oktober 2018]. Tersedia pada: https://www.photonics.com/Article.aspx?AID=54123
[11] Slyusar VI. 2007. 60 Years of Electrically Small Antennas Theory. Proceedings of the 6-th International Conference on Antenna Theory and Techniques. P. 116 – 118
[12] Früngel, Frank BA. 2014. Optical Pulses, Lasers, Measuring Techniques. Academic Press. P.192. ISBN 9781483274317

Acknowledgement
I would like to express my gratitude to the Ministry of Agriculture of the Republic of Indonesia for the awarding of Doctoral Program scholarship, Physics Research Center for the support of spectroscopic lab in the collection of soil LIBS spectrum, Department of Land and Land Resources IPB for the support of soil fertility laboratory in the measurement of soil nitrogen content.

AUTHOR INDEX